1750/EL

Biological and Biochemical Oscillators

Johnson Research Foundation Colloquia

Energy-Linked Functions of Mitochondria
Edited by Britton Chance
1963

Rapid Mixing and Sampling Techniques in Biochemistry
Edited by Britton Chance, Quentin H. Gibson, Rudolph H. Eisenhardt, K. Karl Lonberg-Holm
1964

Control of Energy Metabolism
Edited by Britton Chance, Ronald W. Estabrook, John R. Williamson
1965

Hemes and Hemoproteins
Edited by Britton Chance, Ronald W. Estabrook, Takashi Yonetani
1966

Probes of Structure and Function of Macromolecules and Membranes
Volume I Probes and Membrane Function
Edited by Britton Chance, Chuan-pu Lee, J. Kent Blasie
1971

Probes of Structure and Function of Macromolecules and Membranes
Volume II Probes of Enzymes and Hemoproteins
Edited by Britton Chance, Takashi Yonetani, Albert S. Mildvan
1971

Biological and Biochemical Oscillators
Edited by Britton Chance, E. Kendall Pye, Amal K. Ghosh, Benno Hess
1973

Papers contributed to Conferences on
Biological and Biochemical Oscillators
Held outside Prague, Czechoslovakia
July 19-21, 1968, and at Hangö, Finland
August 16-17, 1969.

ACADEMIC PRESS RAPID MANUSCRIPT REPRODUCTION

Biological and Biochemical Oscillators

Edited by

Britton Chance
The Johnson Research Foundation
School of Medicine
University of Pennsylvania
Philadelphia, Pennsylvania

E. Kendall Pye
Department of Biochemistry
School of Medicine
University of Pennsylvania
Philadelphia, Pennsylvania

Amal K. Ghosh
Harrison Department of Surgical Research
School of Medicine
University of Pennsylvania
Philadelphia, Pennsylvania

Benno Hess
Max-Planck-Institut für Ernährungsphysiologie
Dortmund, West Germany

Academic Press   New York and London   **1973**
A Subsidiary of Harcourt Brace Jovanovich, Publishers

COPYRIGHT © 1973, BY ACADEMIC PRESS, INC.
ALL RIGHTS RESERVED.
NO PART OF THIS PUBLICATION MAY BE REPRODUCED OR
TRANSMITTED IN ANY FORM OR BY ANY MEANS, ELECTRONIC
OR MECHANICAL, INCLUDING PHOTOCOPY, RECORDING, OR ANY
INFORMATION STORAGE AND RETRIEVAL SYSTEM, WITHOUT
PERMISSION IN WRITING FROM THE PUBLISHER.

ACADEMIC PRESS, INC.
111 Fifth Avenue, New York, New York 10003

United Kingdom Edition published by
ACADEMIC PRESS, INC. (LONDON) LTD.
24/28 Oval Road, London NW1

Library of Congress Cataloging in Publication Data

Conference on Biological and Biochemical Oscillators,
    Prague, 1968.
    Biological and biochemical oscillators.

    (Johnson Research Foundation colloquia)
    Held as a satellite of the 5th meeting of the
Federation of European Biochemical Societies.
        1. Biology–Periodicity–Congresses.   2.   Biological chemistry–Congresses.   I.   Chance, Britton, ed.
II.   Federation of European Biochemical Societies.
III.   Title.   IV.   Series: Pennsylvania.   University.
Eldridge Reeves Johnson Foundation for Medical Physics.
Colloquia.
QH527.C66   574.1   69-12281
ISBN 0–12–167872–5

PRINTED IN THE UNITED STATES OF AMERICA

CONTENTS

CONTRIBUTORS ix
PREFACE xv

INTRODUCTION

Introduction 3
    Benno Hess

I. OSCILLATOR THEORY

Oscillatory Behavior, Excitability, and Propagation Phenomena on
Membranes and Membrane-like Interfaces 7
    U.F. Franck

Two-Dimensional Analysis of Chemical Oscillators 31
    Dieter H. Meyer

Stability Properties of Metabolic Pathways with Feedback
Interactions 41
    Gustavo Viniegra-Gonzalez

II. OSCILLATIONS IN DEFINED CHEMICAL AND BIOCHEMICAL SYSTEMS

Some Experiments of a Chemical Periodic Reaction in
Liquid Phase 63
    Heinrich-Gustav Busse

A Study of a Self-Oscillatory Chemical Reaction: I. The
Autonomous System 71
    V.A. Vavilin, A.M. Zhabotinsky, and A.N. Zaikin

A Study of a Self-Oscillatory Chemical Reaction: II. Influence
of Periodic External Force 81
    A.N. Zaikin and A.M. Zhabotinsky

CONTENTS

A Study of a Self-Oscillatory Chemical Reaction: III. Space Behavior 89
    A.M. Zhabotinsky

Chemiluminescence in Oscillatory Oxidation Reactions Catalyzed by Horseradish Peroxidase 97
    Hans Degn

A Siphon Model for Oscillatory Reactions in the Reduced Pyridine Nucleotide, O_2 and Peroxidase System 109
    Isao Yamazaki and Ken-nosuke Yokota

Damping of Mitochondrial Volume Oscillations by Propranolol and Related Compounds 115
    A.J. Seppälä, M.K.F. Wikström, and N.-E.L. Saris

III. GLYCOLYTIC OSCILLATIONS

The Control Theoretic Approach to the Analysis of Glycolytic Oscillators 127
    Joseph Higgins, Rene Frenkel, Edward Hulme, Anne Lucas, and Gus Rangazas

Problems Associated with the Computer Simulation of Oscillating Systems 177
    E.M. Chance

The Effect of Fructose Diphosphate Activation of Pyruvate Kinase on Glycolytic Oscillations in Beef Heart Supernatant: An Experimental and Simulation Study 187
    Rene Frenkel, Murray J. Achs, and David Garfinkel

On the Mechanism of Single-Frequency Glycolytic Oscillations . . . 197
    E.E. Sel'kov and A. Betz

Kinetics of Yeast Phosphofructokinase and the Glycolytic Oscillator 221
    A. Betz

Substrate Control of Glycolytic Oscillations 229
    Benno Hess and Arnold Boiteux

Control Mechanism of Glycolytic Oscillations 243
    Arnold Boiteux and Benno Hess

CONTENTS

Component Structure of Oscillating Glycolysis 253
    B. Hess, H. Kleinhans, and D. Kuschmitz

Glycolytic Oscillations in Cells and Extracts of Yeast - Some
Unsolved Problems 269
    E. Kendall Pye

Synchronization Phenomena in Oscillations of Yeast Cells and
Isolated Mitochondria 285
    B. Chance, Gary Williamson, I.Y. Lee, L. Mela,
    D. DeVault, A. Ghosh, and E.K. Pye

IV. OSCILLATIONS IN TISSUES

Oscillating Contractile Structures from Insect Fibrillar Muscle . . . 303
    J.C. Rüegg

Kinetic Model of Muscle Contraction 311
    V.I. Descherevsky

Excitation Wave Propagation during Heart Fibrillation 329
    V.I. Krinsky

Conformational Oscillations of Protein Macromolecules of
Actomyosin Complex 343
    S.E. Shnoll

Oscillations in Muscle Creatine Kinase Activity 347
    E.P. Chetverikova

Oscillation of Sodium Transport across a Living Epithelium 363
    James E. Allen and Howard Rasmussen

Biochemical Cycle of Excitation 373
    M.N. Kondrashova

Possible Pathways for the Succinate Concentration Burst in the
Active Metabolic State 389
    Y.V. Evtodienko and M.N. Kondrashova

V. OSCILLATIONS IN GROWING CELL POPULATIONS

Undamped Oscillations Occurring in Continuous Cultures of
Bacteria 399
    D.E.F. Harrison

CONTENTS

Stable Synchrony Oscillations in Continuous Cultures of
Saccharomyces cerevisiae under Glucose Limitation 411
   H. Kaspar von Meyenburg

Physiological Rhythms in *Saccharomyces cerevisiae* Populations . . . 419
   G. Kraepelin

Long- and Short-Period Oscillations in a Myxomycete with
Synchronous Nuclear Divisions 429
   W. Sachsenmaier and K. Hansen

Oscillations in the Epigenetic System: Biophysical Model of
the β-Galactosidase Control System 449
   W.A. Knorre

VI. CIRCADIAN OSCILLATIONS

The Investigation of Oscillatory Processes by Perturbation
Experiments: I. The Dynamical Interpretation
of Phase Shifts 461
   Arthur T. Winfree

The Investigation of Oscillatory Processes by Perturbation
Experiments: II. A Singular State in the Clock-Oscillation
of *Drosophila pseudoobscura* 479
   Arthur T. Winfree

The Circadian Oscillation: An Integral and Undissociable Property
of Eukaryotic Gene-Action Systems 503
   C.F. Ehret, J.J. Wille, and E. Trucco

Respiration Dependent Types of Temperature Compensation
in the Circadian Rhythm of *Euglena gracilis* 513
   Klaus Brinkmann

The Role of Actidione in the Temperature Jump Response of the
Circadian Rhythm in *Euglena gracilis* 523
   Klaus Brinkmann

SUBJECT INDEX 531

CONTRIBUTORS

Murray J. Achs,[a] The Johnson Research Foundation, University of Pennsylvania, Philadelphia, Pennsylvania 19174

James E. Allen, Department of Biochemistry, School of Medicine, University of Pennsylvania, Philadelphia, Pennsylvania 19174

A. Betz,[b] Institut für Molekulare Biologie, 3301 Stöckheim/Braunschweig, West Germany

Arnold Boiteux, Max-Planck-Institut für Ernährungsphysiologie, Dortmund, West Germany

Klaus Brinkmann, Institut für Molekulare Biologie, 3301 Stöckheim/Braunschweig, West Germany

Heinrich-Gustav Busse,[c] Institut für Molekulare Biologie, Biochemie und Biophysik, Stöckheim/Braunschweig, West Germany

Britton Chance, The Johnson Research Foundation, University of Pennsylvania, Philadelphia, Pennsylvania 19174

E. M. Chance, Department of Biochemistry, University College, London, England

E. P. Chetverikova, Institute of Biophysics, Academy of Sciences of the USSR, Puschino, Moscow Region, USSR

Hans Degn,[d] The Johnson Research Foundation, University of Pennsylvania, Philadelphia, Pennsylvania 19174

---

[a]Present address: The Moore School of Electrical Engineering, University of Pennsylvania, Philadelphia, Pennsylvania 19174

[b]Present address: Botanical Institute, 53 Bonn, Kirschallé, West Germany

[c]Present address: Max-Planck-Institut für Ernährungsphysiologie, Dortmund, West Germany

[d]Present address: Institute of Biochemistry, University of Odense, Odense, Denmark

CONTRIBUTORS

V. I. Descherevsky, Institute of Biophysics, Academy of Sciences of the USSR, Puschino, Moscow Region, USSR

D. DeVault, The Johnson Research Foundation, University of Pennsylvania, Philadelphia, Pennsylvania 19174

C. F. Ehret, Division of Biological and Medical Research, Argonne National Laboratory, Argonne, Illinois 60439

Y. V. Evtodienko, Institute of Biophysics, Academy of Sciences of the USSR, Puschino, Moscow Region, USSR

U. F. Franck, Institut für Physikalische Chemie der Rhein.-Westf. Techn. Hochschule, Aachen, West Germany

Rene Frenkel,[a] The Johnson Research Foundation, University of Pennsylvania, Philadelphia, Pennsylvania 19174

David Garfinkel,[b] The Johnson Research Foundation, University of Pennsylvania, Philadelphia, Pennsylvania 19174

A. K. Ghosh,[c] The Johnson Research Foundation, University of Pennsylvania, Philadelphia, Pennsylvania 19174

K. Hansen, Zoologisches Institut der Universität Heidelberg, Heidelberg, West Germany

D. E. F. Harrison,[d] The Johnson Research Foundation, University of Pennsylvania, Philadelphia, Pennsylvania 19174

Benno Hess, Max-Planck-Institut für Ernährungsphysiologie, Dortmund, West Germany

Joseph Higgins, The Johnson Research Foundation, University of Pennsylvania, Philadelphia, Pennsylvania 19174

Edward Hulme, The Johnson Research Foundation, University of Pennsylvania, Philadelphia, Pennsylvania 19174

[a]Present address: Department of Biochemistry, University of Texas, Southwestern Medical School, Dallas, Texas 75235

[b]Present address: The Moore School of Electrical Engineering, University of Pennsylvania, Philadelphia, Pennsylvania 19174

[c]Present address: Harrison Department of Surgical Research, University of Pennsylvania, Philadelphia, Pennsylvania 19174

[d]Present address: Shell Research, Ltd., Sittingbourne, Kent, England

CONTRIBUTORS

H. Kleinhans, Max-Planck-Institut für Ernährungsphysiologie, Dortmund, West Germany

W. A. Knorre, Department of Biophysics, Institute for Microbiology and Experimental Therapy, German Academy of Sciences, Berlin, 69 Jena, East Germany

M.N. Kondrashova, Institute of Biophysics, Academy of Sciences of the USSR, Puschino, Moscow Region, USSR

G. Kraepelin, Botanisches Institut der Technischen Universität, Braunschweig, West Germany

V. I. Krinsky, Institute of Biophysics, Academy of Sciences of the USSR, Puschino, Moscow Region, USSR

D. Kuschmitz, Max-Planck-Institut für Ernährungsphysiologie, Dortmund, West Germany

I. Y. Lee,[a] The Johnson Research Foundation, University of Pennsylvania, Philadelphia, Pennsylvania 19174

Anne Lucas, The Johnson Research Foundation, University of Pennsylvania, Philadelphia, Pennsylvania 19174

L. Mela,[b] The Johnson Research Foundation, University of Pennsylvania, Philadelphia, Pennsylvania 19174

Dieter H. Meyer,[c] The Johnson Research Foundation, University of Pennsylvania, Philadelphia, Pennsylvania 19174

E. Kendall Pye, Department of Biochemistry, School of Medicine, University of Pennsylvania, Philadelphia, Pennsylvania 19174

Gus Rangazas, The Johnson Research Foundation, University of Pennsylvania, Philadelphia, Pennsylvania 19174

Howard Rasmussen, Department of Biochemistry, School of Medicine, University of Pennsylvania, Philadelphia, Pennsylvania 19174

---

[a]Present address: B.C.P. Jansen Institute, University of Amsterdam, Amsterdam, The Netherlands

[b]Present address: Harrison Department of Surgical Research, University of Pennsylvania, Philadelphia, Pennsylvania 19174

[c]Present address: Casimirring 60, 675 Kaiserslautern, West Germany

CONTRIBUTORS

J.C. Rüegg, Department of Cell-Physiology, Ruhr University, Bochum, and Max-Planck-Institute for Medical Research, Heidelberg, West Germany

W. Sachsenmaier, Institut für Experimentelle Krebsforschung, Deutsches Krebsforschungszentrum, Heidelberg, West Germany

N.-E.L. Saris, Department of Clinical Chemistry, University of Helsinki, Helsinki, Finland

E.E. Sel'kov, Institute of Biophysics, Academy of Sciences of the USSR, Puschino, Moscow Region, USSR

A.J. Seppälä, Department of Clinical Chemistry, University of Helsinki, Helsinki, Finland

S.E. Shnoll, Moscow State University, Faculty of Physics, Moscow; and Institute of Biophysics, Academy of Sciences of the USSR, Puschino, Moscow Region, USSR

E. Trucco,[a] Division of Biological and Medical Research, Argonne National Laboratory, Argonne, Illinois 60439

V.A. Vavilin, Institute of Biophysics, Academy of Sciences of the USSR, Puschino, Moscow Region, USSR

Gustavo Viniegra-Gonzalez,[b] Cardiovascular Research Institute, University of California, San Francisco Medical Center, San Francisco, California 94122

H. Kaspar von Meyenburg, Department of Microbiology, Federal Institute of Technology, Zurich, Switzerland

M.K.F. Wikström, Department of Clinical Chemistry, University of Helsinki, Helsinki, Finland

J.J. Wille, Division of Biological and Medical Research, Argonne National Laboratory, Argonne, Illinois 60439

Gary Williamson, The Johnson Research Foundation, University of Pennsylvania, Philadelphia, Pennsylvania 19174

---

[a] Deceased

[b] Present address: Instituto de Investigaciones Biomédicas, Ciudad Universitaria, Mexico, D.F.

CONTRIBUTORS

Arthur T. Winfree,[a] Biology Department, Princeton University, Princeton, New Jersey 08540

Isao Yamazaki, Biophysics Division, Research Institute of Applied Electricity, Hokkaido University, Sapporo, Japan

Ken-nosuke Yokota, Biophysics Division, Research Institute of Applied Electricity, Hokkaido University, Sapporo, Japan

A.N. Zaikin, Institute of Biophysics, Academy of Sciences of the USSR, Puschino, Moscow Region, USSR

A.M. Zhabotinsky, Institute of Biophysics, Academy of Sciences of the USSR, Puschino, Moscow Region, USSR

---

[a] Present address: Department of Biological Sciences, Purdue University, West Lafayette, Indiana 47907

PREFACE

The tremendous upsurge of interest in biochemical oscillators has stemmed from the recent realization that enzyme systems, containing appropriate feedback, can generate fundamental rhythms from which, one might speculate, many properties of the cell and organism could be regulated and controlled. Interest, both in the United States and in Europe, developed sufficiently rapidly in this field that during the summers of 1968 and 1969, two international symposia were organized: one in Prague, Czechoslavakia, and the other in Hangö, Finland; the first as a satellite of the 5th meeting of the Federation of European Biochemical Societies. In both cases the colloquia consisted of contributed papers and intensive and enthusiastic discussion. The first meeting brought together those who had worked for some time with purely chemical oscillators and those whose primary interest was in biochemical and biological oscillations. The opportunity was also taken to open up communication on computer simulation and a variety of other theoretical approaches. For the first time the possibility of achieving common points of view between the various fields was established.

The second meeting, with the title Biochemical Oscillations and Chemical Instabilities, continued with the approach initiated in the first meeting and in addition focused upon bistabilities and instabilities from the theoretical standpoint. Many aspects of membrane oscillators, covering the borderline area between various biochemical, biophysical, and physical fields were also considered, together with their relationship to a number of the electrophysiological systems.

The results of these symposia covering, as they did, a wide range of disciplines and a variety of nationalities were painstakingly gathered together after the two symposia and are presented here through the main work of E. Kendall Pye and his associated editors. Especial thanks are due to Mrs. Ann Pye for her generous donation of time in typing the manuscripts.

Britton Chance
E. Kendall Pye
Amal K. Ghosh
Benno Hess

Biological and Biochemical Oscillators

INTRODUCTION

INTRODUCTION

Benno Hess

The great interest in biochemical oscillations derives from the fact that only recently has it become recognized that biochemical systems can generate self-sustained oscillations. Studies over the past decade have clearly shown that relatively simple enzyme reaction systems with appropriate coupling mechanisms for activation and inhibition, as well as suitable input and output rates, can generate oscillations with a wide variety of periods and waveforms. This discovery has revolutionized the study of the biological systems and has opened the way for their understanding at the molecular level. In addition, the availability of technical methods to analyze these instabilities has tremendously stimulated both theoretical and experimental work.

The experimental field of biochemical oscillations began with the measurement of intracellular components in studies of photosynthesis. The observation of cyclic changes in the concentrations of phosphoglycerate and ribulose diphosphate in 1955 led, in 1958, to a detailed kinetic theory of the self-oscillating mode of the dark reactions of photosynthesis. In non-photosynthetic cells the direct readout of intracellular components, such as NADH, showed overshoots (1952, 1954) and oscillations (1957) in glycolysis. In 1964 nearly continuous oscillations of NADH-fluorescence were reported in yeast cells. This observation was followed by the demonstration of continuous glycolytic oscillations in a cell-free system of yeast (1966). The development coincided with the observation of oscillatory ion movements in mitochondria (1965) where different metabolic functions, as well as components of the respiratory chain, displayed sustained periodic behavior.

The frequencies of biological rhythms cover a bandwidth of over ten orders of magnitude which reflects the temporal organization of the living world. It is interesting to see that the biochemical oscillations now fill the gap between the circadian oscillations and the neural frequencies observed in biological systems. Therefore, it was only proper that communication should be opened between those interested in biochemical oscillations and others who focused their attention on the numerous periodic phenomena such as diurnal rhythms, the circadian clock and other periodic activities of biological systems.

This volume of contributed papers is one result of the communication between researchers of various nationalities in the two fields. The papers summarise the major view points from a number of laboratories and clearly illustrate the present situation of the field from both a theoretical and experimental standpoint.

I
OSCILLATOR THEORY

OSCILLATORY BEHAVIOR, EXCITABILITY AND PROPAGATION
PHENOMENA ON MEMBRANES AND MEMBRANE-LIKE INTERFACES.

U. F. Franck

Institut für Physikalische Chemie
der Rhein.-Westf. Techn. Hochschule,
Aachen, Germany

The quantitative treatment, on the basis of primary principles, of oscillating physico-chemical and biological systems is extremely cumbersome. This is on account of their essentially non-linear nature and their complex kinetics involving a set of independent variables. Moreover the general principles which lead to oscillatory behavior are, in this way, hard to recognize because each oscillating system requires, as a rule, a particular treatment.

For these reasons a phenomenological approach is proposed in this paper which considers oscillating systems as "black boxes" whose kinetic properties are studied by directly measurable forces and fluxes which exist between them and their environment.

Sustained periodic transportation processes and/or chemical reactions occur, as we are aware, only in energetically open systems which are in contact with an appropriate environment. Physico-chemical and biological oscillations are therefore always a common result of the kinetic properties of the system and the given environment. They are energetically brought about by two or more independent driving forces existing in the environment, or in the system, or in both of them.

I. <u>One process systems</u>.

Energetically open systems containing only one kind of transportation process are not able to oscillate, but, as shown later, they can exhibit, under certain circumstances, excitation and propagation phenomena.

As already mentioned every energetically open system can be regarded as a "black box" which represents a kind of "dipole"(1-3). At the "poles" (these are the entry and the exit of the flux) the intensity of the flux, I, and the force, X, can be measured without knowing the special inner structure of the dipole.

The system under consideration, as well as its environment, are dipoles. In a given working situation they are connected together forming a flux circuit containing a source of driving forces, X_0, inside of one or both of the dipoles (See Figure 1).

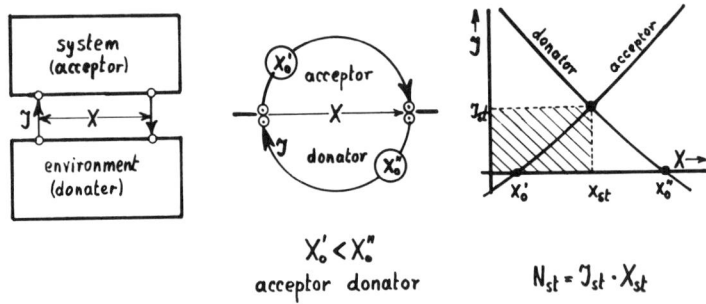

Figure 1. Black-box treatment of physico-chemical systems and their environments.

For all stationary states there exists, for the system as well as for its environment, a defined relationship between the force, X, and the conjugated flux, I. Both relationships can be measured independently of each other at the separated dipoles and can be plotted in the form of "force-flux characteristics". In either case, the characteristic is the geometric locus of all stationary states in which the dipole in question can exist (Figure 1). Accordingly, their intersection represents the common stationary state of the system and the environment. All the other states are, as a function of the force, X, non-stationary and therefore they change with time.

Between the connected dipoles a transfer of energy takes place. Denoting the dipole having the higher strength force source, X_0, as "donator" and the other dipole as "acceptor", then the area of the rectangle formed by the coordinates of state X and I in Figure 1, can be directly interpreted as the "power" transferred from the donator to the acceptor under stationary conditions.

The distinction between acceptor and donator is important with respect to the definition of the sign of the flux and to the criteria of stability of the stationary states.

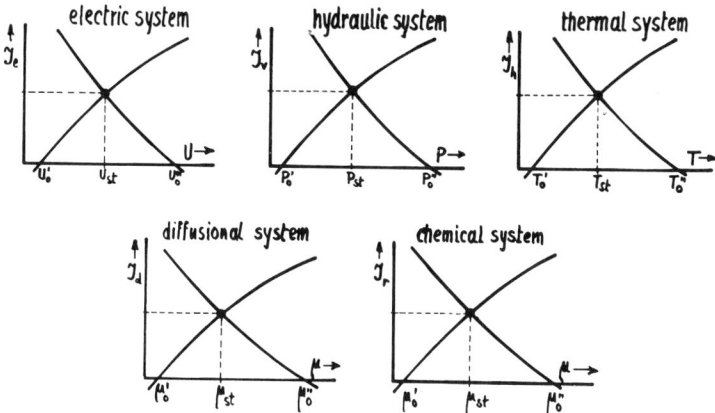

Figure 2. Force-flux characteristics for electric, hydraulic, thermal, diffusional and chemical systems.

For all one-process systems there are valid corresponding graphs. Figure 2 shows examples of electric, hydraulic, thermal, diffusional and chemical systems. In the first case, the characteristic of the electric environment is the well-known "load-line" of electric circuits. This useful term may well be extended to all characteristics of energy transferring environments.

Although the ensemble of characteristics of the system and its environment describes stationary states, direct information concerning the time-dependent behavior of the non-stationary states (as a function of X) can be derived from it. In non-stationary states there exists a defined difference between the stationary state values of the flux of the system and the flux of the environment (see Figure 3). The deviation of the flux supply of the environment, with respect to the stationary state value of the system, necessarily has to be balanced by the "storage elements" inside

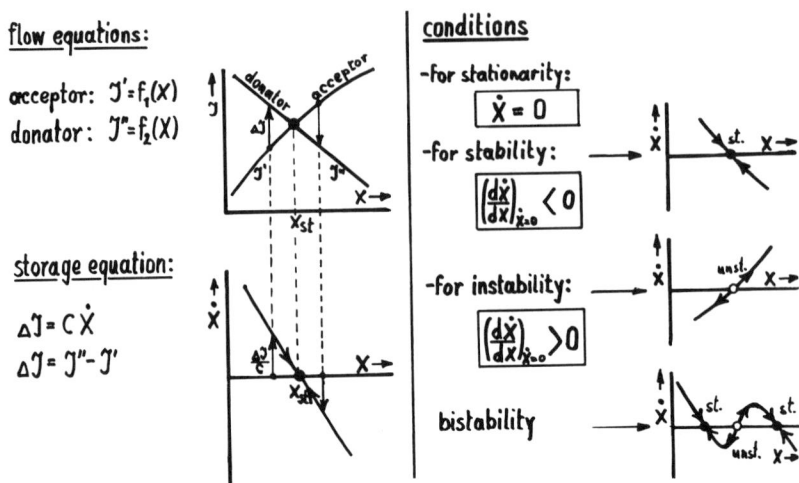

Figure 3. Time-dependence of non-stationary states.

the system giving rise to a change of the force, X, with time, according to the well-known capacity equation (1):

$$\Delta I = C \frac{dX}{dt}$$

Here "C" denotes the "capacity", in a generalized sense, e.g.:

    electric systems — an influx of electric charges into an electric capacitance leads to an increase in voltage;

    chemical systems — an influx of matter into a volume of solvent leads to an increase in concentration;

    thermal systems — an influx of heat into a heat capacitance leads to an increase in temperature.

In simple cases these generalized capacities are constant. They may, however, depend on the forces, but they are always positive as a consequence of the conservation laws.

If we plot the diverging current, $I = I''-I' = C\frac{dX}{dt}$, as a function of X (see Figure 3), then we obtain a relationship between the time derivative of X and X itself. Such a representation is called a "dynamic diagram" because it describes graphically the time behavior of the ensemble of the system and its environment, with respect to the variable of state, X.

The stationary states correspond to the intersections of the dynamic characteristic with the X-axis, according to the condition of stationarity:

$$\dot{X} = 0 \; ; \quad (\dot{X} \equiv \frac{dX}{dt})$$

Stationary states behave in a different manner with respect to disturbances by "stimuli", depending on the dynamic nature of their non-stationary neighborhood in the dynamic diagram. If the dynamic characteristic has a negative slope at the X-axis intersection, then the stationary state is stable, i.e., after disturbances by stimuli the original stationary state is spontaneously restored by the system (see Figure 3). If, however, the slope is positive, then the stationary state is unstable. Here, already, the weakest disturbance leads to an autocatalytically-increasing deviation away from the original stationary state. With respect to the dynamic diagram the stability conditions (4) are:

Stability: $\left(\frac{d\dot{X}}{dX}\right)_{\dot{X}=0} < 0$; Instability: $\left(\frac{d\dot{X}}{dX}\right)_{\dot{X}=0} > 0$

Dynamic instability of one process in one of the dipoles is an essential, but not yet sufficient condition for oscillatory behavior. Therefore, in order to understand oscillation kinetics we have primarily to look for instability constellations in the systems in question. Instability arises, in general, when the dynamic characteristic is non-monotonic, giving the possibility of more than one intersection with the X-axis (4). Then we get an odd number of stationary states with alternating stability and instability as shown in Figure 4.

In the case of "forward inhibition" instability (6), the acceptor dipole has a non-monotonic characteristic. In the case of "backward activation" instability, the donator dipole is non-monotonic. In either case we get the same type of dynamic characteristic as shown in Figure 5.

Figure 6 shows how stable and unstable states are produced by intersections of force-flux characteristics. Stability arises when in the neighborhood to the right of a

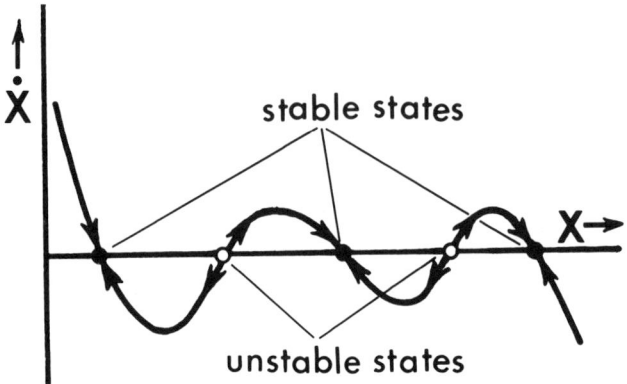

Figure 4. Stable and unstable states given by a non-monotonic dynamic characteristic.

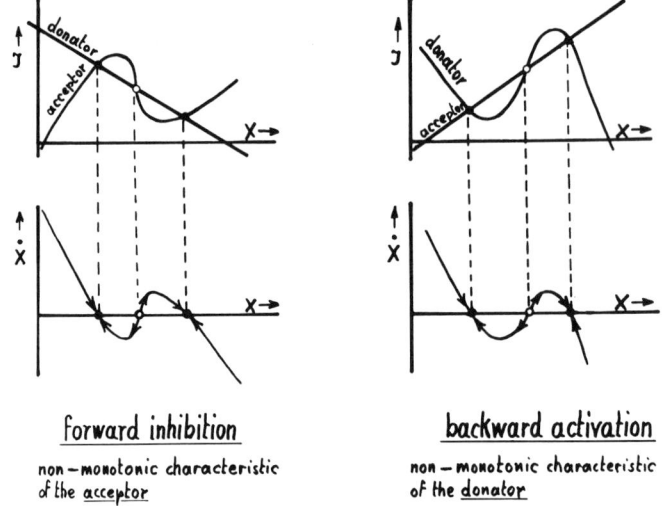

forward inhibition
non-monotonic characteristic of the <u>acceptor</u>

backward activation
non-monotonic characteristic of the <u>donator</u>

Figure 5. Forward inhibition and backward activation represented by force-flux characteristics.

given stationary state the value of the stationary current of the acceptor is greater than that of the donator, or smaller in the neighborhood of the left, respectively. Instability arises under inverse conditions. With respect to

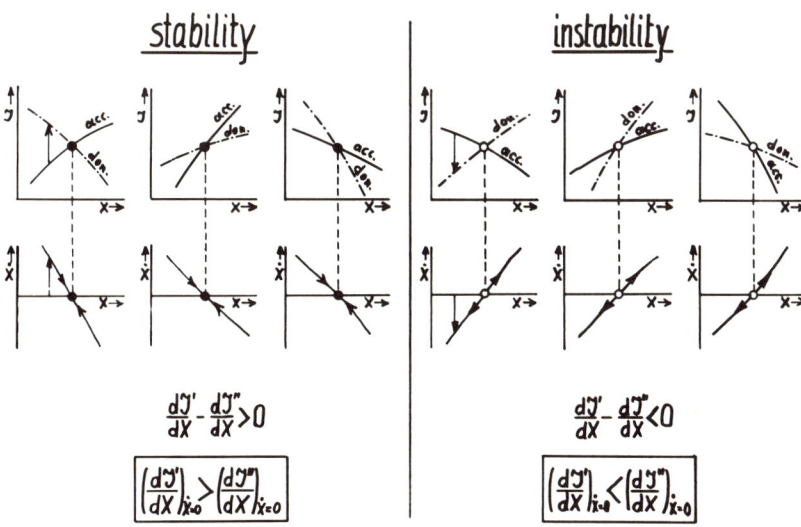

Figure 6. Stability and instability situations given by force-flux characteristics.

the slopes of the force-flux characteristics these conditions are:

Stability: $\left(\dfrac{dI}{dX}\right)_{acceptor} - \left(\dfrac{dI}{dX}\right)_{donator} > 0$; or,

$\left(\dfrac{dI}{dX}\right)_{acceptor} > \left(\dfrac{dI}{dX}\right)_{donator}$ respectively.

Instability: $\left(\dfrac{dI}{dX}\right)_{acceptor} - \left(\dfrac{dI}{dX}\right)_{donator} < 0$; or,

$\left(\dfrac{dI}{dX}\right)_{acceptor} < \left(\dfrac{dI}{dX}\right)_{donator}$ respectively.

In most of the known systems with unstable states the non-monotonic force-flux characteristic is caused by resistances, R, (in a generalized sense) which depend on the driving force in an appropriate manner. Figure 7 shows the condition for the occurrence of non-monotonicity. In the R/X graph the R curve is steeper in a certain range of X than the geometric locus for I = constant. Hence the condition for non-monotonicity with respect for the force dependence of R becomes:

forward inhibition: $\left(\dfrac{dR'}{dX}\right)_{\dot X=0} > \dfrac{1}{\dot I'}$, $\quad (R' \equiv \dfrac{X - X'_o}{\dot I'})$

backward activation: $-\left(\dfrac{dR''}{dX}\right)_{\dot X=0} > \dfrac{1}{\dot I''}$, $\quad (R'' \equiv \dfrac{X''_o - X}{\dot I''})$

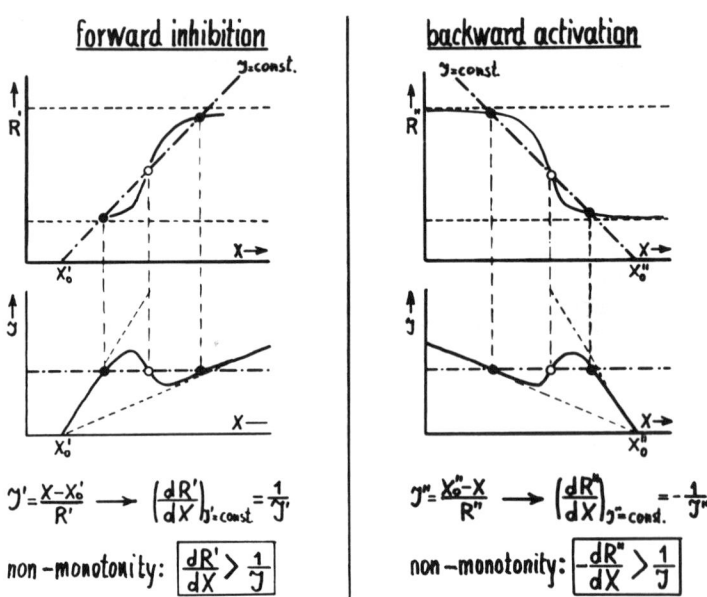

Figure 7. The condition of non-monotonicity.

Force-dependent resistances occur in systems (dipoles) in which high field strengths of forces arise, such as in membranes of all kind, electrolytic layers on electrodes, boundaries of crystal grains, junction zones in semiconductors and all other systems containing thin conducting interfaces. Figure 8 gives examples of such membrane-like interfaces exhibiting strong voltage-dependent electrical resistances. Similar examples are known for pressure-dependent volume fluxes and concentration-dependent fluxes of solutes. In the case of chemical systems, autocatalytic (backward activation) and autoinhibition (forward inhibition) reactions produce the analogous effects.

All of the systems mentioned show non-monotonic force-flux behavior. As it will be explained later, they are able to produce oscillations under suitable additional conditions. In contact with appropriate environments they have one unstable and two stable stationary states. They are called "bistable" because they can exist in two distinct stable

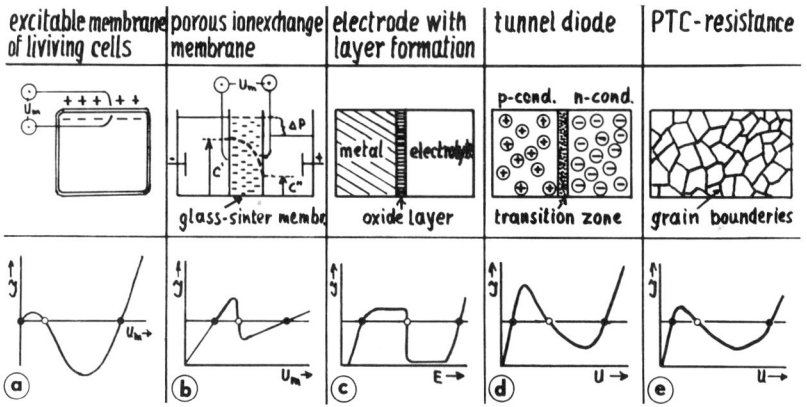

Figure 8. Membrane-like interfaces inhibiting non-monotonic force-flux characteristics.

states. The transition from one stable state to the other is a "trigger" process obeying the "all-or-nothing" law of excitation. This behavior can easily be demonstrated by means of the dynamic diagram shown in Figure 9 (4).

In the case of bistability the dynamic characteristic intersects the X-axis three times giving one unstable state in the middle and two stable states. During the period of the stimulus the situation of the dynamic characteristic is different from the situation under unirritated conditions. As shown in Figure 9, with the beginning of the stimulus the hitherto stable resting state, 1, becomes non-stationary, 2, and changes with time ($\dot{X}>0$), according to the dynamic characteristic of the stimulus condition, toward the unstable state X*. The result of the stimulus now depends on how far the state is shifted to the right. After the stimulus has ceased the system jumps back into the stimulus-free dynamic characteristic. If it happens before reaching the unstable state X* (3'→ 4') then the system returns spontaneously into its initial state (4' → 5; $\dot{X}<0$). But if a state is reached beyond X* (3"→4"), then the total transition into the other stable state takes place (4"→6; $\dot{X}>0$). This kind of all-or-nothing mechanism of triggering is valid for transitions in both directions ($X_1 \rightarrow X_2$ and $X_2 \rightarrow X_1$) (see Figure 9b).

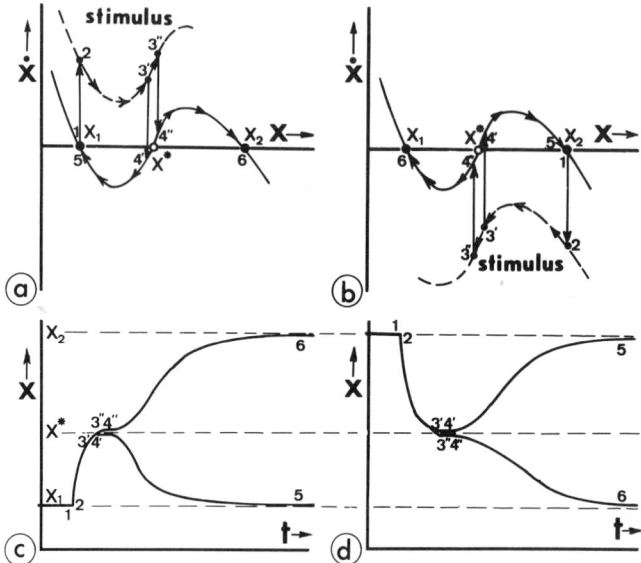

Figure 9. The trigger process of a transition from one stable state into another stable state.

Another important property of bistable systems is their ability to propagate localized triggered state transitions over the entire surface of the triggerable interface, or, in the case of chemical reactions, through the entire volume of triggerable space (7-11). At the boundary between areas of different stable states, local fluxes arise as a consequence of the existing difference of force, which act as stimuli on each area (see Figure 10). The direction in which the propagation actually takes place depends on the quantities of the threshold strengths of both transitions. (12).

According to the nature of the triggerable systems the local fluxes may be electric currents, local diffusion fluxes, local heat fluxes, etc.

In summarizing the essential properties of bistable one-process systems we can state the following facts:

BIOCHEMICAL OSCILLATORS

1. They can exist in two distinct stable stationary states.
2. They contain one unstable stationary state situated between the stable states. With respect to stimuli the unstable state plays a role as a critical state dividing the kinetic ranges of the two stable states.
3. The transition from one stable state to the other is a trigger process which obeys the "all-or-nothing" law.
4. They exhibit propagation phenomena.

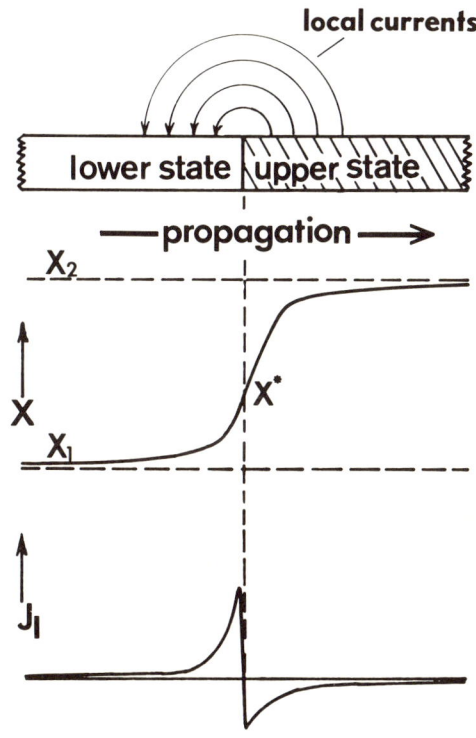

Figure 10. The propagation of state transition in a bistable system.

II. Multi-process Systems

As already mentioned oscillatory behavior occurs only in multi-process systems which contain more than one independent driving force. Because the moving particles which are driven through the interfaces by the driving forces have volume, mass, energy, and, in case of ions, electric charge, the different transportation processes are not independent of each other, (e.g. a driving pressure difference causes not only a volume flux but also a flux of mass and electric charge; a concentration difference causes not only a mass transportation but also a flux of volume and charge, etc.). In this way certain coupling effects exist between the simultaneous transportation processes. As a consequence the dynamic characteristic of the driving force in question also

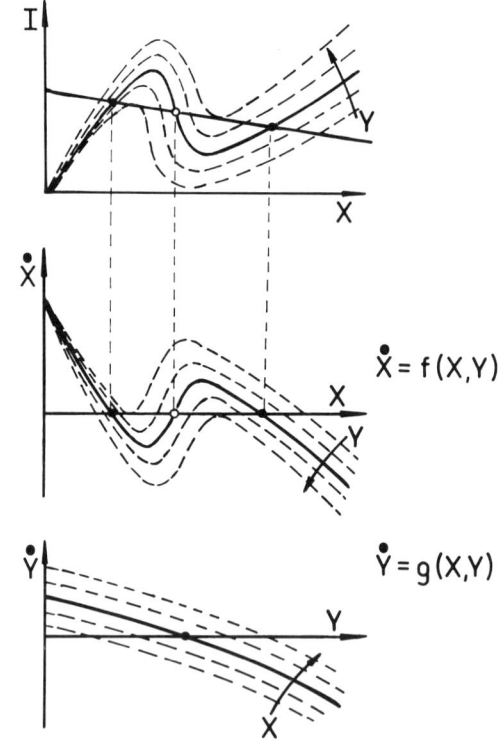

Figure 11. The effects of the trigger and the recovery processes.

BIOCHEMICAL OSCILLATORS

depends on the other driving forces, thus yeilding several varieties of dynamic characteristics according to the specific values of the other forces (4,2).

Figure 11 exemplifies the kinetic situation of a coupled two-process system with respect to the force-flux relationship and the dynamic characteristics of the two variables X and Y. The quantitative description now requires, accordingly, a set of two simultaneous differential equations (3):

$$\dot{X} = f(X, Y).$$
$$\dot{Y} = g(X, Y).$$

Oscillatory behavior, in particular, arises in two-process systems when two essential conditions, among others, are satisfied:

1. There must exist one process which exhibits bistability, as explained above.
2. The other process, which may well have a monotonic dynamic characteristic, must depend on the bistable process in such a way that all occurring "all-or-nothing" state transitions recover spontaneously in a delayed counteraction.

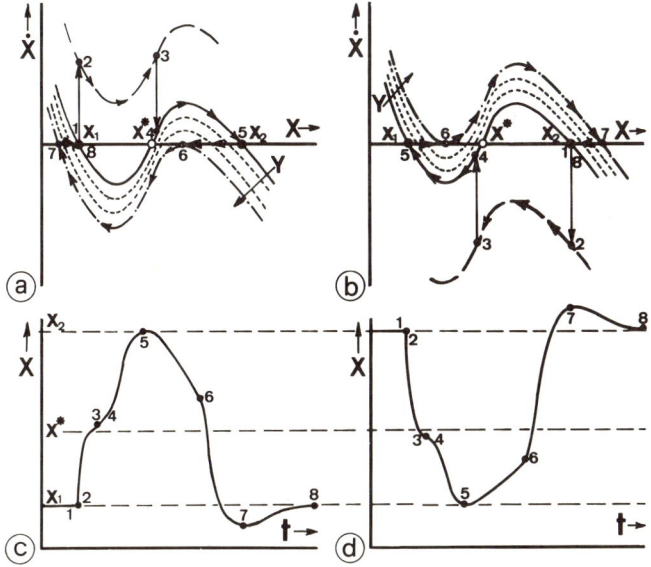

Figure 12. The recovery process of a monostable system.

19

It is reasonable to denote the first process as the "trigger process" and the other as the "recovery process" and, accordingly, the responsible variables of force as the "trigger variable", X, and the "recovery variable", Y. Some important new properties arise in bistable systems due to the action of the recovery process. Figure 12 illustrates the recovery process of a state transition triggered by a superthreshold stimulus. State 1 corresponds to the general resting state, where X as well as Y are stationary stable. At the beginning of the stimulus the system jumps to state 2 and changes with time, reaching state 3 when the stimulus ceases. The system now jumps back to the initial characteristic at state 4 and changes with time to state 5. Here, unlike in Figure 9, this state is not purely stationary since the recovery variable, Y, is simultaneously non-stationary and changes with time, shifting the characteristic of X downward, as in Figure 12a, or upward, as in Figure 12b, until the system again becomes unstable at state 6 and recovers by self-triggering into the original resting state (7→8).

This behavior is called "monostability" because only one state (1 and 8) is actually stationary in X and Y at the same time. The graph of X versus t for a triggered state transition, with self-recovery, is shown in Figure 12c and 12d. It corresponds to the well-known recordings of "action potential spikes" in nerves and muscles. The undershoot or overshoot, points 6 to 7 to 8 in Figures 12c and 12d respectively, is the result of the delayed readjustment of Y to its final resting state after the self-triggering at 6.

It also can happen, as shown in Figure 13, that the recovery variable shifts the characteristic of X so far that state 8 also becomes unstable and self-triggering to state 5 recommences. Then the system exhibits oscillatory behavior and no stable state exists where both the variables X and Y are simultaneously stationary.

In this way, oscillations arise in two-process systems by successive self-triggering and recovery as a result of an appropriate kinetic coupling of both processes. If there are more than two coupled processes effective in the system, then the general feature of the kinetics is not much different from that of the two-process systems. Obviously, in these cases also, instability and recovery may play a similar role, but either of the properties may be a common result of

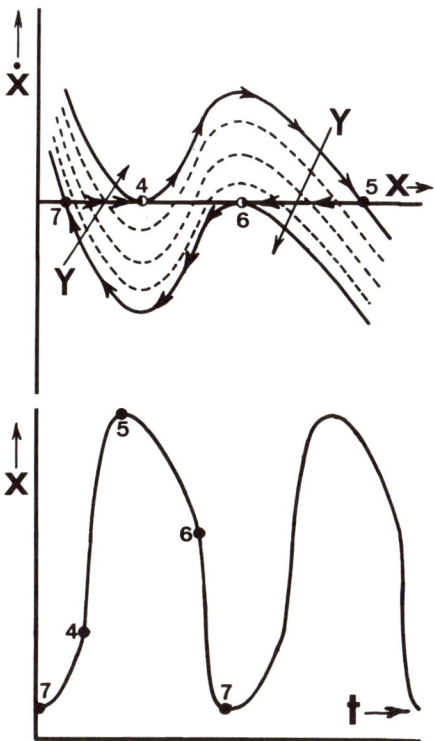

Figure 13. The dynamic diagram of an oscillating system.

several coupled processes.

As already mentioned, the quantitative treatment requires, in case of two-process systems a pair of simultaneous differential equations, one representing the trigger process and the other the recovery process:

$$\dot{X} = f(X, Y) \quad \text{trigger process (Y as parameter)}$$
$$\dot{Y} = g(X, Y) \quad \text{recovery process (X as parameter)}.$$

This set of differential equations describes the simultaneous variation of X and Y with respect to the common independent variable, t(13). Elimination of the time variable by division of the second equation by the first leads to:

$$\frac{\dot{Y}}{\dot{X}} = \frac{dY}{dX} = \frac{g(X, Y)}{f(X, Y)}$$

Using a Y/X diagram this equation describes, for any state given by a pair of instantaneous values of X and Y, the slope of the "path of state" (trajectory). All paths form a "portrait of state" (phase plane) according to the special form of the kinetic equations. (Examples of portraits of state for bistability, monostability and oscillatory behavior are given in Figures 17, 18 and 19).

Recovering triggerable systems exhibit the propagation of "spikes" which can be triggered repeatedly (7,10) as is well-known for action potential propagation in nerves and muscles.

III. Examples of physico-chemical oscillatory systems.

There are several non-living systems which show oscillations and all the essential features of biological excitation phenomena. All of them contain a trigger process and a recovery process, as explained above.

In the following, the phenomenological treatment is exemplified by three typical non-living systems containing triggerable membrane-like interfaces. These systems are "complete nerve models". They simulate excitability, all-or-nothing behavior, propagation, monostability (spike response), refractoriness, accommodation and oscillations (see Figure 15).

1. The "neurodyne" (3,14) is an example of an electronic solid-state nerve model consisting simply of a tunnel diode (TD) in parallel with a NTC-resistance (negative temperature coefficient). The tunnel diode represents, with its tunnelling p/n-junction zone, the triggerable element having a non-monotonic force-flux characteristic. The NTC-resistance represents the recovery process, because a triggered voltage transition causes a change in the voltage-dependent heat production of the NTC-resistance. Thus, as a result of its negative temperature coefficient, a corresponding shift of the dynamic characteristic takes place in just the same

way as explained in the preceding section. The driving forces here are the voltage, U, and the temperature of the NTC-resistance, T. The storage elements are, correspondingly, the electric capacity, C, and the heat capacity of the NTC-resistance.

Figure 14a shows the different characteristics for the I/U, U̇/U, and Ṫ/T diagrams while Figure 16 summarizes the "excitation" phenomenology of the neurodyne, which is similar to Figure 15 for the nerve.

The kinetic equations for both processes are well known, so that the time behavior of the neurodyne can easily be simulated by analog computation. In Figures 17, 18 and 19, typical portraits of state are shown, which were recorded directly from the analog computer (14).

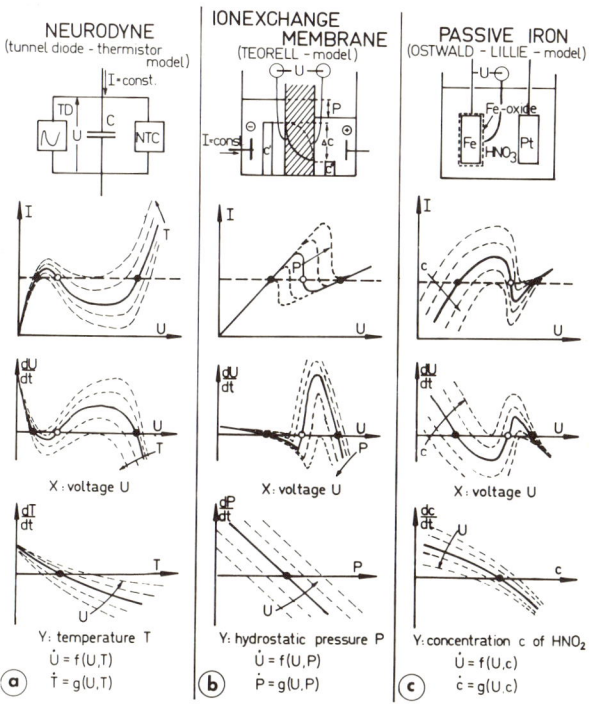

Figure 14. Three examples of recovering triggerable systems with membrane-like interfaces.

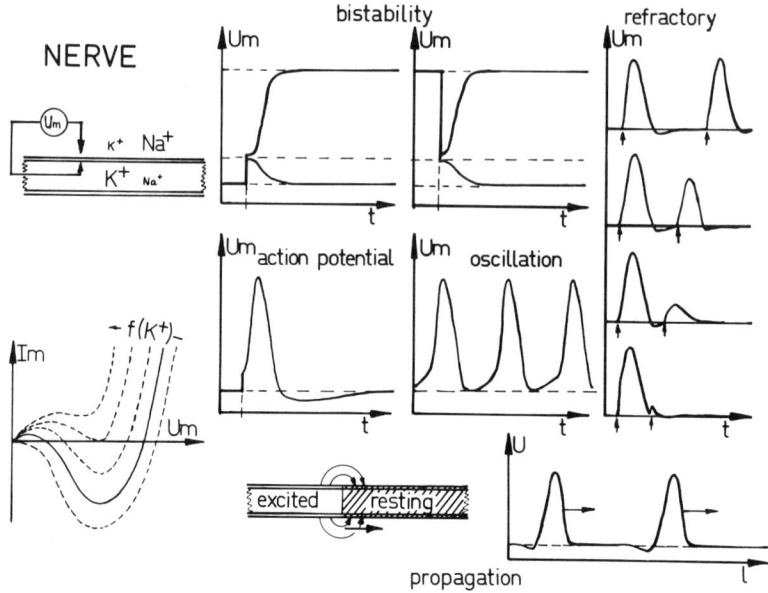

Figure 15. The excitation phenomena of the nerve.

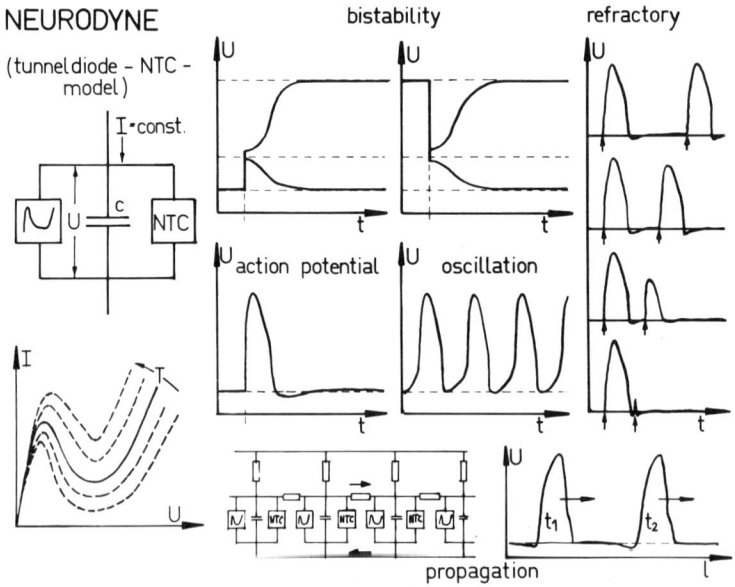

Figure 16. The excitation phenomena of the neurodyne.

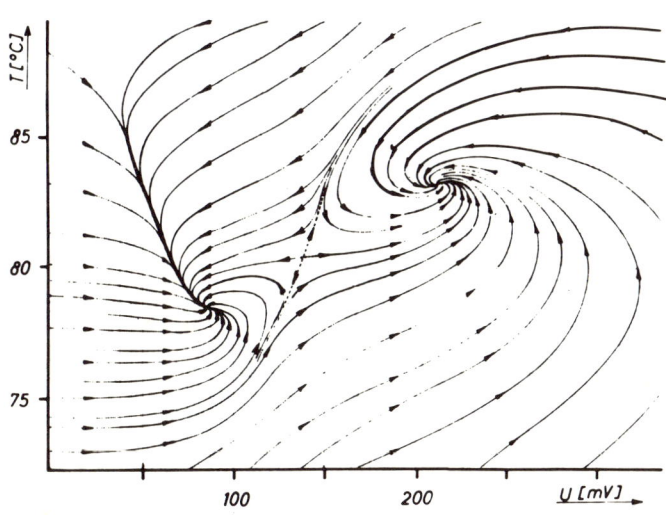

Figure 17. Analog computation of a bistable neurodyne (two stable nodes and one saddle-point).

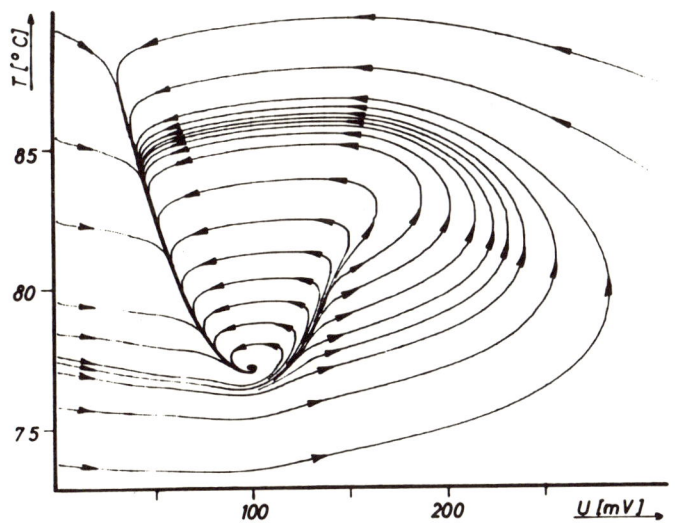

Figure 18. Analog computation of a monostable neurodyne (one stable focus, one separatrix).

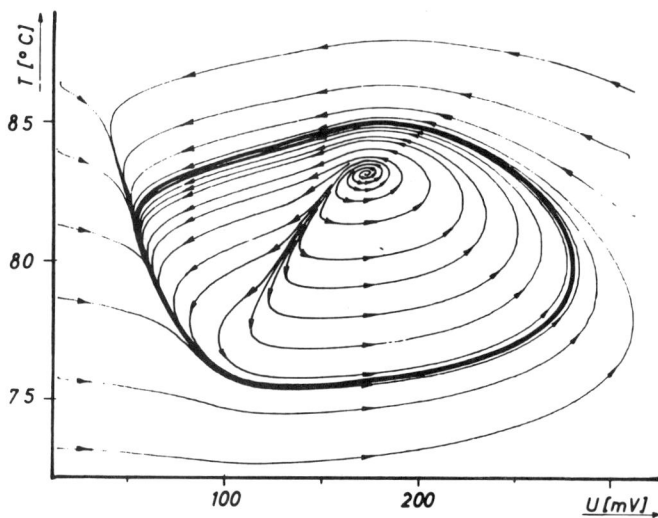

Figure 19. Analog computation of an oscillating neurodyne (one unstable focus, one stable limit cycle).

2. The <u>ion exchange membrane</u> system (Teorell's nerve model) consists of a triggerable porous ion exchange membrane which separates two chambers. These are filled with sodium chloride solutions of different concentrations and hydrostatic pressures and polarized by a constant electric current. Under these conditions the ion exchange membrane has a non-monotonic current-voltage characteristic (20) which is dependent on the hydrostatic pressure difference. Here the driving forces are the transmembrane voltage, U, and the hydrostatic pressure, P. The conjugated storage elements are the membrane itself (storing ions) and the chambers (storing volumes of solution). Figure 14b shows the characteristics and Figure 20 shows the excitation phenomenology, in the same manner as in Figure 15.

3. The "<u>Ostwald-Lillie iron wire</u>" in nitric acid is an example of an electrochemical nerve model which was proposed by Ostwald (21) and thoroughly studied

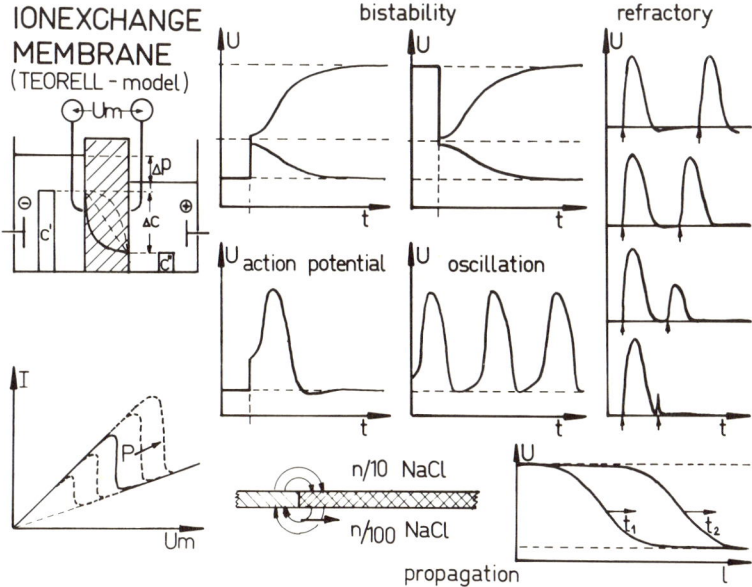

Figure 20. The excitation phenomena of the ion-exchange membrane system (Teorell model).

by Heathcote (22,23), Lillie (24,25) and Bonhoeffer (26), as well as others. There are many other electrochemical systems now known which behave quite similarly.

Oscillations are frequently observed in electrochemical systems (10,27) because non-monotonic current-voltage characteristics readily arise when electrolytic layers are precipitated on electrodes, these being dependent on the potential of the electrode (27,28). Metals which exhibit passivity, like iron, cobalt, nickel, chromium, gold, zinc, cadmium, etc., are particularly susceptible to oscillations and instability phenomena.

In the iron wire nerve model the driving forces are the electrode potential, U, and the concentration, C, of nitrous acid being simultaneously produced. The layer formation (passivation) acts as the recovery process. The kinetics of iron passivation and recovery by nitrous acid are analogous to the neurodyne and Teorell's nerve model

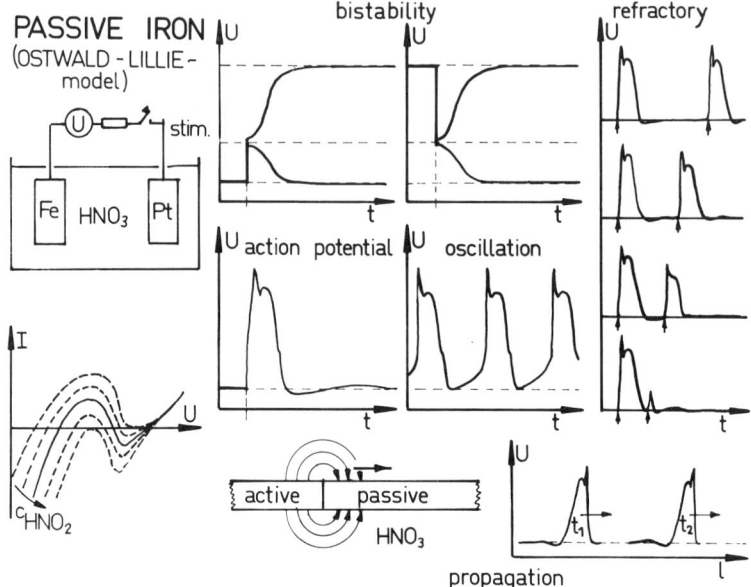

Figure 21. The excitation phenomena of an electrochemical system (Ostwald-Lillie model).

(see Figure 14c and Figure 21).

Experimental studies which examined the kinetics of the iron wire system in more detail (29) have recently shown that four independent variables are actually involved, – one electrical and three chemical variables (the concentrations of NO_2^-, NO_3^- and $Fe(NO)^{++}$), – but with respect to recovery the essential variable is the concentration of nitrous acid.

Concerning the number of variables it is worth noting that the independence of a kinetic variable arises from the existence of a conjugated storage element providing a particular differential equation $X_i = f_i(X_i,...)$. For instance, in systems containing diffusion processes and/or chemical reactions every localized volume of solvent which can take up a solute represents a storage element which introduces an additional chemical variable with its own time-dependence for each chemical component. For this reason a great number of variables are to be expected in such systems. Among the multitude of such variables kinetic analysis should, therefore, be primarily directed at the recognition of the variables which are actually essential for the appearance of oscillations.

REFERENCES

1. Franck, U.F. Zeitschrift f. physik. Chemie N.F. 3, 183 (1955).
2. Franck, U.F. Studium Generale 18, 313 (1965).
3. Franck, U.F. and Kettner, K. Ber. d. Deut. Bunsenges. 68, 875 (1964).
4. Bonhoeffer, K.F. Naturwiss. 31, 270 (1943).
5. Bonhoeffer, K.F. Journ. de Chim. Phys. 51, 521 (1948).
6. Higgins, J. This volume, p.127.
7. Bonhoeffer, K.F. and Renneberg, W. Z. f. Elektrochem. 51, 67 (1954).
8. Franck, U.F. Progress in Biophysics 6, 171 (1956).
9. Maillard, E. and Le Chatelier, H.L. Recherches espérimentales et théoriques sur la combustion des mélanges gazeux explosives. Paris, 1883.
10. Franck, U.F. Z. f. Elektrochem. 62, 649 (1958).
11. Franck, U.F. Z. f. Elektrochem. 55, 154 (1951).
12. Franck, U.F. Ber. d. Deut. Bunsenges. 71, 789 (1967).
13. Bonhoeffer, K.F. Z. f. Elektrochem. 51, 24 (1948).
14. Franck, U.F. and Lammel, E. To be published.
15. Teorell, T. Exp. Cell Res. Suppl. 3, 339 (1955).
16. Teorell, T. Faraday Soc. Discussions 21, 9 (1956).
17. Teorell, T. Acta Soc. med. Upsaliensis 62, 60 (1957).
18. Teorell, T. Exp. Cell Res. Suppl. 5, 83 (1958).
19. Teorell, T. Arkiv Kemie 18, 401 (1961).
20. Franck, U.F. Ber. d. Deut. Bunsenges. 67, 657 (1963).
21. Ostwald, Wi. Z. physik. Chemie 6, 17 (1890).
22. Heathcote, H.L. Z. physik. Chemie 37, 368 (1901).

23. Heathcote, H.L. J. Soc. Chem. Ind. 26, 899 (1907).
24. Lillie, R.S. J. Gen. Physiol. 3, 107 (1920).
25. Lillie, R.S. Biological Review 11, 181 (1936).
26. Bonhoeffer, K.F. Z. f. Elektrochem. 47, 147, 441, 536 (1941).
27. Franck, U.F. Werkstoffe und Korrosion 11, 401 (1941).
28. Franck, U.F. and FitzHugh, R. Z. f. Elektrochem. 65, 156 (1961).
29. Franck, U.F. and Hoppé, K. To be published.

TWO-DIMENSIONAL ANALYSIS OF CHEMICAL OSCILLATORS

Dieter H. Meyer

Johnson Foundation, University of Pennsylvania
Philadelphia, Pennsylvania 19104

Introduction

Common to all biological systems is a complexity in structure and function which exceeds that of most physical and engineering systems by several orders of magnitude (1). As a consequence, the mathematical tools presently available for the study of the dynamics of such systems are far from adequate. On occasion, however, as for example, when dealing with sub-systems on a purely chemical level, mathematical description and analysis become feasible. The dynamical behavior of homogeneous chemical systems in which all constituents are in one phase is completely determined by a generalized form of the law of mass action (2,3). Such systems are therefore describable by a set of first order ordinary differential equations whose right hand sides are polynominals with the system constituents as variables. The total number of equations is equal to the number of constituents involved in the elementary reactions.

The main problem in the analytical study of these equations arises from the fact that, in all but a few cases, no explicit solution other than numerically by digital computer is obtainable. However, this difficulty may be somewhat overcome through techniques due to the researches of Poincaré (4), Liapunov (5) and Bendixon (6) which allow a determination of the qualitative behavior of the solutions without resorting to explicit analytical expressions.

These methods are particularly applicable to two-dimensional systems (two independent variables). In many cases this restriction is not at all critical, since methods are available which allow the reduction of larger systems to

two variables. One such technique which also provides the
mathematical basis for the familiar steady state assumption
is taken from the theory of singular perturbations (3,7,8,9)
of ordinary differential equations and has been successfully
applied to biochemical systems by Prof. Sel'kov (10) and in
our laboratory (3).

The basic feature of this approximation is a separation of motions into two time scales, a fast one and a slow
one. Following a disturbancy, all constituents belonging
to the first group move rapidly to a quasi-stationary state
at which they remain during the slower phase of the motion.
Geometrically this process can be depicted as the motion of
a representative point in a space whose ordinates correspond
to the concentrations of the system-constituents. The motion
of this point itself may be thought of as proceeding along
certain trajectories. Regular stationary states appear as
points, while such representing oscillatory solutions appear
as closed trajectories. For systems involving two variables
only, all these motions are planar and may be depicted on
what is called phase plane.

The investigation of such systems with regard to the
possible existence of oscillatory stationary states is based
on a determination of the conditions under which the topology on the phase plane of the particular system includes a
closed trajectory or limit cycle, and the subsequent location of such. To that purpose the affine coordinates of the
phase plane are transformed to the homogeneous coordinates
of Poincaré's sphere (3,4). This mapping brings about a
closure of the phase plane, so that points which were previously at infinity now have finite coordinates. To determine
the existence of a limit cycle it is then only necessary to
analyze all the stationary states corresponding to the singular points of the differential equations and to determine
the approximate topology of the trajectories remote from
these singular points. Existence and location of a limit
cycle are then found from the fact that none of the trajectories can remain unterminated, but must start and end at
stationary states.

Results

One of the problems which may be studied with the help of these methods is that of generating generalized forms of Lotka's first mechanism (11) capable of sustained oscillations. The individual reactions and associated equations of Lotka's mechanism are shown in Figure 1. This system has one stationary state which is stable for all possible values of rate constants and input flux.

$$[c_0] \xrightarrow{k_1} c_1$$

$$c_1 + c_2 \xrightarrow{k_2} c_2 + c_2$$

$$c_2 \xrightarrow{k_3} \text{Product}$$

$$\dot{c}_1 = k_1 c_0 - k_2 c_1 c_2$$

$$\dot{c}_2 = k_2 c_1 c_2 - k_3 c_2$$

$$\dot{x} = \delta - xy$$

$$\dot{y} = xy - \beta y$$

Figure 1. Reaction mechanism and differential equations of Lotka's first mechanism.

Figure 2 depicts the global phase portrait of the system for the case that the stationary state is a focus. (It should be noted that for chemical systems only the first quadrant is of importance since concentrations are always positive).

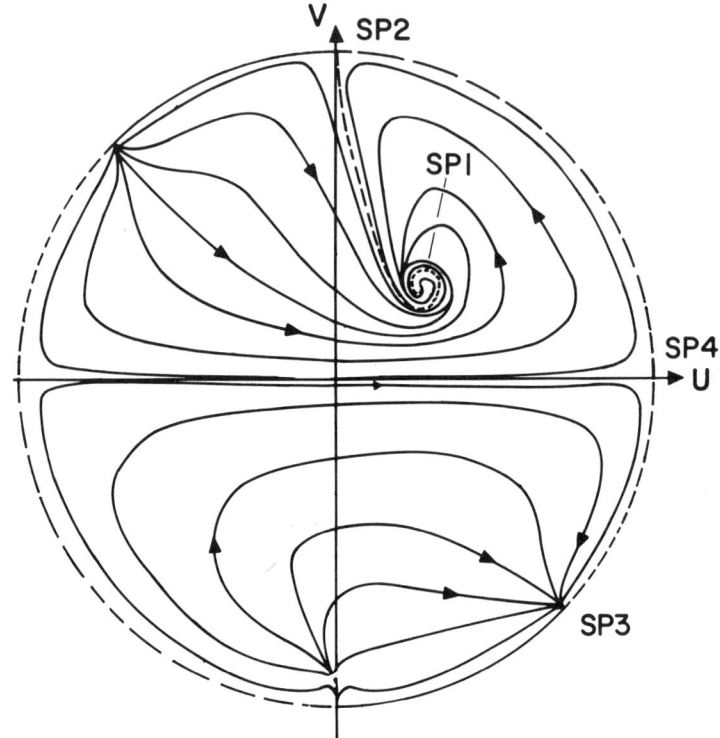

Figure 2. Global phase portrait for Lotka's first mechanism.

One way of modification of this mechanism is clearly that of altering the rate laws of the individual reaction steps. This is demonstrated in Figure 3 which depicts a reaction scheme where the product reaction which was previously of first order is now enzymatic and C_1 is also shunted into a by-product. The number of stationary points is now increased to two and there are values of parameters for which one (SP1) has the form of an unstable node or focus. Analysis of the stationary states at infinity shows furthermore that whenever SP1 is unstable there exists a stable limit cycle. The global phase portrait for this case is given in Figure 4. An interesting simplification occurs when the rate of the shunt reaction is diminished. With decreasing flux SP2 moves along the x-axis towards infinity which it reaches in the limit.

BIOCHEMICAL OSCILLATORS

$$[C_0] \xrightarrow{k_1} C_1$$

$$C_1 \xrightarrow{k_2} \text{By - Products}$$

$$C_1 + C_2 \xrightarrow{k_3} C_2 + C_2$$

$$C_2 + E \xrightarrow{k_4} C_2E$$

$$C_2E \xrightarrow{k_5} E + \text{Product}$$

$$\dot{C}_1 = k_1 C_0 - k_2 C_1 - k_3 C_1 C_2$$

$$\dot{C}_2 = k_3 C_1 C_2 - \frac{C_2 V_m}{C_2 + K_m}$$

$$\dot{x} = (y + b)(\delta - \beta x - xy)$$

$$\dot{y} = y(xy + bx - a)$$

Figure 3. Reaction mechanism and differential equations of Lotka's first mechanism modified by a shunt reaction and an enzymic output step.

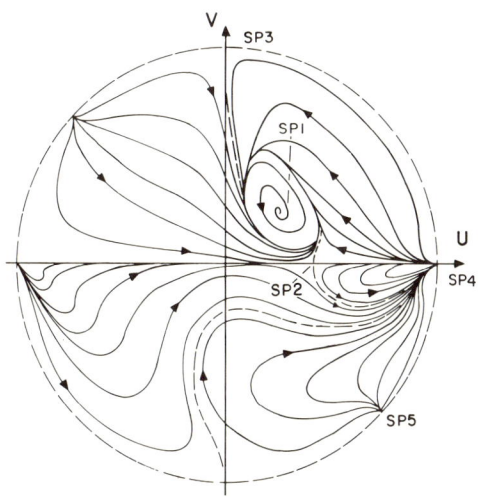

Figure 4. Global phase portrait for the system shown in Figure 3.

DIETER H. MEYER

A second way in which the original system may be modified consists of an alteration of the feedback reaction, in particular an increase of its order. The detailed reaction mechanism is depicted in Figure 5. Here we find three stationary states and it is again possible to obtain conditions under which one of these points is an unstable node or focus. We obtain the global phase portrait given in Figure 6. It should be noted that now, for the first time, we encounter the presence of two stable stationary states, since in the case that SP3 is stable, both it and SP1 are attainable by the system. When SP3 is unstable, a limit cycle exists and we have one oscillatory and one non-oscillatory stationary state.

$$[\,C_0\,] \xrightarrow{k_1} C_1$$

$$C_1 \xrightarrow{k_2} \text{By - Products}$$

$$p\,C_1 + q\,C_2 \xrightarrow{k_3} (p+q)\,C_2$$

$$C_2 \xrightarrow{k_4} \text{Product}$$

$$\dot{C}_1 = k_1 C_0 - k_2 C_1 - k_3 C_1^p C_2^q$$

$$\dot{C}_2 = k_3 C_1^p C_2^q - k_4 C_2$$

$$\dot{x} = \delta - a\,x - x^p y^q$$

$$\dot{y} = x^p y^q - b\,y$$

Figure 5. Reaction mechanism and differential equations for Lotka's first mechanism modified by a shunt reaction and higher order feedback.

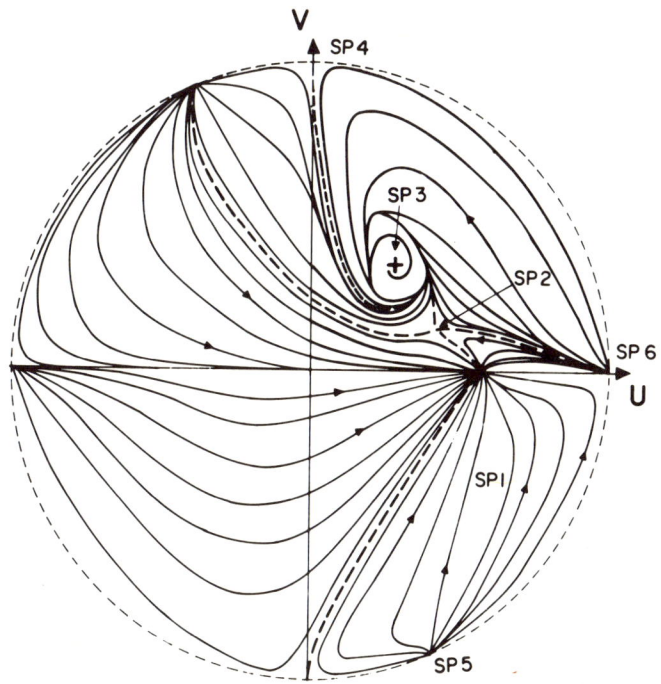

Figure 6. Global phase portrait for the system shown in Figure 5.

Similar to the previous system, elimination of the shunt-reaction moves SP1 to the infinite end of the positive x-axis where it merges with SP2 and SP6. This system was discussed by Sel'kov in a recent publication (10).

A third type of modification was suggested by Lotka himself and is known as his second mechanism (12). The reaction scheme is shown in Figure 7. There are two stationary states, a saddlepoint at the origin and a center within the first quadrant. The global phase portrait is as shown in Figure 8.

$$[C_0] + C_1 \xrightarrow{k_1} C_1 + C_2$$

$$C_1 + C_2 \xrightarrow{k_2} \text{By - Products}$$

$$C_1 + C_2 \xrightarrow{k_3} C_2 + C_2$$

$$C_2 \xrightarrow{k_4} \text{Product}$$

$$\dot{C}_1 = k_1 C_0 C_1 - (k_2 + k_3) C_1 C_2$$
$$\dot{C}_2 = (k_2 + k_3) C_1 C_2 - k_4 C_2$$

$$\dot{x} = x(a - y)$$
$$\dot{y} = y(x - b)$$

Figure 7. Reaction mechanism and differential equation for Lotka's second mechanism.

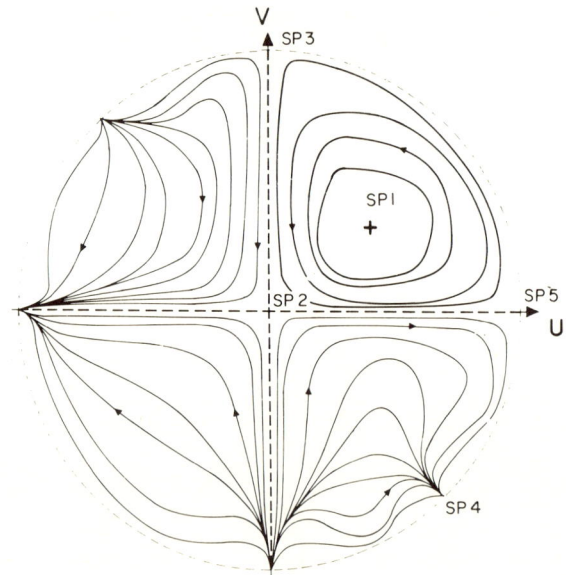

Figure 8. Global phase portrait for Lotka's second mechanism.

Discussion

An excellent basis for a comparative evaluation of the systems just introduced is provided by the input flux, which in each case was assumed to be independent.

In Lotka's first mechanism, both position and character of the single stationary state depend on the amount of flux into the system. We find, in particular, that for small input fluxes the singular point which represents this state in the phase portrait is a stable focus with the associated damped oscillations. For larger values, there are no rotations and the point is a stable node. The character of the singular points at infinity is not affected by variations of input flux.

The situation is slightly more complicated for the system with enzymatic output. As opposed to the previous system, the singular point SP1 is now located inside the first quadrant only for particular values of rate constants and input flux. The requirement that SP1 must have a positive index imposes further restrictions on the values of these parameters. In this case, but depending on the values of V_m and K_m of the output reaction and the rate constant of the feedback step, the singular point is either stable for all permissible values of input flux or stable for low, unstable for intermediate, and stable for high flux values. Elimination of the shunt-reaction step leads to either complete stability or instability for low and stability for larger values of input flux, depending on the other parameters.

A similar situation is found in the third type of mechanism, where the amount of feedback has been increased. Again there are conditions which must be satisfied for the phase portrait to exist in the form given above. In the system with the shunt it is in particular necessary that the input flux exceeds a certain minimum value. For values below that, both SP2 and SP3 disappear while the x-ordinate of SP1 increases with decreasing input. Over the permissible flux range SP3 is unstable for small and stable for large values. Without the shunt-reaction there is one attainable stationary state which exists up to a certain maximum flux. It is unstable for small and stable for large values.

The last system discussed, Lotka's second mechanism, shows a markedly different behavior. As opposed to the previous systems which had only one closed trajectory, the limit cycle, now the entire first quadrant is filled with such. Typical for this type of oscillation is the fact that both frequency and amplitude are intimately related to the initial conditions. Furthermore, following a perturbation, there is no transient time during which the system reaches its stationary state as is the case by limit cycle oscillations. An interesting extension of this mechanism is one in which the product step is made enzymatic. The stationary state then takes the form of an unstable focus and the trajectories spiral outward to infinity which itself consists of separatrixes.

REFERENCES

1. Zadeh, L.A., Proc. I R E , 50, 5, 856 (1962).
2. Guldberg, C.M. and Waage, P., J. Prakt. Chem., 127, 69 (1879).
3. Mayer, D.H., Ph.D. Dissertation, University of Pennsylvania, Philadelphia (1968).
4. Poincaré, H., Oeuvres, Vol. 1, Gauthier-Villars, Paris (1928).
5. Liapunov, M.A., Comm. de la Société Math. de Kharkow (1893).
6. Bendixon, I., Acta Math., 24, 1 (1901).
7. Tikhonov, A.N., Mat. Sb. 22, 64, 193 (1948).
8. Tikhonov, A.N., Mat. Sb. 31, 73, 575 (1952).
9. Vasileva, A.B., Russ. Math. Surveys, 18, 13 (1963).
10. Sel'kov, E.E., European J. Biochem., 4, 79 (1968).
11. Lotka, A., Z. Phys. Chem., 72, 508 (1910).
12. Lotka, A., J. Am. Chem. Soc., 42, 1595 (1920).

STABILITY PROPERTIES OF METABOLIC PATHWAYS WITH FEEDBACK INTERACTIONS

Gustavo Viniegra-Gonzalez

Cardiovascular Research Institute
University of California
San Francisco Medical Center

In the search for a basis of biochemical oscillations, various chemical schemes have already been considered. For example, Spangler and Snell (1961), Higgins (1967) and Pye and Chance (1966) have thought of actually occurring <u>special</u> reactions which might be ultimately describable by Volterra's equations and which therefore may very well lead to oscillations. A somewhat different approach has consisted in very <u>general</u> biochemical patterns which could lead to the existence of sustained oscillations for specific choices of parameters. It has seemed to Goodwin (1963) that a scheme of "genetic control" is such a possible general pattern. But to Sel'kov (1967) and to Morales and McKay (1967) the pattern of "metabolic control," such as originally described by Yates and Pardee (1956) seemed to suffice as a general scheme where oscillations of the self-excited type would occur.

Neither of the previous approaches is exclusive, since eventually every kind of those schemes can exhibit limit cycles if the parameters of a system are conveniently chosen. The first class of special schemes would require the existence of at least two couplings or feedback interactions, which in the biochemical sense would involve the existence of enzymes that might be influenced by the products of the system generated at different steps of the reaction scheme. However, this class of systems has the advantage of being related to the sort of events which are known to occur in the oscillatory reactions of the glycolytic pathway, observed both in intact and broken yeast cells (Pye, 1969).

On the other hand, the class of controlled reactions

with single metabolic or genetic feedback interactions are perhaps of much wider occurrence, since they only require one enzyme or regulatory protein to be sensitive to the end product of a chain of reactions. Furthermore, the existence of damped oscillations of the galactosidase reaction on constitutive strains of E. coli B has been reported by Knorre (1968). This observation shows that oscillatory phenomena can occur in actual metabolic feedback systems. It is the purpose to show here that a simplified scheme of interactions with a single metabolic feedback loop not only can generate sustained oscillations as has been conclusively demonstrated both by Sel'kov (1967) and Walter (1969), but also to show the constraints on the parameters and/or the input level if there is to be a stable stationary point or a stable limit cycle. These constraints may be of interest in the biochemical evolution of a given regulatory system. If a certain stability property has some selective advantages, some of the constraints proposed below suggest situations in which genetical selection may be closely related to the existence of either a steady-state or a limit cycle, so that selection could be achieved by the choice of the proper environmental conditions, e.g., arranging certain prescribed waveforms of the input substrates such as a step or a sinusoidal change of their concentrations in time. Such arrangements would impose advantageous or disadvantageous states as regards the flux of the substrates through a given metabolic pathway, e.g., stable constant fluxes, accumulation of products, or forced or self-excited concentration oscillations.

It is up to the experimentalists to show in which cases, if any, the stability constraints on the parameters of a given regulatory system play a significant role in their natural selection. However, the very existence of inheritable properties in many metabolic pathways, such as the stoichiometry of the enzymes, or the allostericity of the regulatory enzymes, seems to indicate that the choice of such parameters is not necessarily random, and, for the same reason the choice may restrict the possible stability features of a particular biochemical system.

Formulation of the System

A metabolic pathway is usually defined in the biochemical literature as an ordered sequence of chemical reactions catalyzed by the enzymes [E_0, E_1, ..., E_n] (see equation 1), in which the product S_i of the (i-1)th reaction is a substrate in the immediately subsequent ith reaction. Furthermore, in many instances the product, S_n, of this sequence modifies (with a certain degree of cooperativity) the velocity of the zeroth enzyme, E_0, mostly by decreasing it. This is called a negative feedback interaction.

$$\text{SOURCE} \rightarrow S_0 \xrightarrow{E_0} S_1 \xrightarrow{E_1} S_2 \xrightarrow{E_2} \cdots \rightarrow S_n \xrightarrow{E_{n-1}} S_{n+1} \rightarrow \text{SINK} \quad (1)$$

with velocities $v_0[1-f(x_n)], v_1, v_2, \ldots, v_{n-1}, v_n$

where: $x_i = [S_i] > 0$ $\qquad d[E_i]/dt = dv_0/dt = 0$

The unidirectional reaction rate catalyzed by the ith enzyme, called v_i here, is a non-linear function of the concentration of the corresponding substrate, x_i. As a first approximation this function is described by the so-called Henri or, also, Michaelis-Menten equation (see equation 2). If the cofactors and the enzyme concentrations are kept nearly constant this equation will have two fixed measurable parameters: the <u>maximum velocity</u> V_i which is proportional to the enzyme concentration [E_i] and the <u>Michaelis constant</u> K_i which is inversely related (as a limit) to the degree of affinity of the enzyme E_i by its own substrate S_i, i.e., $v_i(K_i) = V_i/2$.

$$v_i = V_i x_i / (K_i + x_i) \qquad V_i \text{ proportional to } [E_i] \quad (2)$$

The regulatory action of S_n on the zeroth enzyme E_0 might be of the noncompetitive and cooperative type, because of a given number of sites on E_0 to which molecules of S_n can bind independently from the binding of the molecules of S_0. The bound small molecules S_n could, in turn, modify the further binding both of molecules of S_0 (in <u>negative</u> feedback cases, decreasing) and of molecules of S_n. It is customary to describe such kind of interactions by the Hill equation, here represented by $f(x_n)$ (see equation 3), in which the power p is called the <u>Hill number</u> and gives a quantity related to the number of <u>binding</u> sites for S_n. The constant α is directly related to the binding affinity of S_n for E_0, since $\alpha^{-1/p}$ would appear as a dissociation constant.

With these equations and noting that the only trans-

such that:
$$f(x_n) = ax_n^p / (1 + ax_n^p) \qquad p \geq 1 \qquad a > 0$$
$$f(0) = 0 \qquad 0 \leq f'(x_n) = pax_n^{p-1}/(1+ax_n^p)^2 \leq \frac{p^2-1}{4p}\left[\frac{(p-1)}{a(p+1)}\right]^{1/p} \quad (3)$$
$$f(\infty) = 1$$
$$f(a^{-1/p}) = 1/2 \qquad \text{If } p > 1$$

formation of each S_i is through the ith reaction, the following set of non-linear differential equations can be written as in (4).

$$\dot{x}_1 = v_0[1 - f(x_n)] - v_1(x_1) = F_1(x_1, x_n)$$
$$\dot{x}_2 = v_1(x_1) - v_2(x_2) = F_2(x_1, x_2) \qquad (4)$$
$$\cdots\cdots\cdots\cdots\cdots\cdots\cdots\cdots\cdots\cdots\cdots$$
$$\dot{x}_n = v_{n-1}(x_{n-1}) - v_n(x_n) = F_n(x_{n-1}, x_n)$$

The following kind of stability conditions for the system of equations in (4) will be considered:

1) Conditions for the <u>existence of a stationary</u> state, wherein every instantaneous rate of change of the concentrations, $\dot{x}_i$, vanishes for a given input v_0(const).

2) Conditions for stability or instability of those stationary states for any arbitrary input level v_0(fixed) when perturbed by small deviations Δx_i (<u>local stability</u>).

3) Conditions for asymptotic stability of the stationary solution when the forward pathway is quasi-linear ($x_i \ll K_i$) and the deviations Δx_i have a finite magnitude (<u>global stability</u>).

4) Some particular conditions in regards to the <u>existence of stable limit cycles</u> and the order of magnitude of the frequencies of these oscillations when the stationary solution lies in the quasi-linear region of the forward path and near the inflection point of the feedback interaction $f(x_n)$.

Conditions for the Existence of the Stationary Solution

From the structure of the equations derived at the stationary condition $[\dot{x}_i = F_i(x_j^*) = 0]$ a finite set of positive roots can be obtained if, and only if, $V_i > v_e$, where $v_e \equiv V_i x_i^*/(x_i^*+K_i) = v_0/(1 + \alpha x_n^* p)$, $i = 1,2,\ldots,n$.

From this necessary and sufficient condition, two simpler sufficient conditions follow:

I) $v_0 < V_{min}$; V_{min} = minimum of $[V_1, V_2, \ldots, V_n]$

or

II) $V_i \geq V_n$; $i = 1, 2, \ldots, n-1$; for the existence of the stationary state.

The first condition is a constraint on the input level v_0. In biological systems this condition could be met by means of selective permeability barriers such as membranes. The second condition requires some sort of programmed synthesis and/or destruction of each enzyme in such a way that the right stoichiometry of all the intermediate enzymes is obtained because each V_i is proportional to the enzyme concentration $[E_i]$. Failure to meet II results in wasteful accumulation of substrates (Fig.1). Thus survival may select input levels and/or enzyme concentrations which do not violate the necessary and sufficient condition for the stationary solution (indicated in the previous paragraph).

Local Stability of the Stationary Solution

If the existence of the stationary solution (x_i^*) is insured by any one of the previous conditions, then its stability can be predicted [for small fluctuations $\Delta x_i^* = (x - x_i^*)$] if each non-linear function, $F_i(x_j)$, representing the instantaneous rate of change of each concentration, $\Delta \dot{x}_i$, is expanded in Taylor's series around the stationary "point" (x_i^*) and only first order terms are considered (see equation 5). Thus, there is obtained an equivalent set of linear differential equations (6) whose characteristic equation has the following form (7).

Viniegra, Martinez and Morales (1969) have been able to prove the following properties of such equations (8 and 9), whose formal derivations will be published elsewhere. Here the conjecture (10) is stated. This conjecture has been verified for an arbitrary set of coefficients $[a_i]$ when $n = 2, 3, 4$, and ∞, and for any n but with all the coefficients

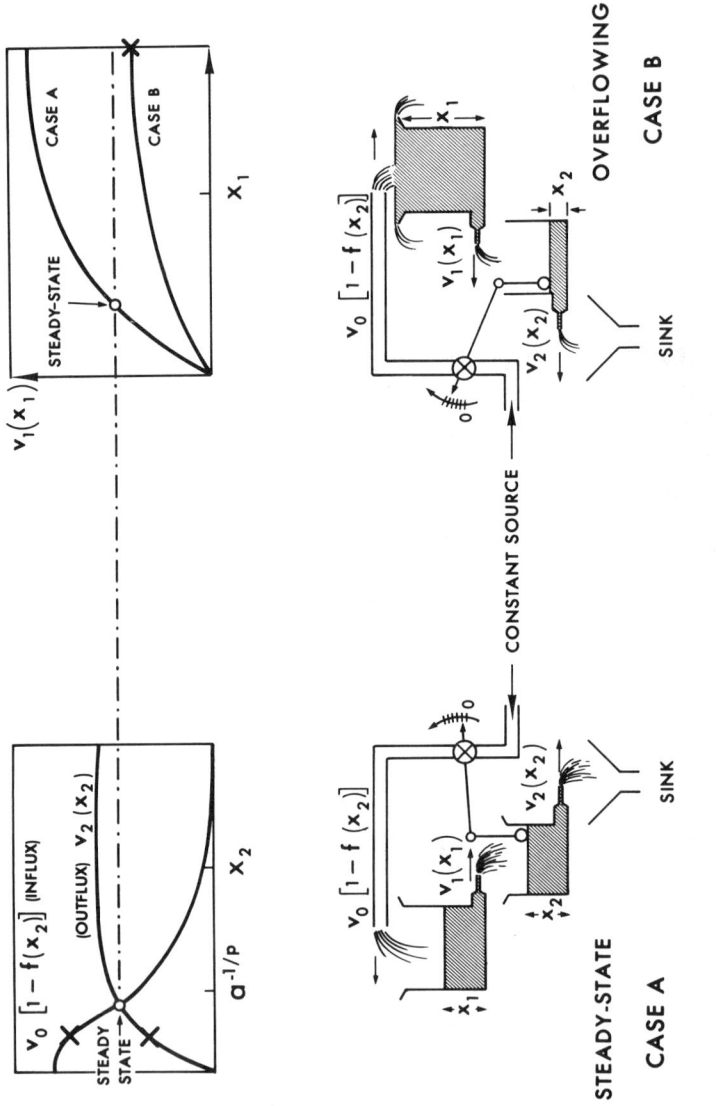

Figure 1.

BIOCHEMICAL OSCILLATORS

$$\Delta x_i \cong F_i(x_j^*) + \sum_{k=1}^{n} \left.\frac{\partial F_i}{\partial x_k}\right|_{x_i^*} \Delta x_k + \ldots \quad (5)$$

where:

$\Delta x_i = x_i(t) - x_i^*$ (with small magnitude)

$F_i(x_j^*) \equiv 0$ (from the definition of stationary point)

$$\frac{d}{dt}\begin{bmatrix} \Delta x_1 \\ \Delta x_2 \\ \ldots \\ \Delta x_n \end{bmatrix} = \begin{bmatrix} -a_1 & 0 & \ldots & 0 & -M \\ a_1 & -a_2 & \ldots & 0 & 0 \\ \ldots & \ldots & \ldots & \ldots & \ldots \\ 0 & 0 & \ldots & a_{n-1} & -a_n \end{bmatrix} \begin{bmatrix} \Delta x_1 \\ \Delta x_2 \\ \ldots \\ \Delta x_n \end{bmatrix} \quad (6)$$

$$a_i = \left.\frac{\partial v_i(x_i)}{\partial x_i}\right|_{x_i^*} = V_i K_i/(K_i + x_i^*)^2$$

$$M = v_0 \left.\frac{\partial f(x_n)}{\partial x_n}\right|_{x_n^*} = \frac{V_n}{(K_n + x_n^*)} \frac{p\, \alpha x_n^{*p}}{(1 + \alpha x_n^{*p})} \;;\; \frac{M}{a_n} = p \frac{\alpha x_n^{*p}}{1 + \alpha x_n^{*p}} \frac{(K_n + x_n^*)}{K_n}$$

$$\prod_{i=1}^{n}(\lambda + a_i) + \frac{M}{a_n}\prod_{i=1}^{n} a_i = 0 \quad (7)$$

a_i equal to each other. A prediction coming from that conjecture (Viniegra and Martinez, 1969) has been supported by a very extensive simulation carried out by Dr. Charles Walter (1970).

(8a)   1) $\dfrac{M}{a_n} < \dfrac{(a_{min})^n}{\prod_{i=1}^{n} } \sec^n(\Pi/n)$

or

(8b)   2) $\dfrac{M}{a_n} < 1$

$\Bigg\}$ imply $R_e(\lambda) < 0$ SUFFICIENT CONDITIONS FOR STABILITY

(9)   3) $\dfrac{M}{a_n} > \dfrac{(a_{aver})^n}{\prod_{i=1}^{n} a_i} \sec^n (\Pi/n)$ implies $R_e(\lambda) \geq 0$ SUFFICIENT CONDITION FOR <u>INSTABILITY</u>

(10)   5) $\dfrac{M}{a_n} < \sec^n(\Pi/n)$ would imply that $R_e(\lambda) < 0$ CONJECTURE

From these results we can conclude that:

1) For a given degree of saturation of the last enzyme (given by the ratio x_n^*/K_n) it is possible to formulate a constraint in terms of "p" and "n" which will insure the local stability of the stationary state. (See Fig.2). Thus, according to the conjecture (10), if $x_n^*/K_n \ll 1$ the constraint $p < \sec^n(\pi/n)$, $n = 3,4,...$ will insure the local stability of the stationary solution, and, according to Walter's simulation, will make impossible the existence of sustained oscillations (see Table I and Fig. 2).

2) A necessary condition for the existence of oscillations in this simple system is that the stationary solution should exist and should be locally instable. Thus, all sufficient conditions for local stability should be violated. The form of the <u>sufficient</u> condition for local instability (9) suggests that making the coefficients a_i unequal makes it more difficult to observe instabilities such as sustained oscillations. This is clearly shown when n = 3 (Fig. 3), since it is obvious that the minimum number which M/a_3 must exceed for instability to occur is obtained when the a_i are equal. The magnitude of this number corresponds to the secant function $[\sec^3(\pi/3)]$ predicted by the conjecture (10). Another

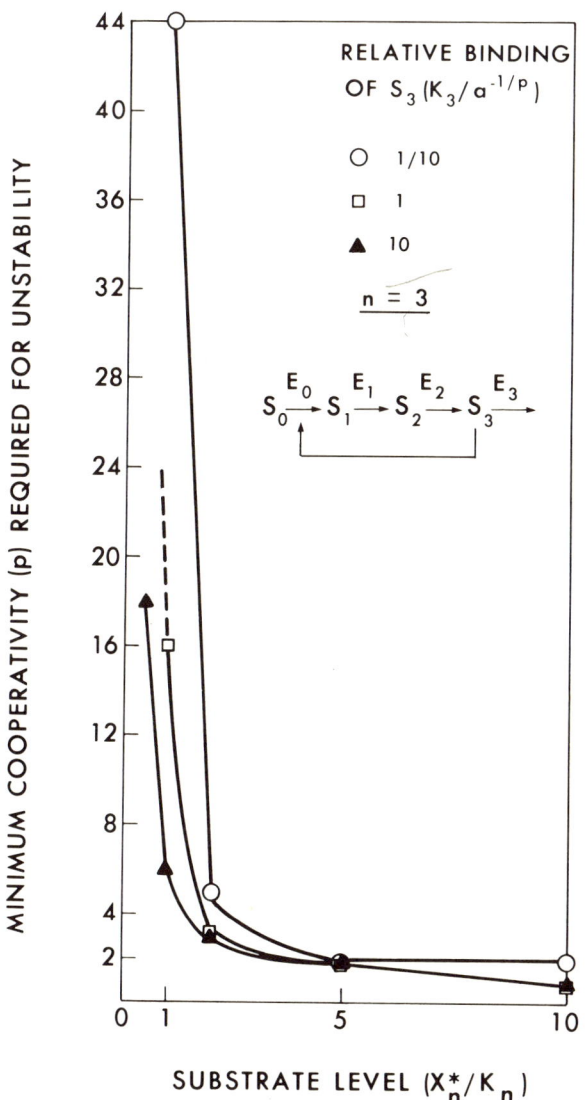

Figure 2.

Table I

COOPERATIVITY (p) REQUIRED FOR OSCILLATIONS ($x_n^* \ll K_n$)

n	$\sec^n (\Pi/n)$	smallest integer $p^+ > \sec^n (\Pi/n)$
2	∞	—
3	8	9
4	4	5
5	2.8	3
6	2.4	3
7	2.1	3
8	1.9	2
9	1.7	2
10	1.65	2

illustrative case can be obtained by plotting the locii of all the possible roots in an equation of the form indicated above (7), in which all the coefficients remain constant and the feedback interaction parameter M/a_n is varied. Such a plot (Fig.4) was obtained with the help of a device called a Spirule (The Spirule Co., 9728 El Venado, Whittier, Calif.). The interrupted lines correspond both to the asymptotes of each locii or to the locii themselves if the coefficients a_i's have the same positive value. For this particular case of equal coefficients, all the possible roots are complex numbers implying the existence of periodic solutions, damped or not. However, if the coefficients are unequal, for very small values of the parameter M, there exists the possibility of finding roots which lie in the negative axis. Such roots would correspond to purely damped responses. It is clear that by separating the distance of the intervals between the coefficients, the region in which purely damped responses are obtained is also increased. Thus, by introducing some inequalities between the values of the ratios (V_i/K_i), which correspond to the initial velocities of the intermediary enzymes, some stabilization is obtained. This effect may

BIOCHEMICAL OSCILLATORS

EXAMPLE (n = 3)

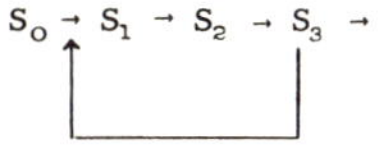

$S_0 \to S_1 \to S_2 \to S_3 \to$

(with feedback from S_3 to S_0)

Necessary and sufficient condition for <u>un</u>stability

$M/a_3 \geq 2 + [a_1/a_2 + a_2/a_1] + [a_2/a_3 + a_3/a_2] + [a_3/a_1 + a_3/a_1] \geq \sec^3(\Pi/3) = 8$

Since $a_i = V_i K_i / (K_i + x_i^*)^2$ and $V_i x_i^* / (K_i + x_i^*) = v_e = \dfrac{v_0}{1 + a x_3^* \bar{p}}$

$a_i / a_j = \dfrac{V_j K_j (V_i - v_e)^2}{V_i K_i (V_j - v_e)^2}$; $\dfrac{\partial v_e}{\partial x_n^*} < 0$; $V_i, V_j > v_e$

also:

$\dfrac{M}{a_3} = p \, \dfrac{a x_3^* \bar{p}}{1 + a x_3^* \bar{p}} \, \dfrac{K_3 + x_n^*}{K_3}$

Thus by making a_i's <u>more unequal</u>, <u>stability</u> <u>increases</u>.

Figure 3.

play some role in the natural selection of these parameters during the biochemical evolution of metabolic pathways. In particular, if for some reason the existence of limit cycles has a selective advantage it might be advantageous not to have very different orders of magnitude of those ratios. But if the existence of oscillations is immaterial or the stability of the stationary solution is more advantageous, the random variation of these parameters will provide some stabilization of the system. This might be of special interest in systems in which the number of intermediary reactions is rather large, e.g., the histidine biosynthetic pathway in Salmonella involves n = 10, and the stoichiometry of the enzymes is believed to be 1 : 1 : ••• : 1 and conserved in that way under a variety of physiological conditions.

ROOT LOCII OF A CHARACTERISTIC EQUATION (n = 5)

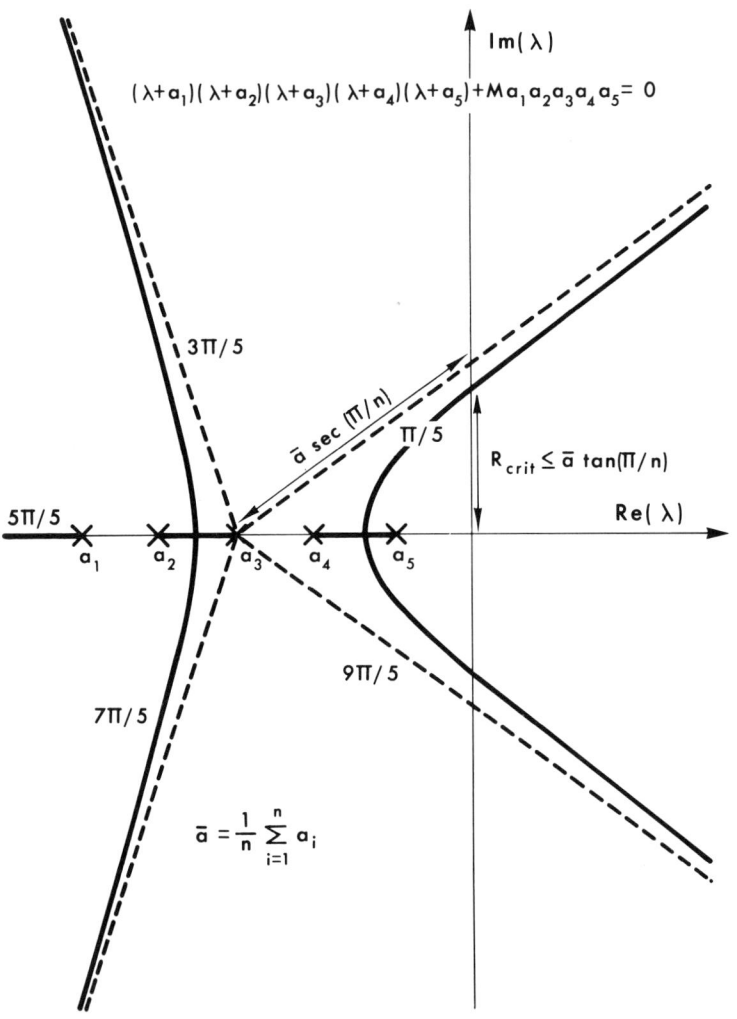

Figure 4.

BIOCHEMICAL OSCILLATORS

Global Stability of the Stationary Solution when the Forward Pathway is Quasi-Linear. $(x_n^* \ll K_n, x_{n-1}^* \ll K_{n-1}, \ldots, x_1^* \ll K_1)$.

If the source input level, v_o, is very low ($v_o \ll V_n$) or if its magnitude is moderate ($v_o \lesssim V_n$), but the affinity of E_o for S_n is much greater than the affinity of E_n for its own substrate $S_n(a^{-1/p} \ll K_n)$, then the degree of saturation of the last enzyme E_n will be very low ($x_n^* \ll K_n$). If the orders of magnitude of the remaining Michaelis constants are also very similar or greater than the last one ($K_i \gtrsim K_n$ for every $i = 1, 2, \ldots, n-1$), then the forward pathway will behave as a quasi-linear subsystem, because, according to (11), each forward velocity will be approximately proportional to each substrate concentration $[v_i \approx (V_i/K_i)x_i]$.

$$\dot{x}_1 = k_1 x_1 + v_o[1-f(x_n)] \quad ; \quad K_n \gg a^{-1/p}$$
$$\dot{x}_2 = k_1 x_1 - k_2 x_2 \quad ; \quad K_i \gtrsim K_n \qquad (11)$$
$$\ldots \ldots \ldots \ldots \quad ; \quad k_i = V_i/K_i$$
$$\dot{x}_n = k_{n-1} x_{n-1} - k_n x_n \quad ; \quad x_i \ll K_i$$

Now the behavior of the system can be approximately described as a set of <u>linear</u> differential equations plus a non-linear feedback interaction whose stability properties can be easily predicted if the non-linear function vanishes together with its independent variable. Since the input v_o is constant by assumption, a transformation of the "coordinates" (x_i) into the new set $\Delta x_i = x_i(t) - x_i^*$ will shift the non-linear interaction $v_o[1-f(x_n)]$ to the new function $v_o g(\Delta x_n) = v_o[f(x_n) - f(x_n^*)]$ which is bounded, vanishes only at the new origin $x_n = x_n^*$ (see 12) and is contained in a certain finite sector $0 < v_o g(\Delta x_n)/\Delta x_n \leq N$ (Fig.5).

SHIFTED SET OF EQUATIONS

where:

$$\dot{\Delta x}_1 = -k_1 \Delta x_1 + v_o g(y)$$
$$\dot{\Delta x}_2 = k_1 \Delta x_1 - k_2 \Delta x_2 \qquad \Delta x_i = (x_i - x_i^*) > -x_i^* \quad i = 1, 2, \ldots, n$$
$$\ldots \ldots \ldots \ldots \qquad g(y) = f(x_n) - f(x_n^*) \quad y = \Delta x_n \quad ; \quad (12)$$
$$\dot{\Delta x}_n = k_{n-1} \Delta x_{n-1} - k_n \Delta x_n \qquad g(y) \begin{cases} g(0) = 0 \\ g(y) \neq 0 \\ \text{if } y \neq 0 \end{cases} \quad ; \quad 0 < \frac{g(y)}{y} < N$$

$$f(x_n) = \frac{a x_n^p}{1 + a x_n^p}$$

53

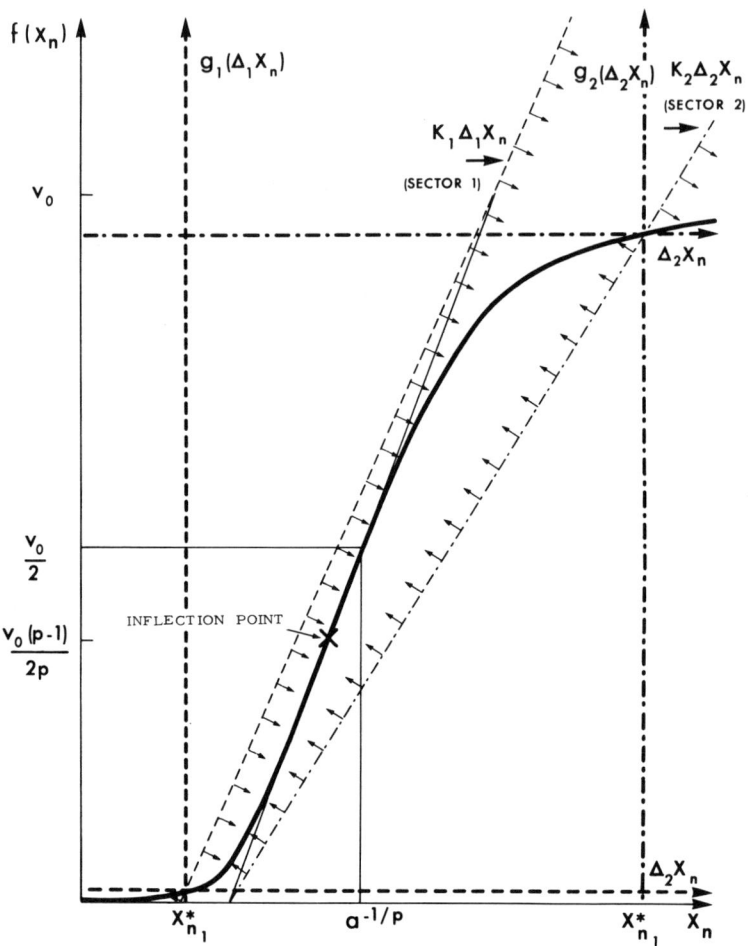

Figure 5.

The minimum value of N can be calculated by solving for x_n at the point at which the partial derivative of the ratio $\partial g(\Delta x_n)/\partial \Delta x_n$ vanishes. However, this procedure yields a rather cumbersome expression. Instead, a simpler evaluation of N can be made by considering the geometrical construction shown in Fig. 5. This construction has a straight line with the maximum slope of $f(x_n)$, which passes through its inflection point with abcissa x_n^{infl}. Now with the help of the new variable $\xi \equiv x_n^*/x_n^{infl}$ it is possible to evaluate N as shown in (13 and 14).

$$0 < v_o g(y)/y \leq N = \frac{k_n (p-1) \xi}{(p+1) - (p-1) \xi} \quad ; \quad \xi \leq 1 \quad (13)$$

$$0 < v_o g(y)/y < N = \frac{k_n (p-1) \xi^{p+1}}{\xi (p+1) - (p-1)} \quad ; \quad \xi > 1 \quad (14)$$

where:

$\xi \equiv x_n^*/x_n^{infl}$ and $\alpha x_n^{infl^p} = p-1/p+1$ (at the inflection point)

$v_o = k_n x_n^* (1 + \alpha x_n^* p)$ (at the stationary point)

On the other hand, the linear subsystem $\Delta x_j = k_{j-1}\Delta x_{j-1} - k_j \Delta x_j$ can be characterized by its corresponding transfer function T(s). This function can be found by solving for the ratio of the Laplace transformation of the last variable $\Delta X_n(s)$ over the Laplace transformation of the first $\Delta X_1(s)$, assuming that all the initial conditions are identically zero $[x_i(0) = 0]$. The resulting expression is shown in (15).

$$T(s) = \frac{\prod_{1}^{n-1} k_i}{\prod_{1}^{n} (s + k_i)} \quad (15)$$

Now the "Popov criterion" for <u>global</u> stability of the stationary solution can be readily applied. This criterion says that a sufficient condition for the unique stationary solution to be asymptotically stable for arbitrary perturbations is that there be a finite real number "q" for which the indicated inequality (16) is satisfied. (Popov, 1961).

$Re[(1 + jq\omega) T (j\omega)] + 1/N > 0$, for any $\omega \geq 0$, and q real, finite (16).

In particular, if the magnitude of T (jω) is a monotonic decreasing function of ω, then the stability of the system can be equally predicted by substituting $v_o g(\Delta x_n)$ by the linear term $N\Delta x_n$ because, for this particular system, the

Popov and Hurtwitz sectors become identical. Therefore, if the value of N is contained within the boundary of stability of the previously studied, linearized system, then the stationary solution can be said to be asymptotically stable in the large.

It should be noted that the crude approximations made in the construction of this model, such as irreversibility of the forward path, absence of branched pathways of reactions, absence of other terms in the feedback interaction but only one to the power "p", absence of accumulation of the very final product S_n+1, negligible effect of mass transport delays (i.e., in heterogeneous systems), etc., could have changed the stability properties from those which would correspond to more realistic models of enzyme behavior. All the previously assumed simplifications, except for the last one on the delays, would in general increase the stability of the system. But the addition of diffusion barriers between the enzymes (i.e., membranes) will decrease it, as has recently been shown by Landahl (1969).

Some Results on the Existence of Stable Limit Cycles.

The system previously described as quasi-linear in the forward pathway can be considered as autonomous (without input) by assuming that v_o is only a fixed parameter which has a value such that the non-linear feedback interaction looks nearly symmetric (steady-state value of $x_n^* \approx a^{-1/p}$) (see Fig.6). In this particular case, especially for large values of n, the stability behavior of the system can be predicted by the so-called "Describing Function Analysis".

This technique essentially consists in obtaining the first harmonic term, in a Fourier series, of the response that the symmetric non-linear characteristic would have if it was driven by a sinusoidal input, $\Delta x_n = E\sin t$. Since the non-linear characteristic shows a saturation effect, the first harmonic of the output-input ratio defined here as $\mu(E)$ (see Fig.6) will be a decreasing function of E if the amplitude of the input "E" is large enough to put in evidence such saturation effects ($E > b$). Otherwise, the ratio $\mu(E)$ will be nearly equal to the value of the slope at the shifted origin (approximately the slope at the inflection point). The graphical test for stability says that a stable limit cycle will occur at the intersection, in the complex plane

BIOCHEMICAL OSCILLATORS

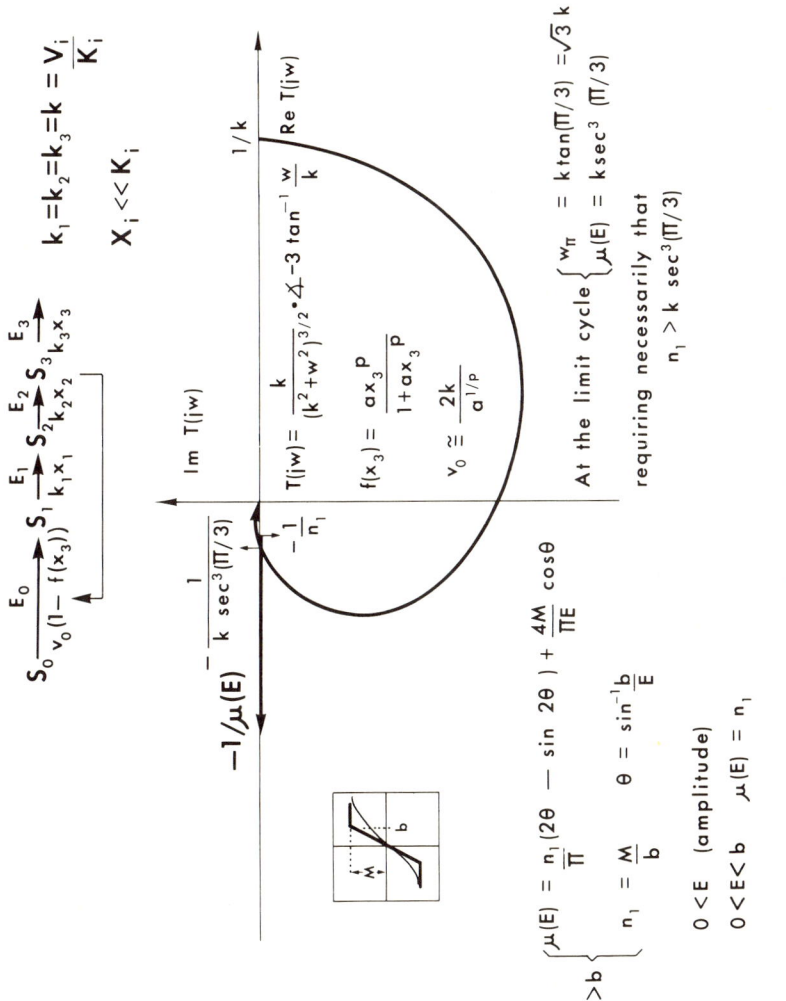

Figure 6.

of the locus $T(j\omega)$ and the locus $\frac{-1}{\mu(E)}$, which here is frequency independent and lies entirely on the real negative axis. The frequency of such stable oscillations will have the frequency of the intersection, which is precisely the corresponding frequency ω_π at which the Nyquist plot $T(j\omega)$ intersects the real negative axis and defines the critical point, for the stability of the corresponding linear system. By using the properties of the function arctan of positive numbers, it can be shown that

$$\sum_{i=1}^{n} \tan^{-1}(\omega_n/k_i) \geq n \tan^{-1}(\omega_\pi/\bar{k})$$

[the sum of arctan of a set of n positive numbers is greater than or equal to the arctan of the harmonic mean times n]. This immediately implies that the frequency ω_π is bounded above by the expression $\bar{k} \tan(\pi/n)$ and only makes sense for $n = 3, 4, \ldots$. It clearly decreases on the number "n" of intermediate reactions. This could provide an estimate of the order of magnitudes of the frequency of stable limit cycles when occurring in biochemical reactions that can be roughly modeled by the present system of equations.

In summary, idealized metabolic pathways exhibiting feedback loops can generate sustained biochemical oscillations, can exist in stable steady-state or can show accumulation of substrates (depending upon analytical conditions expressible in terms of measurable chemical parameters). An approach to predicting the frequency of allowed oscillations also seems possible.

ACKNOWLEDGEMENTS

The author is grateful for the valuable suggestions, help and criticism provided by Dr. Hugo Martinez and for encouragement, support and advice received from Dr. Manuel F. Morales.

This research was supported by grants from American Heart Association (60 CI 8) and U.S. Public Health Service (GM 14076). The author is a Dowdle Fellow of the University of California at San Francisco.

REFERENCES

Goodwin, B. (1963), "Temporal Organization in Cells", London: Academic Press.

Higgins, J. (1967), Ind.Eng.Chem. 59, 18.

Knorre, W.A. (1968), Biochem. Biophys. Res. Comm. 31, 812.

Landahl, H.D. (1969), Bull. Math. Biophys. 31, 775.

Morales, M. and McKay, D. (1967), Biophys. J., 7, 621.

Popov, V.M., (1961) Automatika i Telemekhanika, 22, 857.

Pye, K. (1969) Canadian J. of Botany, 47, 271.

Pye, K. and Chance, B. (1966), Proc. Natl. Acad. Sci. 55, 888.

Sel'kov, E.E. (1967), "Oscillatory Processes in Biological and Chemical Systems," Moscow: "Nauka".

Spangler, R.A. and Snell, F.M. (1961), Nature, 191, 457.

Viniegra-Gonzalez, G. and Martinez, H. (1969), Proc. Biophys. Soc. Abstracts, 13, A210.

Viniegra-Gonzalez, G., Martinez, H., and Morales, M. (1969) unpublished results.

Walter, C. (1969), Biophys. J., 9, 863.

Walter, C. (1970), J. Theoret. Biol., 27, 259.

Yates, P.A. and Pardee, A.B. (1956), J. Biol.Chem. 221, 757.

NOTE ADDED IN PROOF: The proofs of the relations presented in this paper are included in: G. Viniegra-Gonzalez (1971) Ph.D. thesis (Appendix of Part C), University of California at San Francisco.

II

OSCILLATIONS IN DEFINED CHEMICAL AND BIOCHEMICAL SYSTEMS

SOME EXPERIMENTS OF A CHEMICAL PERIODIC REACTION IN LIQUID PHASE

Heinrich-Gustav Busse

Institut für Molekulare Biologie, Biochemie und Biophysik, Stöckheim/Braunschweig, Germany

A solution of bromate ions and malonic acid does not measurably react at 60°C. Cerium ions, however, catalyze a reaction between these two components, as shown in Figure 1. Ce^{3+} is oxidized by bromate ions to Ce^{4+} and the Ce^{4+} is reduced in turn by malonic acid. Looking at the solution one sees a periodic appearance of the yellow of Ce^{4+} ion. Thus this system seems a suitable example of a chemical oscillator.

A spectrophotometric recording of the Ce^{4+} concentration and a recording of the solution's redox potential shows that the amplitude and frequency of the oscillation are stable over several periods.

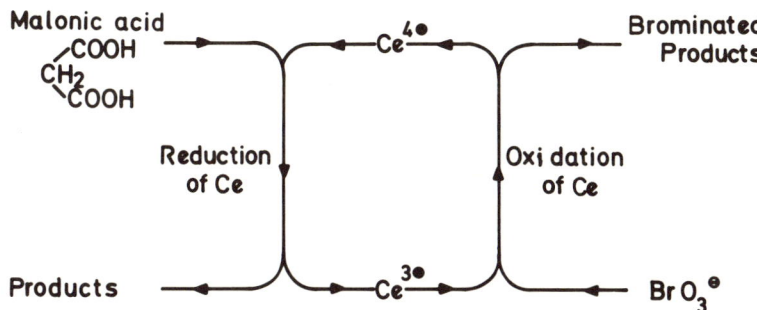

Figure 1. A qualitative picture of the oscillating reaction. The catalyzing cerium ions are reduced by malonic acid and oxidized by bromate ions.

Zhabotinsky (1) described in 1964 a large number of similar systems, the first of which was described by Belousov (2) in 1958. It is possible to replace malonic acid for example by oxaloacetic acid and malic acid. Malic acid is especially interesting because of its optical activity. Measurements of the change in the optical activity are as yet too inaccurate. Cerium can be replaced by manganese and the sulfuric acid of the medium by nitric acid. Until now no adequate replacement has been reported for bromate ions. The shape and frequency of the oscillations depend on the initial concentrations of the four compounds (3) as well as on temperature and the system used. The frequency varies from approximately one second to several minutes. The shape can be rectangular, spike shaped or sinusoidal. Sometimes the oscillations occur after a lag period.

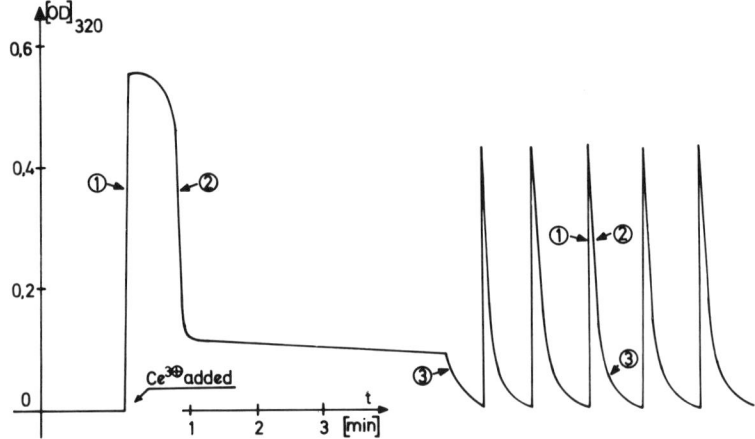

Figure 2. Time dependence of the optical density at 320 mµ. Concentrations: 1M malonic acid, 0.06M $KBrO_3$ and 5×10^{-3}M Ce^{3+} in 2N H_2SO_4 medium. Temperature: 20°C. After a lag period of about 6 minutes the oscillation starts. The numbers refer to the steps described in the text.

Figure 2 shows a plot of the optical density against time. At the marked time Ce^{3+} ions are added to the solution of malonic acid and bromate ions. Ce^{4+} ions are produced in the first autocatalytic step, which then disappear in an autocatalytic like manner in a second process. After a few minutes, the rest of Ce^{4+} ions disappear exponentially in a third process. Then the oscillation starts. If one looks at the form of the spikes, it seems that they are built by the three steps just described. The autocatalytic one, the second one and the third one.

The S-shaped curve of Figure 3 indicates the autocatalytic behavior of the first step. At time t equal zero, bromate ions and Ce^{3+} ions are mixed.

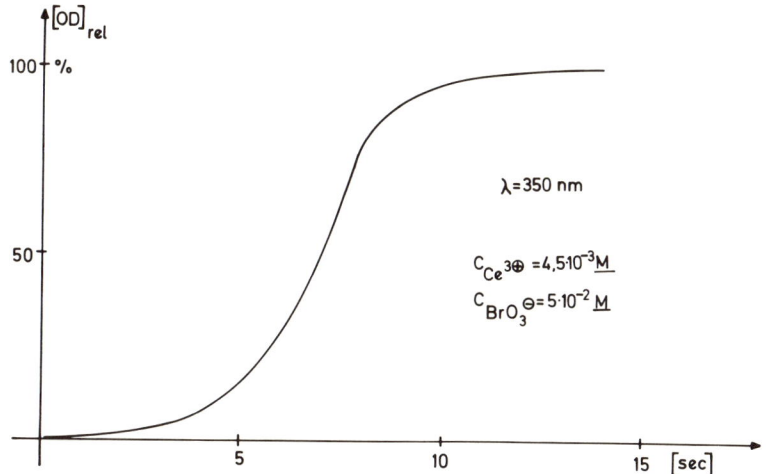

Figure 3. Bromate and Ce^{3+} ions are mixed in a stop flow device and the percentage of optical absorption at 350 mμ is measured as a function of time.

A simultaneous recording of the variation of the optical density during one period (upper diagram) and the solution's redox potential measured with a Platinum electrode (lower diagram) is given in Figure 4. The middle one shows the potential measured on a silver electrode referred to a silver, silver bromide cell. The electromotive force of this electrode combination is sensitive to bromide ions.

In all these three diagrams one can distinguish the three main steps. But the middle one indicates that during the reaction the concentration of bromide ions probably also oscillates.

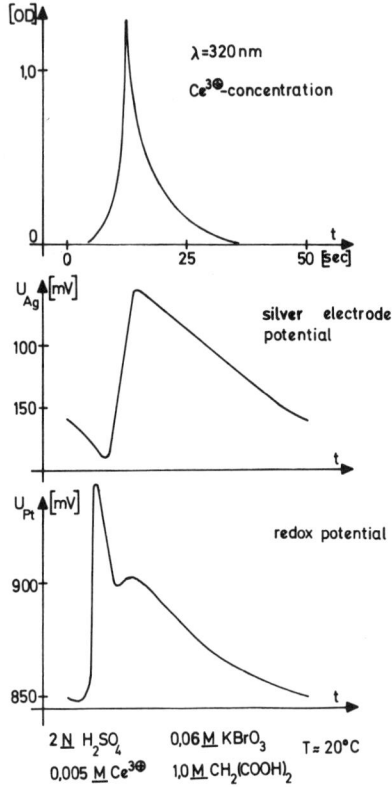

Figure 4. Optical density at 320 mµ (upper graph), potential on a Ag/AgBr electrode (middle graph) and redox potential on a platinum electrode (lower graph) within an oscillation period.

Now what will happen if we add bromide ions to the solution during the oscillation? If we add a certain amount at the time when the concentration of the bromide ions is high (this means at low potential in Fig.5) and if we choose the amount so that it increases the total concentration only by a negligible amount, nothing will happen. But if we add the same amount at a later time when the concentration of the bromide ions is low, this amount is critical and will change the reaction course, because now the bromide ions have to be removed before the next step of oscillation can start.

In Figures 5 and 6 we see the experimental result.

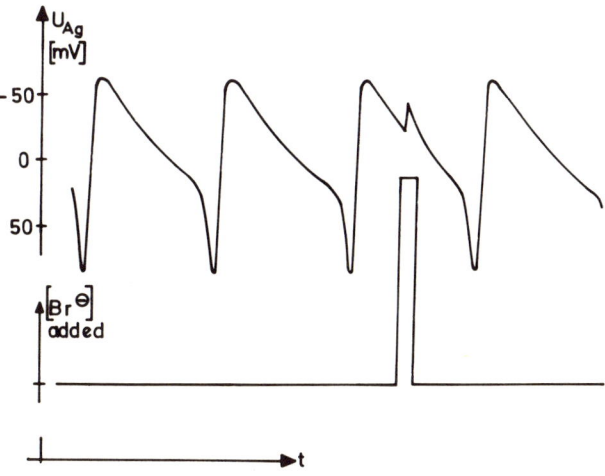

Figure 5. The potential U_{Ag} on a silver, silver bromide electrode oscillates (upper part). The time of the addition of bromide ions to the oscillating solution is shown below this recording. The added bromide ions do not alter the frequency of the oscillation.

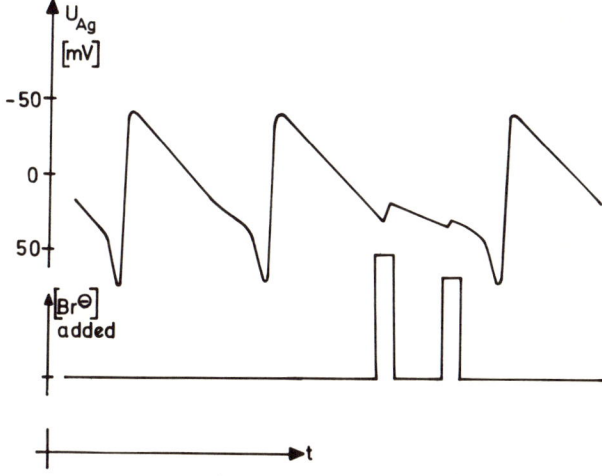

Figure 6. The same plot as in Fig. 5. Now bromide ions are added at a critical position. Then the addition of bromide ions lengthens the period. Here, an amount of bromide ions is added twice.

What can be concluded from this experiment? In Figure 7 one sees what occurs if bromide ions are added periodically with a frequency a little bit slower than the eigenfrequency of the oscillating chemical system. After a few cycles the oscillating chemical system is forced to oscillate with the frequency given by the added bromide ions.

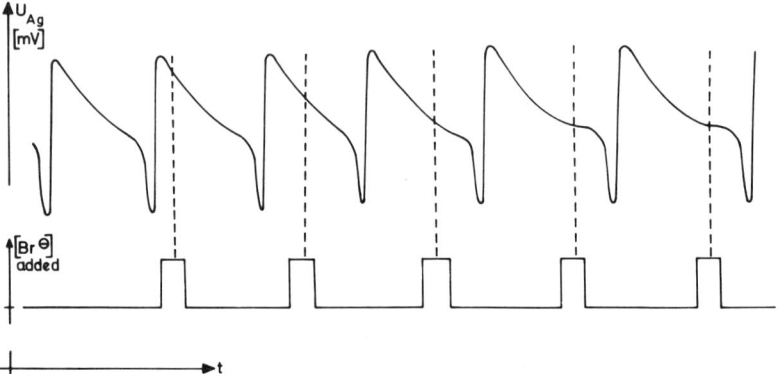

Figure 7. The same plot as in Fig. 5. Now bromide ions are added with a frequency slower than the eigenfrequency of the oscillating system. Thus, it is possible to force the system to oscillate with the frequency given by the addition of bromide ions.

I have to mention that in this particular case there is experimentally superimposed an additional effect in which the period of oscillation is lengthened by the addition of bromide ions. But nevertheless this may possibly be a model system for a regulated biological rhythm.

Now I want to mention some other points of view according to this phenomenon:

First: this oscillation generator gives out a voltage of 100 mV and can be built up for a few dollars in a volume so small that one cannot see it.

Second: the appearance of periodic oscillations divides the time axis and gives it a higher symmetry, like the periodicity in crystals.

Third: such reactions are interesting for cybernetics or information theory because one has a regulation input by the concentration of the bromide ions. Furthermore, it is possible to transcribe external information, that is the frequency of Br^-, into the periodical oscillation.

Now if one chooses a space coordinate x, y or z instead of the time axis, then it should be possible to put information into the chemical reaction, as shown in the case of the time axis with the addition of Br^-. It should also be possible to construct a chemical memory. Perhaps nature has used such processes.

Finally, I want to mention that the theory behind this type of behavior is that of chemical reactions far away from the chemical equilibrium of a set of specially coupled reactions.

REFERENCES

1. Zhabotinsky, A.M., Dokl. Akad. Nauk SSSR, <u>157</u>, 392 (1964).
2. Belousov, B.P., Sborn. referat. radiats. med. za 1958, Medgiz, Moscow, 1959.
3. Zhabotinsky, A.M. In <u>Oscillatory Processes in Biological and Chemical Systems</u>, Nauka Publ., Moscow, 1967.

A STUDY OF A SELF-OSCILLATORY CHEMICAL REACTION
I. THE AUTONOMOUS SYSTEM

V.A. Vavilin, A.M. Zhabotinsky and A.N.Zaikin

Institute of Biophysics of the
USSR Academy of Science
Puschino, Moscow Region, USSR

Catalytic reactions in the liquid phase are the most adequate models of biochemical systems. Their experimental investigation is much simpler and the reproducibility of the results is considerably better than that of biochemical systems. All essential features of concentrational oscillatory systems may be investigated on such a model. A survey of investigations on chemical and biochemical oscillatory systems is given elsewhere (1,2).

The present paper deals with the results of an experimental study of one-type oscillating reactions. In these reactions, cerium (or manganese) ions catalyze the oxidation of substances with active methylene groups by bromate. The reactions take place in an acid medium. Under certain conditions the concentrations of oxidized and reduced forms of the catalyst change periodically. These oscillations may be monitored by measuring the optical density of the solution or rH-potential of the system. The oscillations were monitored with the following reducers being used: malonic (MA), bromomalonic (BMA), citric, malic, acetone dicarboxylic and oxaloacetic acids, acetylacetone and acetoacetic ether. These compounds are easily brominated and reduce Ce^{4+}. The mechanism of the oscillatory reactions was investigated in the systems using MA and BMA as reducer and cerium ions as catalyst. The BMA system is the simpler.

The results given below concern both systems. The results have been previously published, in part (3-5).

1. <u>Reaction Scheme.</u>

Oscillations of Ce^{4+} concentration during the reaction are shown in Figure 1. They are relaxation oscillations, the period (T) of which is clearly divided into two parts: T_1 is the phase of $[Ce^{4+}]$ increase while T_2 is the phase of $[Ce^{4+}]$ decrease. Correspondingly, (in a simplified scheme) the reaction has two stages. During stage I cerous ion is being oxidized by the following reaction:

$$Ce^{3+} \xrightarrow[H^+]{BrO_3^-} Ce^{4+} \qquad (1)$$

In stage II reduction takes place as follows:

$$Ce^{4+} \xrightarrow{MA} Ce^{3+} \qquad (2)$$

The products of bromate reduction, having been formed in the course of stage I, brominate MA. Bromide ion (Br^-) is formed as a result of the decomposition of the MA-bromine derivatives induced by ceric ion oxidation of MA and BMA. Direct oxidation of BMA by Ce^{4+} also produces Br^-. Bromide is a strong inhibitor of reaction (1).

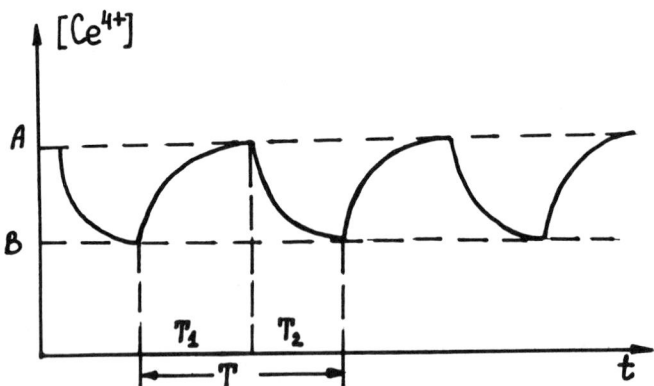

Figure 1. Change of Ce^{4+} concentration during the oscillatory chemical reaction (refer to the text).

BIOCHEMICAL OSCILLATORS

An oscillatory cycle may be qualitatively described in the following way. Suppose there is some Ce^{4+} concentration in the system. Then Br^-, formed in the reactions of stage II, interacts with active intermediates of reaction (1) and disappears from the system at a certain rate. If the $[Br^-]$ is high enough, reaction (1) is completely retarded. When $[Ce^{4+}]$, being removed in reaction (2), reaches a lower threshold, the $[Br^-]$ drops abruptly. Reaction (1) then starts at a high rate and $[Ce^{4+}]$ begins increasing. When $[Ce^{4+}]$ reaches an upper threshold, the $[Br^-]$ sharply increases, stopping reaction (1). Then the cycle is repeated.

The scheme has been confirmed by the following experiments:

1. Small additions of Br^-, introduced into the system at the phase of $[Ce^{4+}]$ increase, cause phase-shifting (Figure 2a). The minimum added bromide concentration required for phase-shifting decreases towards the end of phase T_1.

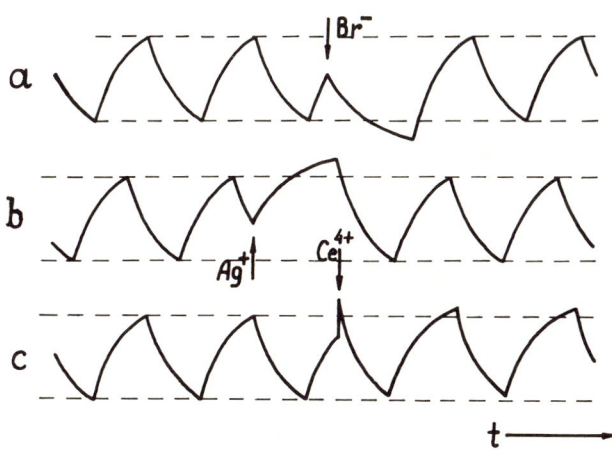

Figure 2. Phase-shifting during oscillations, caused by additions of Br^- (a), Ag^+ (b), and Ce^{4+} (c).

2. Br^- added during phase T_2 prolongs that phase. The larger the addition and the closer it is to the end of phase T_2, the greater the effect.

3. Ag^+ binds Br^-, and additions of Ag^+ have an opposite effect. Ag^+ added during phase T_2 causes phase-shifting (Figure 2b). When added during phase T_1 Ag^+ prolongs this phase.

4. If Br^- were introduced into the system at a low constant rate, we might expect the oscillations to cease if the rate of introduction exceeded a certain value. In this case $[Ce^{4+}]$ remained stable at a level near to B (see Figure 1). Oscillations resumed after stopping Br^- addition.

5. A result analogous to that described in the previous paragraph (4) may be obtained by adding Ce^{4+} at a low rate into the system.

6. Ce^{4+} added in phase T_1 causes phase-shifting (Figure 2b).

2. **Dependence of Reaction Behavior on Parameters.**

The behavior of the oscillating chemical system depends on a number of parameters, such as reagent concentrations, acidity, temperature and others. The parametric space may be divided into some regions within which the behavior of the system does not qualitatively change. Qualitative changes take place at the boundaries between the regions (surfaces of bifurcation). A study of these critical surfaces gives a complete picture of the behavior of the system (6). In the systems described, the essential variables are the concentrations of ceric and cerous ions, and of bromide and the active intermediates directly oxidizing Ce^{3+}. Basic parameters which reveal the system behavior are the concentrations of bromate, BMA, MA and the total concentration of cerium in the system. Such parameters as temperature, acidity and stirring intensity were kept constant in all the experiments. The significance of some other parameters and variables will be considered below.

The region in which oscillations exist. The region in which the oscillations exist was determined in the three-

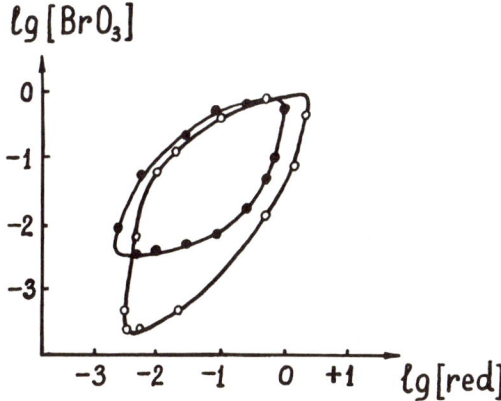

Figure 3. The plane of the oscillation region where the total concentration of cerium in the system is equal to 0.001 M. ●-● = BMA is reducer; o-o = MA is reducer.

dimensional space by the concentrations of bromate, BMA (MA) and the total concentration of cerium (4,5). The plane section of the region where [Ce] = $1 \cdot 10^{-3}$ M is shown in Figure 3. The sectional area is less with the BMA system than it is with the MA system.

Oscillation shape. The main types of oscillatory modes observed in reactions with BMA and MA are shown in Figure 4. Within the oscillation region, the oscillatory mode changes from harmonic to relaxation. It is mainly established by the concentrations of bromate and BMA (MA), the influence of Ce concentration being less important. For a further description it is convenient to use the parameter α.

For the BMA system, $\alpha = \dfrac{[BrO_3^-]}{[CHBr(COOH)_2]}$

For the MA system, $\alpha = \dfrac{[BrO_3^-]}{[CH_2(COOH)_2]}$

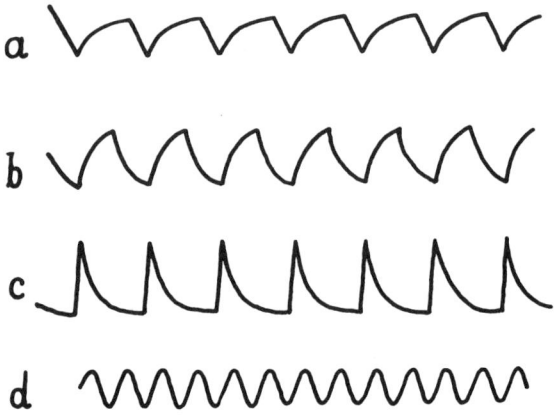

Figure 4. Types of kinetic curves observed at different points in the oscillation region. a) $\alpha > 1$; b) $\alpha \approx 0.4$; c) $\alpha \approx 0.2$; d) only in MA-system, if $\alpha < 0.1$.

The MA system displays a richer variety of kinetics than does the BMA system (4,5). In the system with MA the harmonic mode takes place at small values of α (Figure 4d). The BMA system does not exhibit this mode.

Amplitude and period of the oscillations. The amplitude of the oscillations in the BMA system is maximum at the points of the oscillation region located near the border of the region, where α is small. The amplitude approaches zero at high α. For the MA system, the amplitude is maximum in the center of the oscillation region and approaches zero at the borders. The value of the oscillation amplitude for the MA system varies from 1% (and less) to 60% of total cerium concentration. In the BMA system, the amplitude varies from 1 to 20%. The maximum oscillation amplitude of rH potential reaches 300 mv.

The oscillation period diminishes when the concentrations of bromate and BMA (MA) increase and rises when the total cerium concentration increases. By variation of the three basic parameters the oscillation period may be changed from 3 to 500 sec in the BMA system and from 0.3 to 1000 sec in the MA system (at 40°C and in 3N H_2SO_4). A rise of temperature leads to a quick decrease of oscillation period.

3. Evolution of the Oscillation Behavior in a Closed System.

The results cited above were obtained when the reaction was conducted in a closed system (without flow). Due to the irreversible decrease of bromate and BMA (MA) concentrations, the oscillatory conditions are not strictly stationary. The amplitude and the period of the oscillations gradually change with time. After some time elapses the oscillations cease and the system changes to a monotonous kinetics. The total number of oscillation cycles observed is usually much greater for the MA system than for the BMA system. For quasi-harmonic conditions corresponding to small values of α (i.e., the MA system), the total number of cycles approaches 500-800. Such oscillatory conditions are characterized by a gradual decrease of the amplitude, the period being constant. The oscillations, after having stopped, may be restarted by adding bromate to the system.

Oscillatory conditions with a small value of α, for the BMA system, are notable for a considerable increase of the period during this evolution, while the amplitude remains almost constant until the oscillations stop. At the border of the oscillation region the total number of cycles observed is small, about 5-10. In the MA system such oscillatory conditions are found in the center of the oscillation region ($\alpha \approx 0.3$). In this case the stopping of the oscillations is probably caused by a rapid accumulation of bromine derivatives of organic acids in the system. Indeed, oscillations resume after addition of Ag^+ to the system.

For oscillatory conditions with high values of α, the total number of cycles observed approaches 200. In the points close to the borders of the oscillation region the total number of cycles diminishes to 10-20. In this case, a gradual decrease of amplitude takes place, with the period being slightly increased. After oscillations have died away, they can be restarted by adding BMA (MA).

In a certain region of concentrational space of the MA system, a double frequency mode is established after some dozens of cycles from the beginning of the oscillations (Figure 5a). In the BMA system the double frequency mode may appear just after mixing the reagents (Figure 5b).

The absence of oscillations at the beginning of the

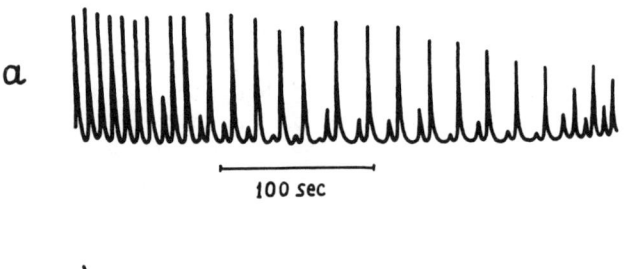

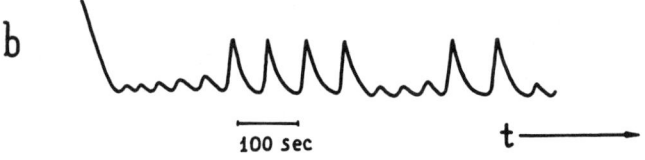

Figure 5. The double-frequence mode. a) The MA system; initial concentrations (M) : [Ce] = 0.003; $[CH_2(COOH)_2]$ = 1; $[NaBrO_3]$ = 0.93. b) The BMA system; initial concentrations (M) : [Ce] = 0.001; $[CHBr(COOH)_2]$ = 0.032; $[KBrO_3]$ = 0.007. In all systems $[H_2SO_4]$ = 3Ñ, T = 40°C.

reaction is typical for the MA system. During this nonoscillatory phase the accumulation of bromine derivatives takes place. Addition of these derivatives shortens the phase (7). The bromine derivatives have different influences on the system. The influence of dibromomalonic and tribromoacetic acids is the strongest. Probably the concentrations of dibromomalonic and tribromoacetic acids are essential variables in the MA system as well as in the BMA system. At the same time, the concentrations of bromomalonic and dibromoacetic acids may be considered as parameters of the system.

For the BMA system, the time of transition to the stationary oscillatory conditions approximates one period of oscillation. During this time there is a negligible change in the concentrations of bromate and BMA and thus during the first few cycles the concentrations are strictly equal to the initial values. For the MA system the time of the initial non-oscillating phase is often much more than a single oscillation period. During this phase, MA and bromate concentrations undergo considerable changes. Therefore the MA system oscillation region, determined by the authors, is the region of initial concentrations from which the system changes to the oscillatory conditions.

REFERENCES

1. Жаботинский, А.М., Колебателъные процессы в биологицеских и химицеских системах, Наука, Москва, стр. 149.
2. Селъков, Е.Е., там же, стр. 7.
3. Жаботинский, А.М., Докл. АН СССР, 157, 392 (1964).
4. Вавилин, В.А., Жаботинский, А.М., Ягужинский, Л.С., Колебателъные процессы в биологицеских и химицеских системах, Наука, Москва, стр. 181.
5. Вавилин, В.А., Жаботинский, А.М., Крупянко, В.И., там же стр. 199.
6. Андронов, А.А., Витт, А.А., Хайкин, С.Э., Теория колебаний, физматгиз, Москва, 1959.
7. Degn, H., Nature, 213, 589 (1967).

Note added in proof by editor: Readers may wish to refer to a recent paper on this subject by A. Winfree, Science, 175, 634 (1972).

A STUDY OF A SELF-OSCILLATORY CHEMICAL REACTION
II. INFLUENCE OF PERIODIC EXTERNAL FORCE

A.N. Zaikin and A.M. Zhabotinsky

Institute of Biophysics of the
USSR Academy of Science
Puschino, Moscow Region, USSR

Introduction

Ultra-violet (UV) radiation, with $\lambda < 300$ nm, effects the oscillatory reaction described previously (1) in the same manner as does the addition of bromide to the system from outside (2). A short UV pulse acts analogously to a momentary addition of Br^-; — it prolongs the time of $[Ce^{4+}]$ decrease and shortens the time of $[Ce^{4+}]$ increase during the oscillation cycle. During the transition of the system to the oscillatory condition, a short UV pulse provokes the appearance of a single impulse of rH potential in the system. UV-irradiation at constant intensity influences the system in a way similar to the continuous influx of Br^-.

An analysis of the experimental data shows that the action of UV-irradiation on the oscillatory system is a result of bromide formation due to the photo-destruction of bromomalonic and bromoacetic acids.

For the quantitative study of the action of UV-irradiation under different oscillatory conditions, some points in the concentration space were chosen. The continuous flow system was employed: it provided long-term stability of both amplitude and frequency of the oscillations of about 1 percent. The results are shown in Figure 1 (using UV-light wavelengths from 220 to 380 nm filtered by $NiSO_4$ solution).

In the closed system (without flow) UV-light produces the same effects as in the open system. However, in this

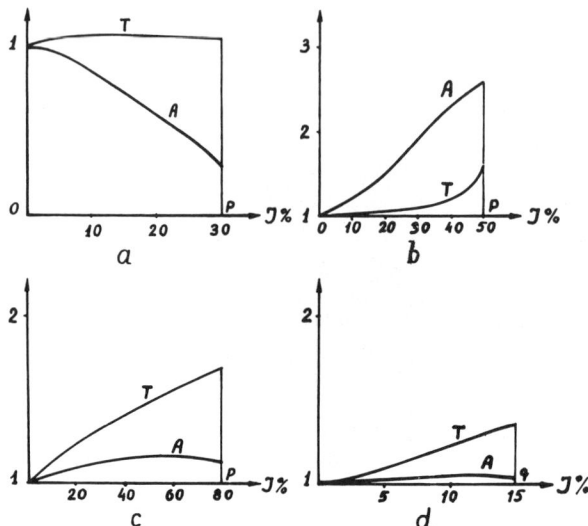

Figure 1. Oscillation amplitude (A) and period (T) functions of UV-irradiation intensity (A and T in relative units). Light intensity equivalent to 10 milliwatt/cm^2 (220-300 nm range) is taken as 100 percent. p - the break-point for oscillations; q - the point to transition to the interrupted oscillation condition.

Initial concentrations (moles per liter).

	$[Ce(SO_4)_2]$	$[NaBrO_3]$	$[CH_2(COOH)_2]$
a)	1×10^{-4}	1×10^{-1}	1×10^{-1}
b)	5×10^{-4}	1.5×10^{-1}	5×10^{-2}
c)	5×10^{-4}	1×10^{-1}	1×10^{-1}
d)	5×10^{-4}	5×10^{-2}	2×10^{-1}

case the photosensitivity of the system increases considerably during the course of the reaction. This is probably associated with the accumulation in the system of bromine derivatives of organic acids.

Some complicated oscillatory modes were obtained by means of a fine control of UV intensity. One of the most interesting modes is that of the interrupted oscillations shown in Figure 2.

The work reported here is concerned with a study of the dynamics of an oscillatory chemical system influenced by a periodic external force. In this case, the external force employed is periodic short-wavelength UV-irradiation of the system.

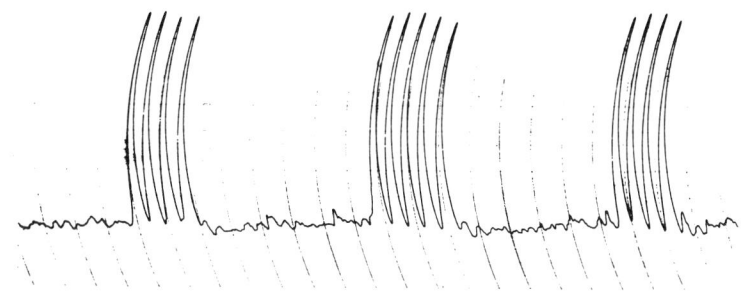

Figure 2. The interrupted oscillation mode.

Experimental Procedure

Stationary oscillatory conditions were maintained by a continuous flow system described elsewhere (2). The rH potential of the system was measured by a platinum electrode with continuous recording. A high-pressure mercury lamp, SVD-120A, was used as a UV-light source. Modulation of the light flux was carried out by a mechanical shutter driven by a rectangular pulse generator. The light flux had the shape of square wave, the frequency of which could be changed from 0.005 to 10 cps. The maximum light intensity was equal to 20 mw/cm^2 in the wavelength range 220-300 nm. This value is taken as 100% throughout. The light intensity between pulses was zero. The volume of the solution was 3 ml, the irradiated surface area being 6.2 cm^2. The solution was stirred by a magnetic mixer. In all the experiments the temperature was kept at 40°C by a water ultra-thermostat. The reagents were of analytical grade.

Results and Discussion.

The system has a large number of oscillatory modes within the region of auto-oscillation. The oscillations may be of sinusoidal, triangular and other shapes, with all intervening versions. The response of the system to external disturbance is different under various conditions.

The behavior of quasi-harmonic and strongly pronounced relaxation modes was studied. In quasi-harmonic conditions, synchronization takes place if the frequency of the UV pulses is close enough to the autonomous frequency of the system. The dependence of oscillation amplitude on the influencing period, within a synchronization band, is shown in Figure 3. Beats are observed near the borders of the synchronization band (Figure 4).

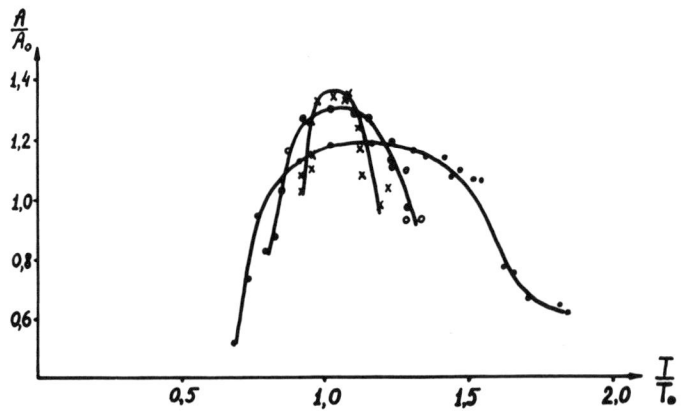

Figure 3. Oscillation amplitude as a function of the period of the external force, within a synchronization band, for quasi-harmonic conditions.
A - amplitude of synchronized oscillations;
A_o and T_o - amplitude and period of the oscillation in the absence of an external influence;
T - period of external force;
P - relative amplitude of the external force
($P = k \cdot I/I_{max}$, where $I_{max} = 20$ mw/cm^2)

●-●-●, for P=1; o-o-o, for P=0.5; x-x-x, for P=0.25.

Initial concentrations of reagents (M): [Ce] = 0.0002; [NaBrO$_3$] = 0.062; [CH$_2$(COOH)$_2$] = 0.75; [H$_2$SO$_4$] = 3 N.

For quasi-harmonic auto-oscillating systems it is possible to calculate the shape of the synchronization zone if the relative amplitude of the influencing force (P) is not too large (3). A qualitative picture of zone arrangement for this case is shown in Figure 5. A comparison with the experimental curves (Figure 3) shows that their shapes

correspond to the theoretical ones. On increasing P, the
experimentally observed synchronization band expands, in
correspondance with the theory. However, the dependence of
A on P is in contradiction to the theory. In our minds
there are two reasons for it. Firstly, the theory does not
take into account the influence of a constant component in
the external force. In the experiments mentioned, the external force has a constant component which has quite a
considerable influence on the system (see the introduction).

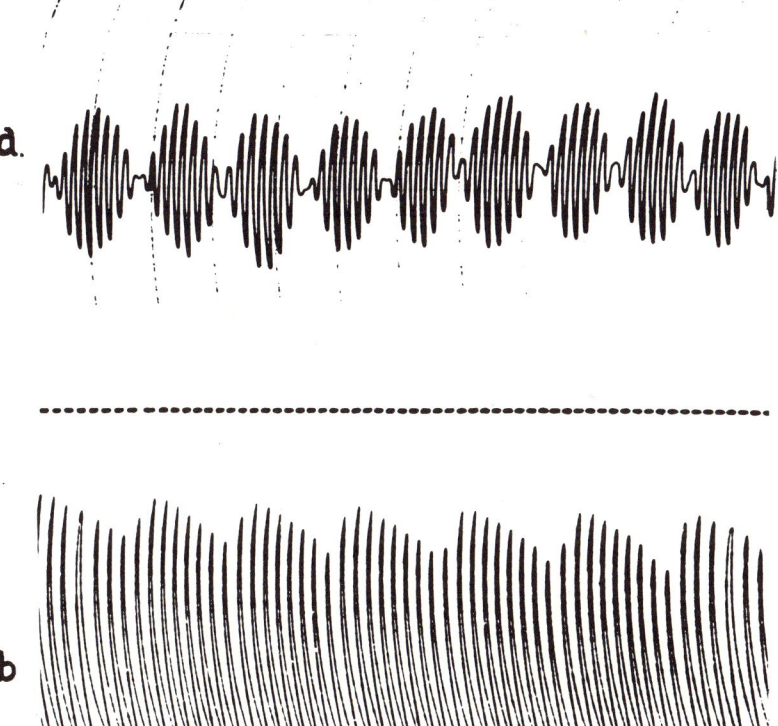

Figure 4. Beats arising near the synchronization band
boundaries. a) quasi-harmonic conditions; b) relaxation
conditions. The spaces in the dashed line correspond
to irradiation time.

Secondly, under the conditions chosen for the experiments, the system probably is still too nonlinear for calculations by this method (3) to be accurate. The decrease of A with increasing P can be phenomenologically explained by a large drop of the systems "quality", due to the increase of P.

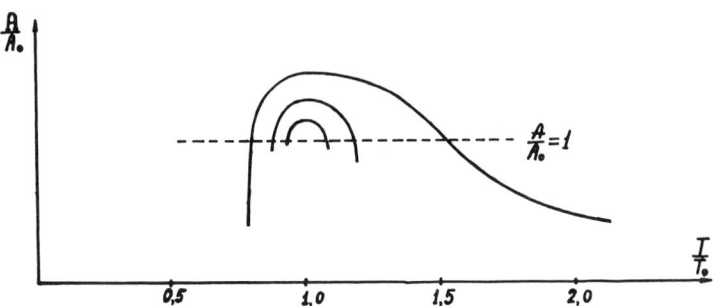

Figure 5. Synchronization bands for quasi-harmonic conditions, calculated by method given previously (3).

Under relaxation conditions, the synchronization band is much wider than with the quasi-harmonic ones. The amplitude of the oscillations does not change within a synchronization band. Synchronization takes place not only at the basic frequency but also at higher harmonics of the auto-oscillation frequency. Synchronization zones for the first, second and third harmonics of the system are shown in Figure 6.

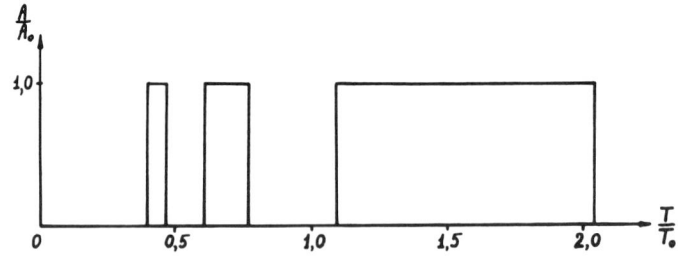

Figure 6. Synchronization bands for relaxation conditions. Initial concentrations of reagents (M): [Ce] = 0.0005; [$NaBrO_3$] = 0.05; [$CH_2(COOH)_2$] = 0.2; [H_2SO_4] = 3 N; I = 15%.

The synchronization regions represented in Figure 7b are calculated for the oscillation shape depicted in Figure 7a, by the method given previously (3). The most significant feature is a shift of the high frequency boundaries of the synchronization regions to the right, from the points where T/T_o is 1, 0.5 and 0.33, for the first, second and third harmonics respectively.

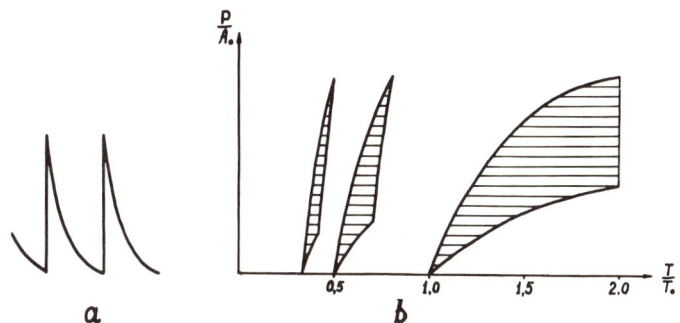

Figure 7. Synchronization regions for the first three harmonics (relaxation conditions) calculated by method given previously (3). a - oscillation shape for which calculations were performed; b - regions of synchronization.

Hence, it appears that for synchronization of the system at a given frequency (1/T), it is necessary that its autonomous frequency be higher than 1/T.

A comparison of the bands in Figure 6 with the section of synchronization regions in Figure 7b, where P/A_o is constant, shows a qualitative conformity of experimental and theoretical data. Quantitative results will be published later.

In conclusion it is necessary to note that external synchronization of oscillating chemical systems allows us to stabilize frequency and to carry out the detection of synchrony by the signal accumulation over many cycles. It also gives an opportunity to study processes which take place during time intervals that are small compared with the oscillation period. In such a way these processes may be investigated in any phase of the oscillation cycle.

REFERENCES

1. Vavilin, V.A., Zhabotinsky, A.M. and Zaikin, A.N. This volume, p. 71.

2. Вавилин, В.А., Жаботинский, А.М., Заикин, А.Н. Журн.физ.химии (в пецати).

3. Теодорцик, К.Ф., Автоколебателъные системы, ГИТТЛ, Москва, 1952.

A STUDY OF A SELF-OSCILLATORY CHEMICAL REACTION
III. SPACE BEHAVIOR

A. M. Zhabotinsky

Institute of Biophysics of the
USSR Academy of Sciences,
Puschino, Moscow Region, USSR

The mechanisms of oscillatory chemical reactions are usually considered only in terms of formal kinetics in which only ordinary differential equations are examined and a complete homogeneity of the system in space is implied. Such treatment does not permit a consideration of questions concerning the means of synchronizing concentration oscillations throughout a finite volume, i.e., concerning the existence of macrooscillations. Such an approach is quite valid if averaging by space, either through diffusion or convection, occurs much more rapidly than the process being investigated, that is if averaging time is much less than the oscillation period. Otherwise, space effects should necessarily be taken into account. In this case the most important question concerns the possibility of space synchronization and the existence of oscillations in the macrovolume.

In chemical systems, the velocity of diffusional averaging may be greatly increased owing to kinetic features of the separate reactions. For instance, autocatalytic reactions are known to spread in space at a rate greatly exceeding the diffusion rate (1). The velocity of the reaction front propagation is given by $V \approx \sqrt{KD}$, where K is the velocity constant of the autocatalytic reaction and D is the diffusion coefficient. That is the reason why the reaction, having started at one point, rapidly occupies the whole volume, like flame propagation.

The presence of autocatalytic stages may provide synchronization of the macrovolume in those cases where diffusion alone is insufficient, with convection being absent.

Experimental Data

The investigated self-oscillatory chemical reaction (2-6) is usually studied in volumes in the range of milliliters; averaging such a volume through pure diffusion takes many hours. The oscillation period in the system in question varies from several seconds to hundreds of seconds depending on the parameters of the system. Even so, under such conditions, macrooscillations are observed (within a definite parameter region) both with and without stirring of the solution.

The oscillatory system being considered has an autocatalytic stage (3); thus the system could be expected to have waves spreading over it.

a. <u>Wave propagation</u>. Indeed, one can observe a periodical wave propagation of chemical activity when carrying out this reaction in long tubes. For this purpose local non-uniformity should be produced at one end of the tube, e.g. by increasing the oxidizer concentration. Waves then spread from this part over the whole tube. In the case where no artificial non-uniformities are produced and a solution-gas interface is present, the wave spreads from it. At a temperature of 25°C the velocity of wave propagation may be varied in the range from 1 to 30 cm/sec by changing the reagent concentrations within standard limits. The wavelength under these conditions is about 10 cm. so, in a rather long tube, several waves can be observed simultaneously. An increase in temperature decreases the wavelength. In the case where there are non-uniformities at both ends of the tube, from which waves are spreading, the waves run first from the ends towards each other. Somewhere in the middle of the reaction system the waves collide and cancel each other. Since the oscillation periods are not precisely equal at both ends, the point of wave collision gradually shifts towards the low-frequency end. Wave propagation under stationary conditions may be observed only from the part having the shorter oscillation period.

b. <u>Influence of reagent concentrations on stability of the macrooscillations</u>. The possibility of realizing macrooscillations in this system under various conditions of stirring depends on how the oxidizer concentration relates to the reducer concentration. Let this ratio be α_1. For the system with malonic acid (MA):

$$\alpha_1 = \frac{[BrO_3^-]}{[CH_2(COOH)_2]}$$

The macrooscillations were monitored by measuring changes in the optical density of the solution. When $\alpha_1 \geqslant 1$ and there is no stirring, macrooscillations are not detected. Nevertheless, by using platinum microelectrodes, it is possible to show that in this case oscillations do occur in each point in space but their phases are not connected with others at different points. If stirring is started in such a system, oscillations at different points in the volume become practically synphase and macrooscillations are observed. After the stirring is stopped the macrooscillations rapidly damp. The damping rate increases with a rise in α_1. The dependence of the macrooscillations on stirring, with $\alpha_1 \approx 3$, is shown in Figure 1a. When $\alpha_1 \approx 0.05$, macrooscillations in the system do not depend on stirring conditions (Figure 1b). An increase in catalyst concentration is similar to an increase in α, with other conditions being unchangeable.

For the system with bromomalonic acid, the relation between oxidizer and reducer concentrations will be designated as α_2.

$$\alpha_2 = \frac{[BrO_3^-]}{[CHBr(COOH)_2]}$$

At $\alpha_2 \geqslant 1$ the dependence of the macrooscillations on stirring is the same as in the case with the MA system when $\alpha_1 \geqslant 1$.

At $\alpha_2 \leqslant 0.3$, macrooscillations are observed both in the presence and in the absence of stirring. However, cessation of stirring causes a decrease of the period and a smoothing of the oscillation shapes as shown in Figure 2a.

The region where oscillations exist in the BMA system is shown in paper II of this series.* Beyond this region

* See this Volume, p.81 , Figure 3a.

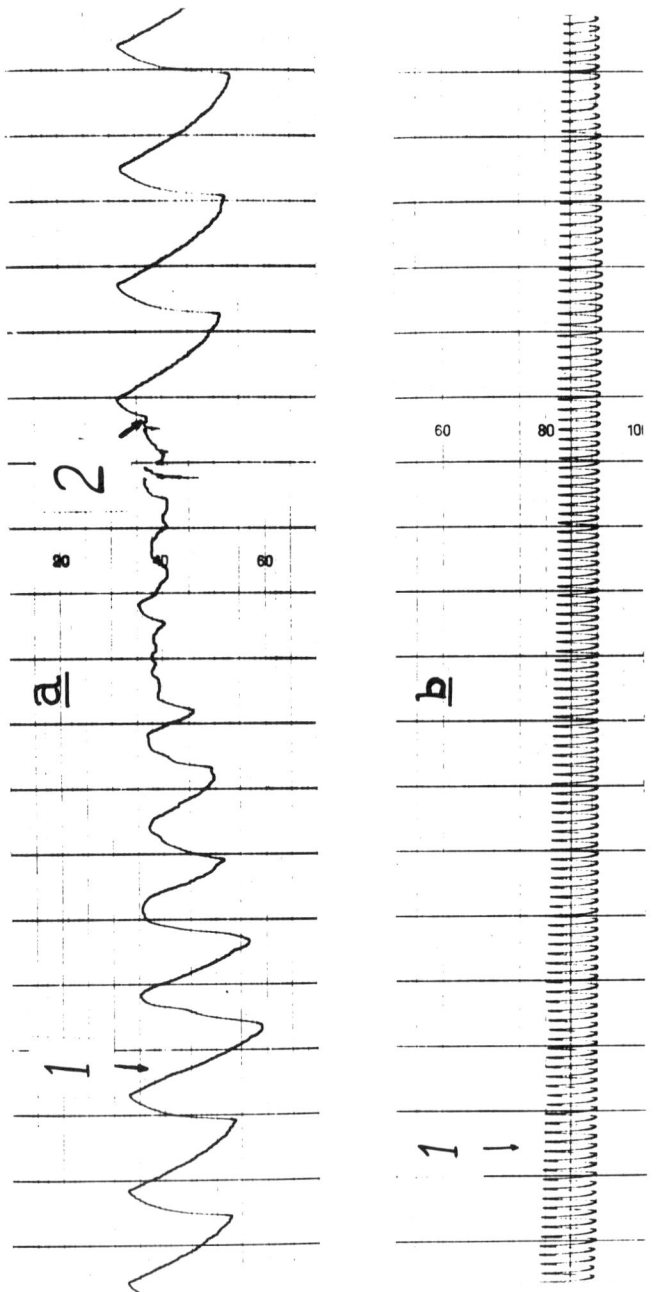

Figure 1. Influence of stirring on the stability of macrooscillations at different values of α_1. 1 – stirring off; 2 – stirring on. Initial concentrations (M) a) $\alpha_1 \approx 3$; [Ce] = 0.0007; [KBrO$_3$] = 0.07; [CH$_2$(COOH)$_2$] = 0.02. b) $\alpha_1 \approx 0.05$; [Ce] = 0.0007; [KBrO$_3$] = 0.07; [CH$_2$(COOH)$_2$] = 1.5. Time scale – distance between vertical lines corresponds to 100 sec.

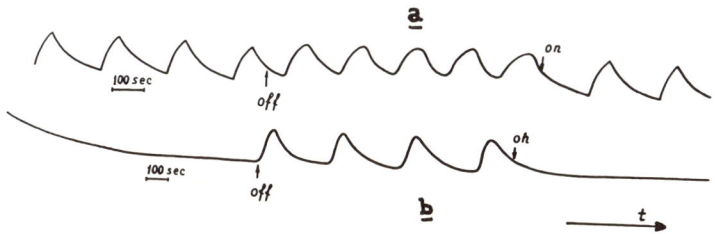

Figure 2. Influence of stirring on macrooscillations in the system with bromomalonic acid. Initial concentrations (M); [Ce] = 0.001; [CHBr(COOH)$_2$] = 0.01; a) [KBrO$_3$] = 0.003; b) [KBrO$_3$] = 0.007.

macrooscillations are absent when the solution is being stirred. However, near the oscillation region (where the concentrations of bromate and BMA acid are about 0.01M) macrooscillations arise when stirring is absent. On switching on the stirrer the oscillations disappear (Figure 2b).

c. <u>Evolution of the macrooscillation conditions with time</u>. It is known (3,4) that when MA is used as the initial reducer, oscillations are not observed at the beginning of the reaction. The system becomes self-oscillatory only after a threshold concentration of CHBr(COOH)$_2$ has been reached. When $\alpha_1 \approx 0.05$, the evolution of the system conditions does not depend on whether stirring is present or not.

When $\alpha_1 \approx 3$, the duration of the initial non-oscillatory state and the transition to the macrooscillation conditions are also independent of stirring. However, if mixing is absent, the initial macrooscillations damp rapidly (Figure 3), as was noted earlier.

Discussion

In the case where wave propagation takes place in the system described, the possibility of macrooscillation observation depends on wave-length. If the wave-length (λ) is much greater than the length of the reaction vessel, the oscillations in different points of space are practically synphase.

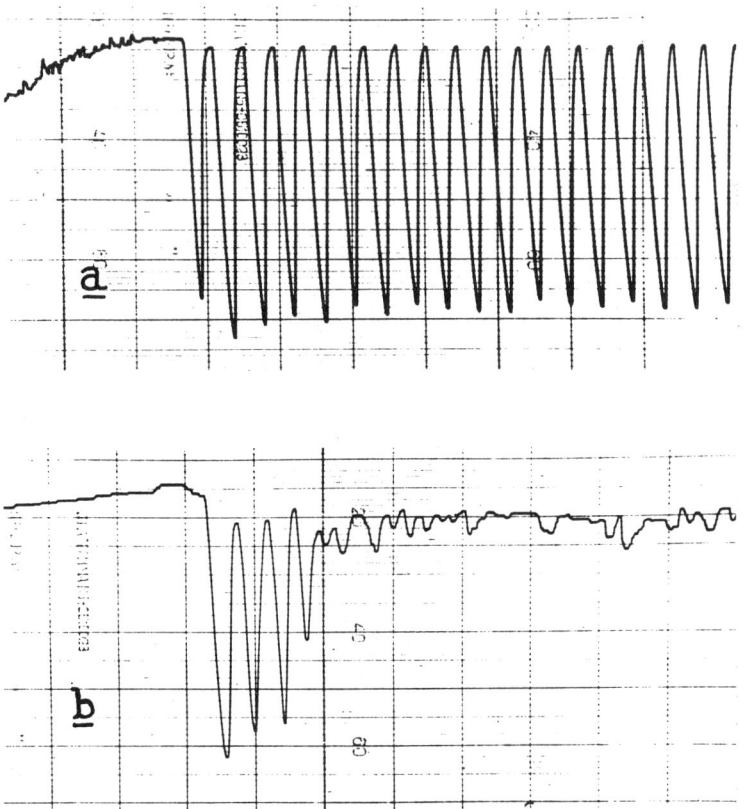

Figure 3. Evolution of macrooscillation conditions depending on stirring. a) reaction course with stirring; b) reaction course without stirring. ($\alpha_1 \approx 3$). Initial concentrations (M); [Ce] = 0.001; [$KBrO_3$] = 0.2; [$CH_2(COOH)_2$] = 0.07. Time scale – distance between vertical lines corresponds to 100 sec.

If λ does not differ very much from the length of the reaction vessel the oscillation phases at different points are shifted. Nevertheless, macrooscillations obtained by integration over the whole length of the vessel still have a considerable amplitude.

If λ is much less than the length of the reaction

vessel then the macrooscillation amplitude is small. Indeed, integration over the length λ gives a time-independent constant. Therefore (in the one-dimensional case) only that part with length, L = λN, where N is a whole number of waves in the reactor length, gives a non-zero contribution to the macrooscillations.

At a constant rate of propagation, V, the wave-length, λ, is given by:

$$\lambda = VT$$

where T is the oscillation period. As was previously stated, the propagation rate of an autocatalytic reaction is $V \approx \sqrt{KD}$. In a complex system, V depends on several diffusion coefficients (D_i). It can be shown that, in this case as well, an m-fold increase of all D_i and K_i always leads to a rise in V by m. It is not necessary to have the conditions of macrooscillation registration improved. These conditions depend on λ but not on V.

REFERENCES

1. Франк-Каменецкий, Д.А., Диффузия и теплопередаца в химицеской кинетике, Наука, Москва, 1967.

2. Vavilin, V.A., Zhabotinsky, A.M., Zaikin, A.N., This issue, p.71.

3. Жаботинский, А. М., Докл. АН СССР 157, 392 (1964).

4. Вавилин В.А., Жаботинский, А.М., Ягужинский, Л.С., Колебательные процессы в биологицеских и химических системах, Наука Москва, 1967, стр.181.

5. Вавилин, В.А., Жаботинский, А.М., Крупянко, В.И., там же, стр.199.

6. Жаботинский, А.М., там же, стр.252.

CHEMILUMINESCENCE IN OSCILLATORY OXIDATION REACTIONS CATALYZED BY HORSERADISH PEROXIDASE*

Hans Degn

Johnson Foundation, University of Pennsylvania
Philadelphia, Pennsylvania 19104

Yamazaki and his collaborators have found that a damped oscillation of the oxygen concentration in the solution may occur when the peroxidase catalyzed aerobic oxidation of NADH takes place in an open system, where oxygen is continuously supplied by bubbling with an N_2-O_2 mixture (1,2). They also found that the damped oscillation in the oxygen concentration was accompanied by a damped oscillation in the concentration of complex III, a form of the enzyme originally found as a product of the reaction between horseradish peroxidase (HRP) and excess hydrogen peroxide (3). In the experiments to be reported here, it was found that damped oscillations of the type reported by Yamazaki et al. also occur when dihydroxy fumaric acid (DHF) or indole-3-acetic acid (IAA) is used as a donor instead of NADH (or NADPH).

The oxidatic as well as peroxidatic reactions catalyzed by HRP are presently assumed to be free radical reactions. Free radicals originating from substrate have been found by electron paramagnetic resonance measurements during some HRP catalyzed oxidatic and peroxidatic reactions (4-6). The occurrence of chemiluminescence during the peroxidation of different substrates has been reported and this is also taken as an indication of a free radical mechanism (7,8). In the experiments to be reported here, chemiluminescence was found in different oscillatory oxidation reactions catalyzed by HRP and some new features of the behavior of complex III in such reactions were found.

---

* This work supported by USPHS grant GM 12202

Experimental

Both chemiluminescence and light absorption measurements were carried out in a dual wavelength spectrophotometer, using a 4x4x1 cm glass cuvette in a thermostated holder (25°C). When chemiluminescence was measured, the light was off, the photomultiplier tube was cooled with solid carbon dioxide, and a stabilized high voltage supply was used. An N_2-O_2 mixture whose composition could be controlled by means of gas flow meters was blown on the surface of the solution in the cuvette. It was not bubbled into the solution. The oxygen concentration in the solution was measured polarographically by a vibrating platinum electrode which also stirred the solution efficiently. The peroxidase used in the experiments was from Boehringer Mannheim Corp. The purity number specified by the manufacturer was 2.8.

Because light absorption and chemiluminescence could not be measured at the same time, all experiments were performed twice using the oxygen measurement as a common reference.

Results and Discussion

When NADH was oxidized, a very weak chemiluminescence was found. At the high sensitivity required to measure the light, the response time was so long that meaningful kinetic curves could not be obtained. However, when DHF or IAA was oxidized, the chemiluminescence was strong enough to allow the recording of kinetic curves. When IAA was oxidized, the intensity of chemiluminescence was about ten times higher than when DHF was oxidized at the same rate of oxygen consumption.

Figure 1 shows experiments where increasing amounts of DHF were added to solutions of HRP in phosphate buffer at pH 5.1, initially in equilibrium with the N_2-O_2 mixture. In all cases the oxygen concentration in the beginning falls with an increasing rate indicating that the oxygen consuming reaction is auto-catalytic. At a certain initial concentration of DHF, a damped oscillation of the oxygen concentration is observed (Fig. 1C).

Long ago it was predicted by Lotka (9) that damped

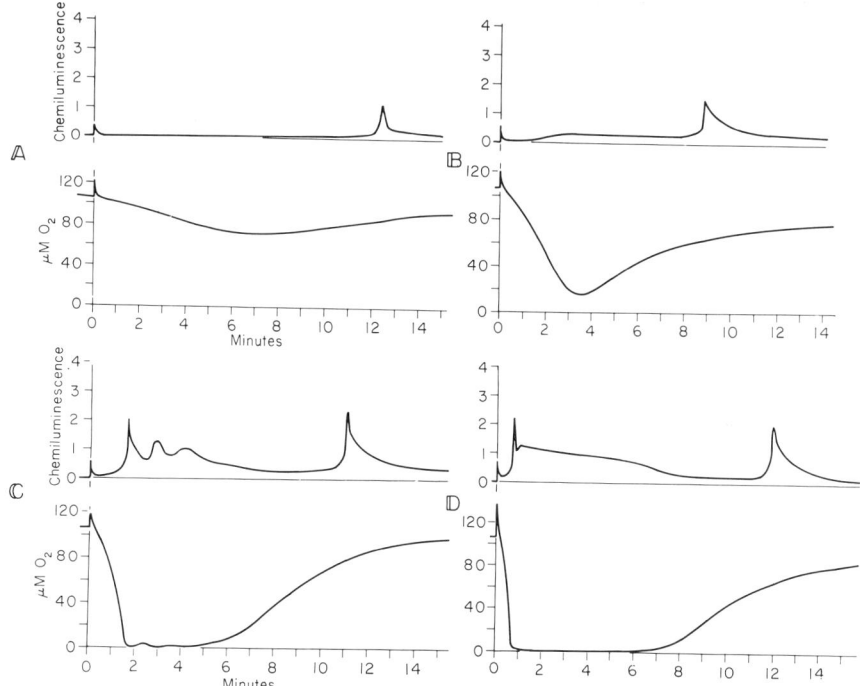

Figure 1. Simultaneous measurement of oxygen concentration and chemiluminescence during HRP catalyzed aerobic oxidation of DHF in system open to oxygen. The concentrations are in A: 0.1 M phosphate buffer, pH 5.1, 0.8 µM HRP and 0.425 mM DHF. In B, C and D the concentrations are as in A except that the DHF concentration is 2, 3 and 4 times higher respectively. DHF dissolved in ethanol is added at time zero.

oscillations may occur in an open system where a reactant, which enters at a constant rate, is consumed by an autocatalytic reaction. Apart from the minor deviation that oxygen enters the reation solution at a rate proportional to the difference between the equilibrium concentration and the actual concentration of oxygen in the liquid, the present experimental system is a realization of Lotka's theoretical system, and the damped oscillation found in the oxygen concentration verifies Lotka's prediction. The chemiluminescence measurements show a maximum about the same time as the oxygen curve has its steepest negative slope, which is approximately when the reaction rate is maximal. The damped oscillation of oxygen concentration in Figure 1C is accompanied by a damped oscillation in the chemiluminescence. The maxima in this oscillation seem to correspond to maxima in the reaction rate. In all the chemiluminescence curves there is a final peak which occurs when the oxygen concentration has nearly returned to its initial value. This maximum in chemiluminescence does not coincide with a high rate of oxygen consumption as in the previous phases of the reaction. A very small upward inflection can sometimes be seen at the time when the final peak of chemiluminescence is observed.

When chemiluminescence is used as a kinetic tool, it is usually assumed that the intensity of the chemiluminescence at any time is a measure of the reaction rate at that time (10). If the reaction rate is taken to be the rate of consumption of oxygen the above experimental results show that this assumption is not valid for the present reaction.

In Figure 2 the experiments from Figure 1 are repeated except that the increase in light absorption at 418 mμ is measured instead of the chemiluminescence. It is seen that the addition of DHF to the enzyme solution causes an immediate formation of complex III. Later on, when the oxygen concentration begins to return to its original value, the concentration of complex III begins to decrease slowly. At a certain time, when the oxygen concentration is close to its initial value, complex III suddenly disappears rapidly. By comparing Figures 1 and 2, it is seen that the final peak of chemiluminescence occurs at the time when the rapid disappearance of complex III takes place.

If DHF was added once more to a reaction solution after the oxygen concentration had returned to its original value, the course of reaction described above was found to repeat itself. This indicates that the oxidation reaction stops because of exhaustion of donor.

Since complex III is formed rapidly when the reaction is starting and disappears rapidly when the reaction is coming to an end, it might be assumed that its concentration during the overall reaction is a steady state concentration

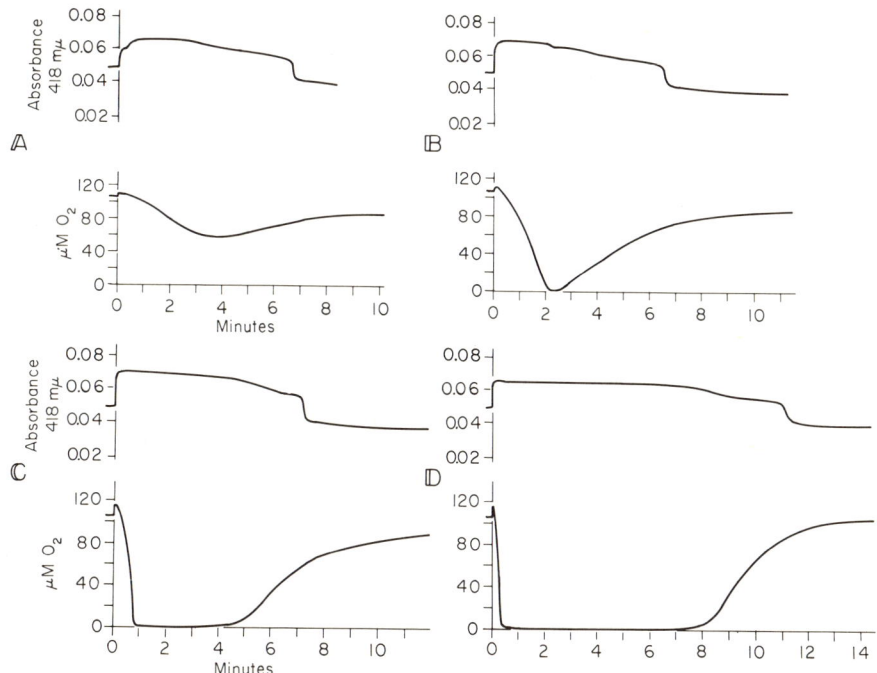

Figure 2. Simultaneous measurement of oxygen concentration and increase in light absorption at 418 mµ (absorption maximum of complex III). Concentrations as in Figure 1.

determined by a rapid reaction between HRP and an intermediate in the overall reaction to form complex III, and a rapid spontaneous breakdown of complex III to form HRP again. However, if the final peak of chemiluminescence during the rapid breakdown of complex III is due to an intermediate in the breakdown reaction, it must be concluded that this reaction is slow during the overall reaction until little donor is left. Then it accelerates strongly. This acceleration may be due to autocatalysis (back-activation) or relief of forward inhibition exerted by the donor, or both.

The effect of turning the oxygen in the gas off and on during the DHF oxidation was studied in experiments (Figure 3) where the conditions were the same as in the experiments of Figures 1C and 2C. It is seen that turning off the oxygen in the gas is followed by a rapid disappearance of the chemiliminescence whereas the complex III concentration decreases slowly. When the oxygen is turned on again, an overshoot in the oxygen concentration followed by a strongly damped oscillation occurs. After the turning on of the oxygen, a lag time before the reappearance of the complex III and the chemiluminescence is observed. This lag time is shorter for complex III than for the chemiluminescence.

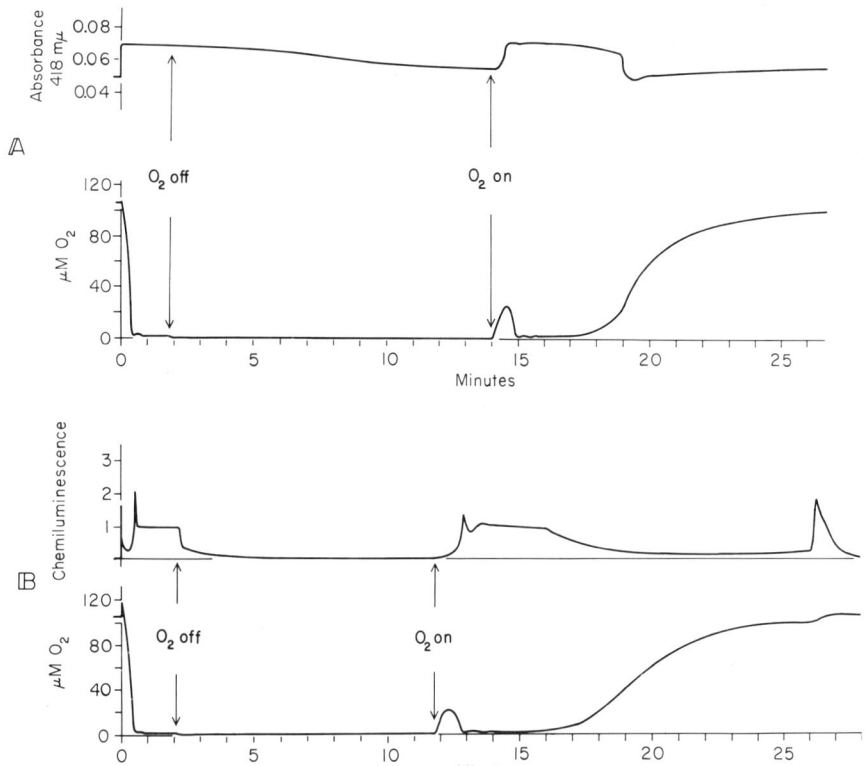

Figure 3. Experiments as in Figures 1C and 2C except that the oxygen in the gas is turned off and on during the reaction.

This experiment shows that the spontaneous degradation of complex III is a slow process, at least when oxygen is not present. The overshoot and damped oscillation when the oxygen is turned on again conforms with computer solutions of the Lotka model (11).

It has often been reported that certain phenols are required, or strongly increase the reaction rate, in oxidation reactions catalyzed by HRP. For this reason the influence of 2,4-dichlorophenol on the DHF oxidation was investigated. The experiments shown in Figures 1 and 2 were repeated with 2,4-dichlorophenol added. The results are shown in Figures 4 and 5. It is seen that the most significant effect of adding 2,4-dichlorophenol is that the final disappearance of complex III occurs at a lower oxygen concentration and is much faster. It is accompanied by a sharp break in the oxygen concentration curve. At the same time the final peak of chemiluminescence is much narrower and more intense. The intensity of chemiluminescence during the first phases of the reaction is not significantly influenced by the addition of 2,4-dichlorophenol. Also, when 2,4-dichlorophenol was present, the course of reaction was repeatable when the addition of DHF was repeated.

The sharp break in the oxygen concentration curve occurring when complex III disappears may indicate that the oxygen consuming reaction suddenly stops when there is still some donor left. A possible interpretation of the coincidence of the stop of the overall reaction and the disappearance of complex III is that complex III is an active intermediate and the reaction stops because complex III disappears. However, it can also be argued that complex III is not an active intermediate and that it disappears as a consequence of the stop of the overall reaction. Most authors, except Yamazaki et al., believe that complex III is not an active intermediate. It has been shown by the present author that the conversion to complex III causes an inhibition of the enzyme activity when NADH is the donor (12). However, such an effect could not be demonstrated when DHF was used as the donor.

The oxygen content of the gas used in the previous experiments corresponded to an equilibrium concentration of 105 µM O_2 in the solution. This concentration was close to the upper limit of oxygen concentration where oscillations could be obtained under the present conditions. It was chosen because it gave a convenient intensity of chemilumin-

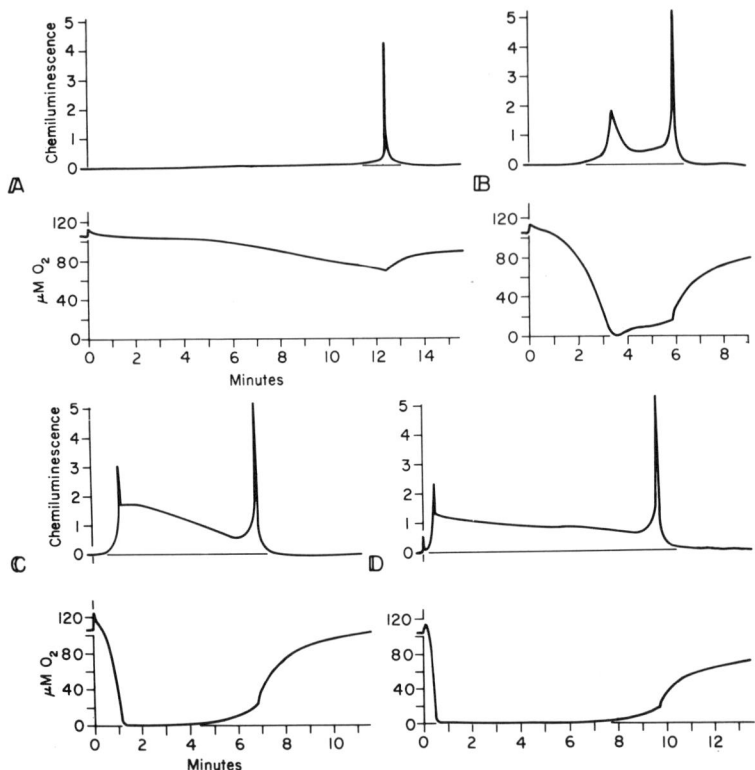

Figure 4. Experiment as in Figure 1 except that 2,4-dichlorophenol was added to the DHF solution. A: 7.7 µM DCP. In B, C and D the DCP concentration is 2, 3 and 4 times higher, respectively.

escence. The optimal equilibrium concentration of oxygen for oscillations was about one third of the concentration used in the above experiments. Figure 6 shows chemiluminescence and light absorption measurements in experiments similar to those of Figures 1C and 2C except that the oxygen concentration of the gas corresponds to an equilibrium concentration in the solution of 35µM O_2. It is noted that the light absorption at 418 mµ shows an oscillation of a very small amplitude synchronous with the oscillations of the oxygen concentration.

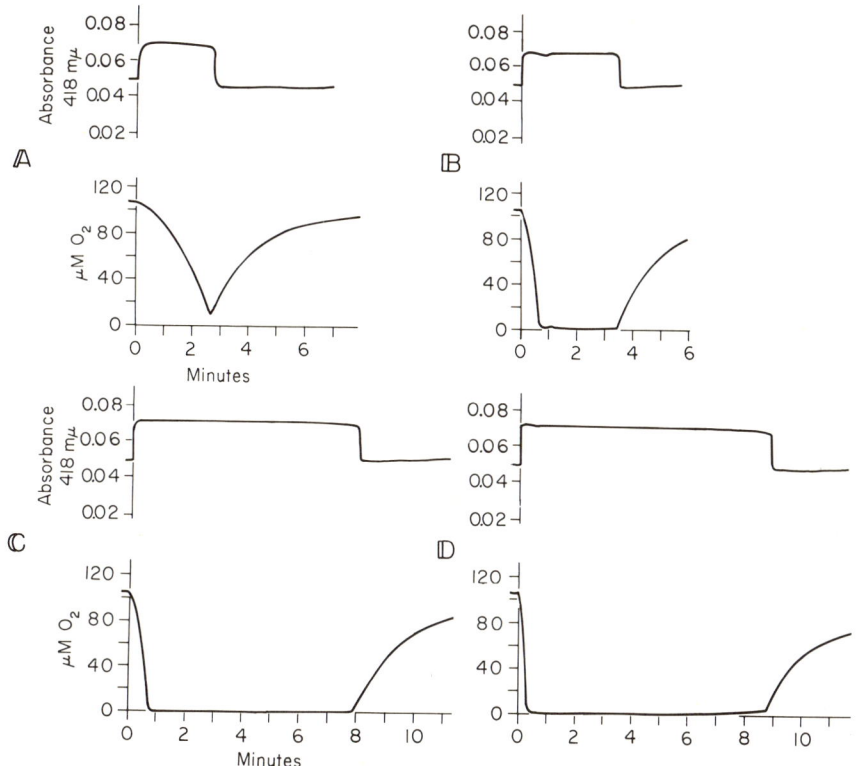

Figure 5. Experiment as in Figure 2 except that DCP is added. Concentrations as in Figure 4.

The curve in Figure 6B resembles closely a curve obtained by Yamazaki et al.(Ref.2, Fig.4) using NADH as the donor. One important difference is that high amplitude oscillations of complex III concentration were found in the NADH case. The absence of a significant oscillation in the complex III concentration during the oscillatory oxidation of DHF contradicts the hypothesis by Yamazaki et al. (2) that complex III is a regulatory intermediate causing the oscillation.

Figure 7 shows a damped oscillation in oxygen concentration and chemiluminescence in an experiment where IAA was used as the donor. The oxygen in the gas was turned

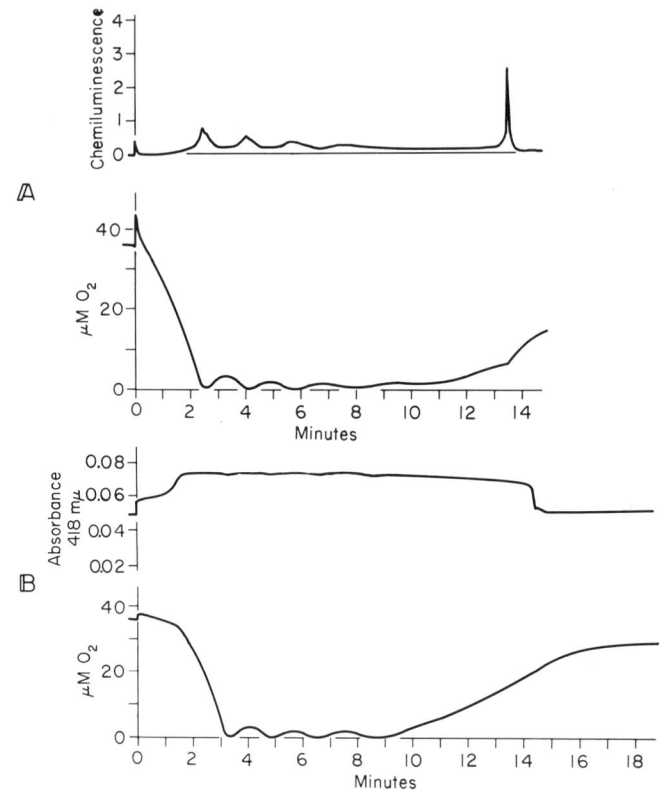

Figure 6. Typical samples of oscillations obtained at optimal O_2 content in the gas. The concentrations are in A: 0.1 M phosphate buffer, pH 5.1, 0.8 µM HRP and 0.85 mM DHF. In B the same except 0.68 mM DHF.

on at zero time. Besides the recording of the oxygen concentration and the chemiluminescence as time functions these two variables were also recorded on an x-y recorder yielding an inward spiralling trajectory.

Conclusion

It was found that the damped oscillations discovered by Yamazaki et al. in the peroxidase system also occur when dihydroxy fumaric acid or indole-3-acetic acid are

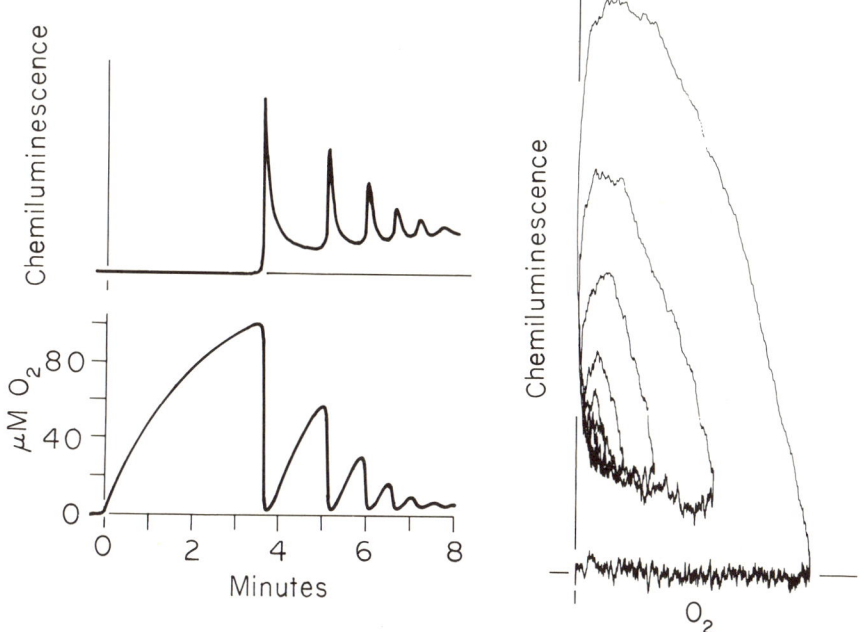

Figure 7. Simultaneous measurement of oxygen concentration and chemiluminescence during HRP catalyzed aerobic oxidation of IAA. The curve to the right is a phase trajectory obtained by recording the chemiluminescence as a function of the O_2 concentration on an x-y recorder. The concentrations are: 0.1 M phosphate buffer, pH 6.1, 0.8 µM HRP and 2.9 mM IAA. The oxygen in the gas is turned on at time zero.

substituted for NADH as the donor.

The HRP catalyzed aerobic oxidation reactions are accompanied by chemiluminescence whose intensity increases in the order NADH, DHF and IAA. Because of this intensity difference it is assumed that the chemiluminescence originates from intermediates derived from donor molecules. These intermediates are presumably free radicals. The measurement of chemiluminescence in the DHF oxidation reveals that the breakdown of complex III by a strongly chemiluminescent reaction takes place during a short interval of time when the donor is nearly exhausted. The breakdown of complex III is accelerated by 2,4-dichlorophenol.

The very weak oscillation in the complex III concentration during the oscillatory oxidation of DHF makes it unlikely that complex III has any essential role in the mechanism of the oscillation. The autocatalytic character of the reaction is sufficient to explain the damped oscillation. The experimental system is a realization of a theoretical system shown by Lotka to yield damped oscillations.

REFERENCES

1. Yamazaki, I., Yokota, K., and Nakajima, R., Biochem. Biophys. Res. Comm., 21, 6 (1965).
2. Yamazaki, I. and Yokota, K., Biochim. Biophys. Acta, 132, 310 (1967).
3. Keilin, D. and Mann, T., Proc. Roy. Soc. B 122, 119 (1937).
4. Yamazaki, I., Mason, H.S. and Piette, L.H., J. Biol. Chem., 235, 2444 (1960).
5. Yamazaki, I. and Piette, L.H., Biochim. Biophys. Acta, 50, 62 (1961).
6. Yamazaki, I. and Piette, L.H., Biochim. Biophys. Acta, 77, 47 (1963).
7. Ahnström, G. and Nilsson, R., Acta Chem. Scand., 19, 2 (1965).
8. Cormier, M.J. and Prichard, P.M., J. Biol. Chem., 243, 4706 (1968).
9. Lotka, A., J. Phys. Chem., 14, 271 (1910).
10. Vassil'ev, R.F., In *Progress in Reaction Kinetics* (Ed. G. Porter), Vol.4, Pergamon Press, New York, 1967, p.305.
11. Lindblad, P. and Degn, H., Acta Chem. Scand., 21, 791 (1967).
12. Degn, H., Nature, 217, 1047 (1968).

A SIPHON MODEL FOR OSCILLATORY REACTIONS IN THE
REDUCED PYRIDINE NUCLEOTIDE, O_2 AND PEROXIDASE SYSTEM

Isao Yamazaki and Ken-nosuke Yokota

Biophysics Division, Research Institute of Applied
Electricity, Hokkaido University
Sapporo, Japan

Since clear oscillations at the level of enzymic reactions were reported in the glycolytic system (1-3), chemical mechanisms for producing oscillatory kinetics have been presented by Higgins (4) and Sel'kov (5). The phosphofructokinase reaction is thought to be a possible source of oscillations in glycolysis and the mechanism is based on the feedback control such as substrate inhibition and product activation.

Oscillatory reactions have been observed also in an open system containing reduced pyridine nucleotide (NADH), O_2 and peroxidase (6,7). The mechanism of oscillations in the peroxidase system seems entirely different from those in glycolysis. A chain reaction which has been suggested for the mechanism of NADH oxidation catalyzed by peroxidase (8) seems to be involved in the mechanism of oscillatory reactions in the peroxidase system. This mechanism might well be explained on the basis of the siphon trick which is involved in a pipette washer.

An oscillatory cycle in the peroxidase system is composed of 3 phases: inductive (State I), active (State II) and terminating (State III). State I corresponds to the lag phase of NADH oxidation caused by the presence of excess peroxidase. An active intermediate, O_2^- is captured by ferriperoxidase and produces an inactive product of the enzyme, called Compound III. Consequently, NADH oxidation is retarded during this period. When the conversion from ferriperoxidase to Compound III reaches a critical level, the NADH oxidation suddenly commences. This critical moment can be seen in Figure 1 as sudden decrease in the O_2 concentration

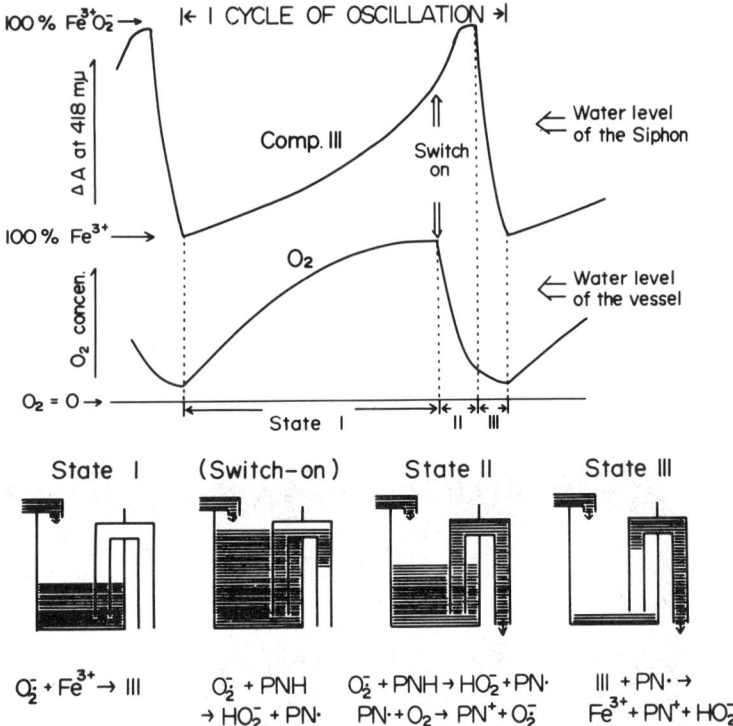

Figure 1. Schematic representation of 1 cycle of oscillation in the NADH, O_2 and peroxidase reaction. Peroxidase oscillates between ferriperoxidase (Fe^{3+}_p) and Compound III ($Fe^{3+}_p O_2^-$) in a similar oscillatory pattern of O_2 concentration in the reaction mixture. Water level in the vessel of a pipette washer will correspond to O_2 concentration in the peroxidase system and water level in the siphon, on the other hand, may be compared to O_2 level in the enzyme which can be measured as the accumulation of Compound III. Elementary reactions characteristic of the corresponding stages of a pipette washer are described below the siphon model.

of the reaction solution. The rapid NADH oxidation (State II) continues until the O_2 concentration falls below a certain limit and is followed by a characteristic phase, State III, which switches off the reaction. Compound III decomposes during the period of State III and the enzyme returns to the original state. When oxygen is being supplied continuously the reaction system comes back to the original state and will repeat the cycle.

This mechanism is now compared with the model of a pipette washer (Fig.1). Water influx to the vessel will take the place of O_2 supply in the peroxidase system. Water flows from the duct (siphon) just after it is filled with water. The siphon empties of water when the water level of the main vessel decreases to the level of the siphon edge. In Figure 1, the characteristic elementary reactions which take place dominantly in the peroxidase system during the period corresponding to each stage of the pipette washer are described.

It should be noticed here that Compound III is a sluggish intermediate for NADH, but its slow decomposition (Reaction c' and h) provides little by little active intermediates, monodehydro-NADH (NAD·) via Reaction a to the chain reaction which is shown in Figure 2 (9). Peroxidase captures O_2 as Compound III after O_2 accepts an electron from the NAD· radical. Propagation of the chain reaction (Fig.2) will start at the time when most of the enzyme is saturated with O_2. At this moment, the gain of the chain reaction will exceed unity, as was discussed previously (7) and it seems desirable to term this moment "switch-on". During the fast oxidation of NADH (State II) the gain of the chain reaction may be kept at a maximum. A particularly ingenious mechanism in the peroxidase oscillatory system is that the enzyme becomes empty of O_2 just after the O_2 concentration in the solution decreases below the certain level. It is easily understood that this trick is very similar to that involved in a pipette washer, and the most important mechanism for oscillation. This phase (State III) indicates the transition from the active state to the inactive one and may be not inaptly termed "switch-off". The switch-off step seems to be a kind of reductive decomposition of Compound III (7,9) but is the most ambiguous reaction in this oscillatory system which needs further investigation.

Phenol derivatives are known to promote the peroxidase-oxidase reaction by stimulating Reaction a and have been

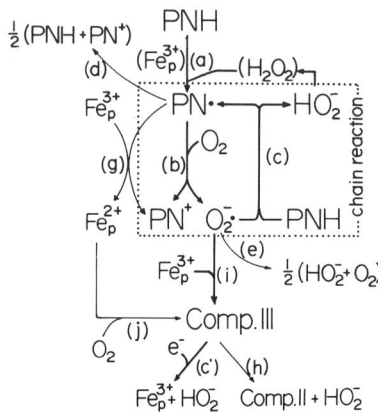

Figure 2. The mechanism of the aerobic oxidation of NADH by peroxidase (Fe_p^{3+}). Reaction <u>a</u> indicates one-electron oxidation of NADH catalyzed by peroxidase in the presence of H_2O_2. Thick lines indicate the reactions which will participate positively in the oscillatory mechanism. As H_2O_2 gives 2 moles of NAD· through Reaction <u>a</u>, the gain of the chain reaction will be defined in the following equation (7).

$$\text{gain} = \frac{v_b}{(v_b+v_d+v_g)} \times \frac{(3v_c+v_e)}{(v_c+v_e+v_i)}$$

recently found to accelerate the switch-off reaction. Methylene blue, on the other hand, has a slightly inhibitory effect on the oxidation of NADH by peroxidase but exhibits an elaborate effect on the switch-off reaction. It appears to compete with oxygen for NAD· and to promote the reductive decomposition of Compound III in the presence of a critical amount of oxygen. Consequently, the addition of both phenol and methylene blue greatly stabilizes the oscillation in the peroxidase system. It is generally accepted that the fundamental condition necessary for sustained oscillation is to keep the reaction system open. From this point of view there is a possibility of improving the stability of oscillation in the NADH-O_2-peroxidase system by coupling it with a dehydrogenase system. The NADH oxidation by peroxidase is active at acidic pH and the choice of dehydrogenase system to be coupled with this reaction is therefore rather limited. Glucose-6-phosphate dehydrogenase has been the most successful and one of the results is shown in Figure 3.

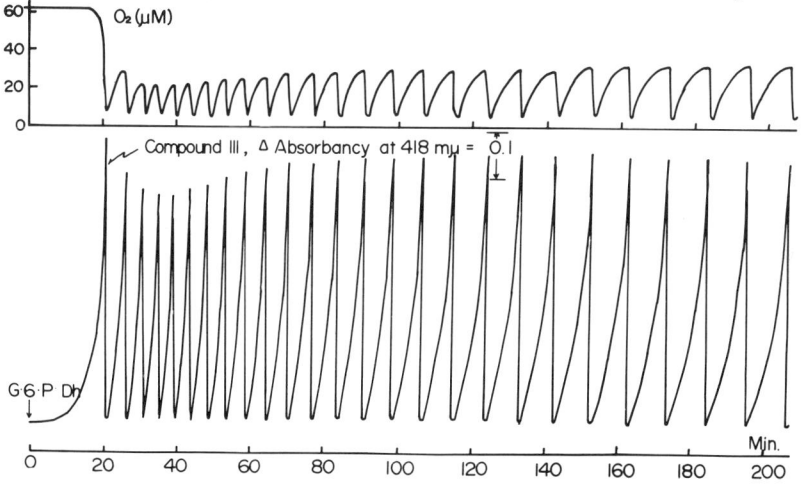

Figure 3. Sustained oscillation in the peroxidase system coupled with glucose-6-phosphate dehydrogenase. Reaction was started by the addition of dehydrogenase. 10 µM horse radish peroxidase, 2 mM $NADP^+$, 20 mM glucose-6-phosphate, 20 µM 2,4-dichlorophenol, 0.6 µM methylene blue and 0.05 M acetate (pH 5.6) at $25°$. Five percent O_2 diluted with nitrogen was bubbled at a constant rate (10 ml per min) into the reaction solution (6 ml) through a capillary.

Elongation of the oscillation pattern as the reaction goes on is due to the decrease in the NADH level caused by the transformation from glucose-6-phosphate to 6-phospho-gluconate in the solution. This oscillatory pattern will probably be modified by many other biochemical substances or biochemical reaction systems and such modification is now under investigation in our laboratory.

REFERENCES

1. Chance, B., Schoener, B., and Elsaesser, S., J. Biol. Chem., 240, 3170 (1965).
2. Frenkel, R. Biochem. Biophys. Res. Commun., 21, 497 (1965).
3. Hess, B., Brand, K., and Pye, K., Biochem. Biophys. Res. Commun., 23, 102 (1966).
4. Higgins, J., Proc. Nat. Acad. Sci. U.S., 51, 989 (1964).
5. Sel'kov, E.E., Euro. J. Biochem., 4, 79 (1968).
6. Yamazaki, I., Yokota, K., and Nakajima, R., Biochem. Biophys. Res. Commun., 21, 582 (1965).
7. Yamazaki, I., and Yokota, K., Biochim. Biophys. Acta, 132, 310 (1967).
8. Yokota, K., and Yamazaki, I., Biochim. Biophys. Acta, 105, 301 (1965).
9. Yamazaki, I., Yokota, K., and Tamura, M. In Hemes and Hemoproteins, (B. Chance, R.W. Estabrook and T. Yonetani, eds.), Academic Press, New York, 1966, p.319.

DAMPING OF MITOCHONDRIAL VOLUME OSCILLATIONS
BY PROPRANOLOL AND RELATED COMPOUNDS

A.J. Seppälä, M.K.F. Wikström and N.-E.L. Saris

Department of Clinical Chemistry,
University of Helsinki,
Helsinki, Finland

Isolated mitochondria can be induced to exhibit damped oscillations of volume, ion movements, oxidation-reduction states of the respiratory carriers, the respiratory rate and fluorescence of membrane-bound ANS (1-6). It has been shown that there is a phase difference between the different parameters, the changes in the oxidation-reduction state of the respiratory carriers and the changes in ANS-fluorescence preceding the energy-linked functions of ion transport (3,6). The two extreme conditions during the oscillations are the contracted state, - in which the respiratory carriers are largely oxidized, the respiratory rate or ATPase activity are high, the cation content of the mitochondria is low and protons have been taken up, - and the swollen state, in which these parameters are reversed, cations being accumulated and protons extruded. Swelling can be prevented and contraction induced by the addition of electron transport inhibitors when ion transport is driven by substrate oxidation, or by oligomycin when it is driven by ATP, or by uncouplers of oxidative phosphorylation in both cases.

In this study we have used agents which possess membrane effects. They all contain nitrogen atoms that may become positively charged and they all exhibit lipophilic characteristics. Propranolol is a β-adrenergic blocking agent which in erythrocytes induces loss of potassium ions and shrinkage (7). Butacaine is a highly potent local anesthetic. Chlorpromazine is a so-called major tranquilizer and the chemically closely-related promethazine is a potent antihistaminic drug. Mela (9) has found an increased uptake of divalent cations into mitochondria and a concomitant enhancement of bromothymol blue absorbance at 618 nm in the presence of butacaine and phenthiazines. Laurylamine is known to uncouple oxidative phosphorylation (8).

Methods

Rat liver mitochondria were prepared as described previously (10). The experimental conditions were similar to those described by Packer et al. (5), with a few modifications.

Changes in optical density were followed with an Aminco-Chance dual wavelength spectrophotometer set at 520 nm, with the other wavelength excluded. Oxygen consumption was assayed with a Gilson Oxygraph and ANS-fluorometer Model A-3 equipped with a primary filter 7-60 and a secondary filter 3-72. Mitochondrial protein was estimated according to Lowry et al. (11).

Results

Figure 1 (A,B and C) shows the effects of propranolol added at various points in the swelling-shrinkage cycle. Propranolol caused damping of the mitochondrial volume oscillations, with the swollen state being preserved in the presence of this agent and the spontaneous contraction phase apparently being inhibited. The swollen state is still respiration-dependent as may be seen in Figure 1A. Shrinkage is induced when oxygen is depleted or when an uncoupler (FCCP) is added (Figure 1C).

The swelling in the presence of propranolol is accompanied by an increas in the initial rate of respiration, as shown in Figure 2. Propranolol also abolishes the oscillations in the respiration rate.

The effect of propranolol on volume oscillations in the presence of valinomycin is shown in Figure 3. The oscillations become gradually more damped with increasing concentrations of propranolol and a steady-state is reached at a more highly expanded mitochondrial volume. From this figure it is clearly seen that the main effect of propranolol is an inhibition of the spontaneous shrinkage phase, while the rate of swelling is not significantly affected.

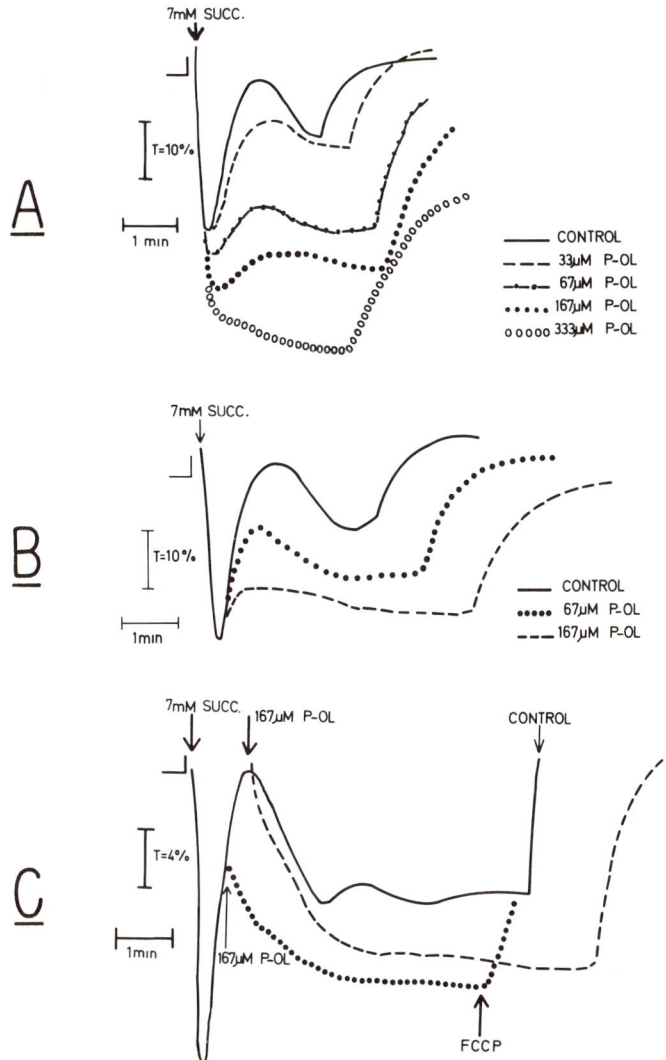

Figure 1. Effect of propranolol (P-ol) on mitochondrial volume oscillations. Experimental conditions: Medium, 40 mM NaPi, 30 mM Tris, 100 mM Sucrose, 0.5 mM EDTA, 5 µM rotenone, pH 8.2. Addition of FCCP in C was 1 µM. In A propranolol was added before succinate. Mitochondrial protein in A, 1.25 mg/ml, B, 0.89 mg/ml, C, 0.81 mg/ml.

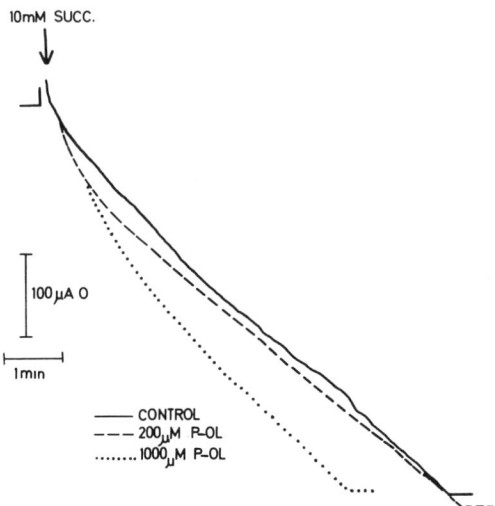

Figure 2. Effect of propranolol on succinate oxidation. Experimental conditions as in Figure 1. Mitochondrial protein, 1.31 mg/ml.

Propranolol causes the same general effect in the presence of gramicidin (Figure 4). The period of oscillation is shortened in the presence of both antibiotics, probably due to the increased permeability of the membrane to monovalent cations. In the presence of propranolol the addition of gramicidin induces further swelling, but when ammonium ions are also present gramicidin causes contraction (Figure 5). This effect is presumably due to the fact that gramicidin renders the membrane permeable to ammonium ions which, when accumulating, exchange for sodium ions according to the mechanism suggested by Chappell and Crofts (12). This latter effect may thus contribute to the observed contraction. The contraction induced by gramicidin plus ammonium ions is unaffected by propranolol (Figure 5).

Figure 6 shows oscillations in the ANS-fluorescence under similar conditions. Propranolol was found to damp these oscillations in the "high-energy state" (low fluorescence), in accordance with the previous data on volume oscillations.

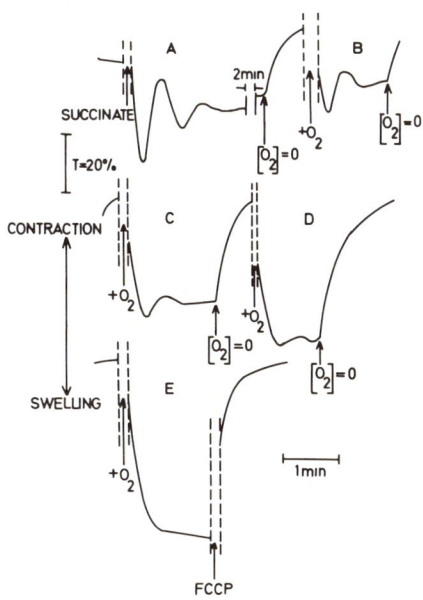

Figure 3. The effect of propranolol on volume oscillations in the presence of valinomycin. Standard reaction mixture as in Figure 1, except that 20 mM KH_2PO_4 was used. Valinomycin concentration was 17 µg/ml, mitochondrial protein, 0.63 mg/ml The experiments (traces A-E) were performed on the same suspension. Every new trace (B-C) was induced by reoxygenation. In the beginning of the experiment (trace A), 7 mM succinate was added and the volume oscillations were recorded. Trace B is the control in the absence of propranolol. Traces C, D and E are in the presence of 167 µM, 333 µM and 500 µM propranolol, respectively. At the end of the experiment (trace E), 1 µM FCCP was added to induce uncoupling.

The effect of chlorpromazine in damping oscillations is very similar to that of propranolol. In this case the rate of swelling seems to be a little slower than in the control and the amplitude is somewhat larger. The damped oscillations in the presence of chlorpromazine are also readily reversed by the addition of uncouplers, as shown in Figure 7. The effects of chlorpromazine and propranolol are the same when the respiratory substrate is TMPD plus ascorbate, instead of succinate.

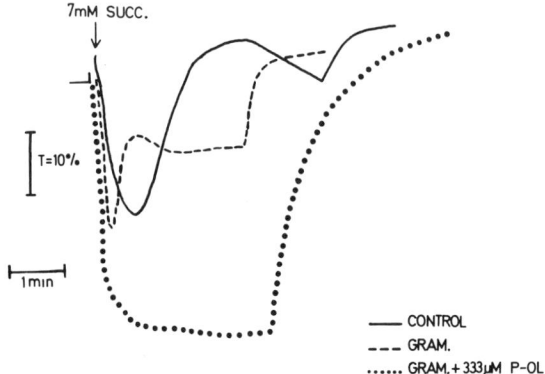

Figure 4. Effect of propranolol on volume oscillations in the presence of gramicidin. Experimental conditions as in Figure 1. Gramicidin concentration was 170 nM and mitochondrial protein, 1.42 mg/ml.

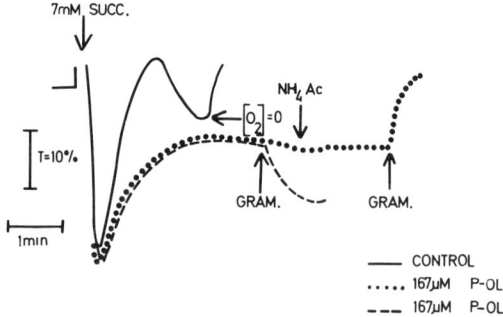

Figure 5. Effects of gramicidin and ammonium ions on the propranolol-damped state. Experimental conditions as in Figure 4. Ammonium acetate (7mM) was added where indicated. Mitochondrial protein, 1.6 mg/ml.

In general the effects on the volume oscillations of butacaine, promethazine and laurylamine very much resemble those of propranolol (Figure 8).

Discussion

Our main finding is that the compounds studied caused an inhibition of the rate of the oscillatory contraction phase and a simultaneous damping of the oscillations in

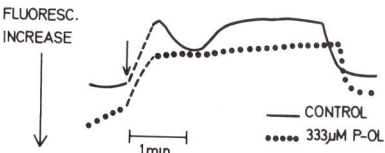

Figure 6. Effect of propranolol on oscillations in ANS-fluorescence. Experimental conditions as in Figure 1. 33 μM ANS was used. Mitochondrial protein, 1.26 mg/ml. Tris-succinate (7 mM) added at arrow.

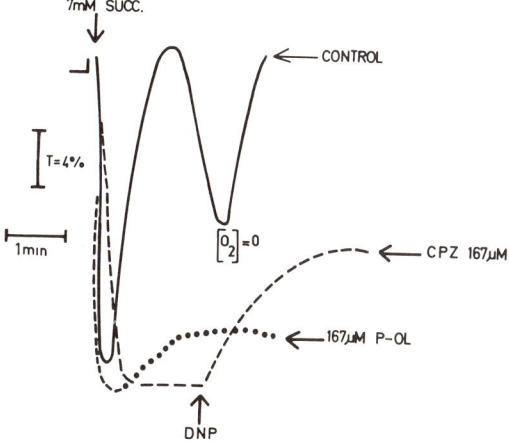

Figure 7. Effect of chlorpromazine (CPZ) on volume oscillations. Experimental conditions as in Figure 1. 2,4-dinitrophenol (DNP) concentration was 67 μM and mitochondrial protein, 1.6 mg/ml.

mitochondria. Contraction induced by other means (e.g. by the expenditure of oxygen, the addition of uncouplers or electron transport inhibitors or gramicidin plus ammonium ions) was not found to be significantly affected. The initial rate of swelling was also little affected. Thus the higher steady-state extent of swelling in the presence of the various agents is more likely to be due to inhibition of contraction than to acceleration of swelling. This very specific action of the compounds posed the question as to what could be their most likely mechanism of action.

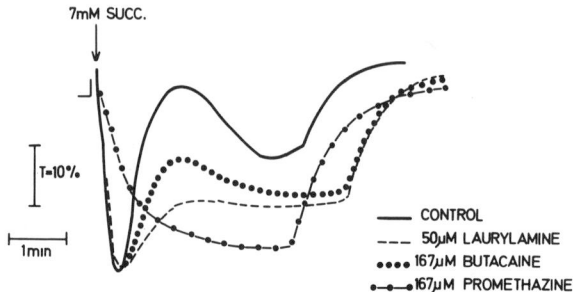

Figure 8. Effects of butacaine, promethazine and laurylamine on volume oscillations. Experimental conditions as in Figure 1. Mitochondrial protein, 0.97 mg/ml.

A specific interaction with ion carriers of the membrane seems very unlikely in view of the fact that the agents inhibited contraction only and even this was under highly specific conditions. Furthermore, the effect was found with a great variety of compounds, which supports this conclusion. The increased initial rate of respiration in the presence of propranolol is probably not due to uncoupling but rather to the simultaneously-occurring increased capacity for ion uptake. In other words it is unlikely that these agents affect the primary energy conserving processes - cf. "classical" uncouplers: 2,4-dinitrophenol, FCCP etc. (13). This is in contrast with previously reported interpretations of the action of chlorpromazine and laurylamine (8).

It seems to us quite plausible that the agents tested are acting by complexing with some components of the membrane, presumably with phospholipids. It is known that most of these compounds can be bound by phospholipids (14,15), e.g. phenothiazines strongly interact with cardiolipin (16), one of the major components of mitochondrial inner membranes. The distribution and density of fixed negative charges in the membranes and/or in the pores of the membranes are probably of great importance in determining their conformational and permeability properties (17). Chance et al. have shown evidence for a rearrangement of the charge distribution in the mitochondrial membrane in the transition from the non-energized to the energized state (18). The finding that propranolol affects the ANS-fluorescence signal in the same fashion as it affects the volume changes shows that the configuration of the membrane is stabilized in the "high-energy" state. A mechanical stabilization, such as an unspecific increase in membrane viscosity or rigidity, by propranolol is unlikely because only oscillatory contraction is inhibited. This focuses interest upon the elasticity forces which would become important when swelling has proceeded to the point where the foldings of the membrane (<u>cristae mitochondriales</u>) have vanished. The agents tested which bind to the membrane may well change its elasticity properties, for instance by combining with fixed negatively charged groups of phospholipids. The specific inhibition of oscillatory contraction may in this context be viewed as conforming to the "overflow" or leakage mechanism introduced by Teorell for nerve cell membranes (17). Damping of the oscillations may thus result from inhibition of this leak, e.g. by a decrease in the elasticity forces of the expanded membrane, with an inhibition of outwardly directed flows.

With regard to the pharmacological effects of these agents it is interesting to note that most of them have local anesthetic properties. In addition, propranolol inhibits contraction-relaxation cycles of the gravid uterus and also stops rhythm distrubances of the heart. The latter effect may also be brought about by some local anesthetics. All these examples more or less represent damping effects on oscillations. Whether the strong tranquilizing action of chlorpromazine is due to a similar stabilizing effect on the nerve cell membrane is not known, but this would seem to be an interesting possibility.

Acknowledgements

This study was supported by a grant from the Sigrid Juselius Foundation. The authors are grateful to Mrs. A. Sarasjoki for technical assistance.

References

1. Pressman, B.C. Fed. Proc. <u>24</u>, 425 (1965).
2. Graven, S.N., Lardy, H.A., and Rutter, A. Biochemistry, <u>5</u>, 1735 (1966).
3. Utsumi, K. and Packer, L. Arch. Biochem. Biophys. <u>120</u>, 404 (1967).

4. Deamer, D.W., Utsumi, K., and Packer, L. Arch Biochem. Biophys. 121, 641 (1967).

5. Packer, L. and Wrigglesworth, J.M. In The Energy Level and Metabolic Control in Mitochondria (Ed. S.Papa, J.M. Tager, E. Quagliarello and E.C.Slater) Adriatica Editrice, Bari, 1969, p.125.

6. Packer, L., Donovan, M.P. and Wrigglesworth, J.M. Biochem. Biophys. Res. Commun. 35, 832 (1969).

7. Ekman, A., Manninen, V. and Salminen, S. Acta Physiol. Scand. 75, 333 (1969).

8. Lees, H. Biochim. Biophys. Acta 131, 310 (1967).

9. Mela, L. Arch. Biochem. Biophys. 123, 286 (1968).

10. Wikström, M.K.F. and Saris, N.-E.L., Europ. J. Biochem. 9, 160 (1969).

11. Lowry, O.H. Rosenbrough, O.H., Farr, N.J. and Randall,R. J. Biol. Chem. 193, 265 (1951).

12. Chappell, J.B. and Crofts, A.R. In Regulation of Metabolic Processes in Mitochondria (Ed. J.M. Tager, S.Papa, E. Quagliarello, and E.C. Slater) BBA Library, Vol.7, Elsevier, Amsterdam, 1966. p.293.

13. Ernster, L. and Lee, C.P. Ann. Rev. Biochem. 33, 729 (1964).

14. Seeman, P.M. Intern. Rev. Neurobiol. 9, 145 (1966).

15. Feinstein, M.B. J. Gen Physiol. 48, 357 (1961).

16. Demel, R.A. and van Deenen, L.L.M. Chem. Phys. Lipids 1, 68 (1966).

17. Teorell, T. Ann. N.Y. Acad. Sci. 137, 950 (1966).

18. Chance, B., Azzi, A., Radda, G.K. and Lee, C.P. In Electron Transport and Energy Conservation (Ed. S.Papa, J.M. Tager, E.C. Slater and E. Quagliarello) Adriatica Editrice, Bari, 1970.

III
GLYCOLYTIC OSCILLATIONS

THE CONTROL THEORETIC APPROACH TO
THE ANALYSIS OF GLYCOLYTIC OSCILLATORS

Joseph Higgins, Rene Frenkel, Edward Hulme,
Anne Lucas, Gus Rangazas

Johnson Research Foundation, University of Pennsylvania
Philadelphia, Pennsylvania 19174

I. INTRODUCTION

As witnessed by this symposium, considerable progress has been made in the understanding and recognition of oscillating biochemical reactions. Since the earlier discovery of oscillations in anaerobic yeast cell preparations (1,2) a great number of experimental studies (e.g. this volume) have provided data on the effects of many substances on the oscillations as well as detailed kinetics for the glycolytic intermediates in both cells and cell extracts. Contrary to our initial expectations that rare and relatively unique chemical mechanisms would be associated with such oscillations, early theoretical developments (3) demonstrated that such mechanisms required simple properties typical of many enzyme systems and suggested that limit cycle oscillatory (and other non-linear) (4,5) behavior should probably be commonplace in biochemical kinetics. Further theoretical studies (6,7) have only enlarged the class of mechanisms which can exhibit these phenomena.

For the glycolytic pathway, based on the <u>in vitro</u> control character of the enzymes, there are a dozen mechanisms, corresponding to distinct control situations, which could account for the oscillatory dynamics. Thus, at this stage the problem is no longer one of providing a basic understanding of biochemical oscillations, but rather the determination of the details which account for the oscillations in specific circumstances. The analysis and formulation of this detailed aspect is not limited to oscillatory dynamics <u>per se</u>; the approach is just as applicable to the analysis of any dynamic response and the oscillatory kinetics provides just one example, though it is of particular interest at this time. In order to contrast the approach we have used with other methods, we shall call it the "control theoretic" approach. The objectives of this paper will be to demonstrate its utility as a basis for theoretical deductions, for the formulation of computer models and for the direct analysis of kinetic data for multi-enzyme systems in terms of the underlying properties of the individual reactions. We shall direct our examples to studies of the oscillating glycolytic system.

II. THE CONTROL THEORETIC APPROACH

It has long been recognized that the expression of enzyme and multienzyme system kinetics finds its formulation in the simultaneous set of non-linear differential equations which derive from the application of the Guildberg-Waage-Van't Hoff Law to the underlying elementary reaction steps. Though fundamentally correct, that formulation is of little value in practice since the equations are analytically intractable and generally involve too many parameters for even rational computer study. Such a circumstance virtually demands that other approaches be formulated which can provide answers as to the properties of the system without the need for detailed integration of the equations. Although some results of a control theoretic nature such as the Poincare-Bendixon Theorems and the Liapounov (8) method were developed long ago, most of the concepts, methods and terminology appear to derive from the more recent (over the past five decades) studies of electronic systems where a similar formulation in terms of non-linear differential equations prevailed. The analogy between electronic circuits and biochemical pathways is very deep. Both systems possess linear and non-linear control elements (9,10) (i.e. resistors, capacitors, and vacuum tubes or transistors as compared to first order reactions and enzyme reactions). In electronics, currents (fluxes) are determined by concentrations. And in both cases the control elements are connected together into a topological mesh called a circuit or pathway which can and often does involve feedback. The dynamics of the system is determined by the nature of each control element and the topology of the connections. Finally, the control characteristics of the non-linear elements are similar in both cases, usually being monotonically increasing or decreasing functions of the variables and typically displaying saturation effects. The earlier days of these electronic developments provided much in the way of graphical methods of analysis as well as a broad understanding of the effect of feedback on the overall dynamics. Unfortunately for the general understanding of non-linear systems, these non-linear methods soon gave way to the more tractable "linear systems" and "small signal analysis" (11), perhaps because would-be mathematicians prefer linear systems or because electronics is a synthetic science and once reliable systems of interest could be constructed there was little reason to investigate more complicated systems. In biochemistry and biology, the system is provided by nature and the problem is to analyze and understand the properties which it can or does exhibit. As a result, some of our developments and methods in the control theoretic approach are new and have been motivated by the study of biochemical systems.

Our formulation of the control theoretic approach is based on the tenet that the qualitative features of the system dynamics are determined by the qualitative features of the individual reaction control characteristics, the relative relations between these control characteristics and the topology of the pathways. By qualitative features we refer to directions of slopes, maxima, minima, inflections, changes of curvature, etc. while relative relations between control characteristics refers to the existence of intersections (when appropriate) and the qualitative size of slopes (i.e. slope 1 greater than slope 2, etc.). In accord with this tenet, changes in individual control characteristics which preserve the qualitative and relative properties of the individual control characteristics will also preserve the qualitative features of the system dynamics. We have no direct proof that this tenet is true nor do we know what limitations should be placed on the functional forms describing the control characteristics. We anticipate that exceptions can be found in the domain of non-linear differential equations and that these would delineate the bounds of the tenet and perhaps further limit or extend the meaning of "qualitative features". Nevertheless, the tenet has provided motivation and found <u>a posteriori</u> justification from both theoretical (12) and computer studies (13,14). Thus far we know of no exceptions in the domain of non-linear dynamics encountered in multi-enzyme systems.

While the basic elements and simple applications of the control theoretic approach have been discussed in detail elsewhere (4,5,7) it is helpful to review some of these concepts here. In most cases of cellular dynamics, the individual enzymes can be treated as obeying their stationary state equations since the characteristic time for exciting the enzymatic intermediates is much smaller than that for changes in the cellular intermediates (e.g. the glycolytic intermediates). The stationary state equations, which may find an algebraic expression (e.g. the Michaelis-Menten equation) or may be given simply as a graphical plot of rate versus concentrations, are referred to as the control characteristics and designated as $v = v(x)$ where x represents the concentration of some substrate or allosteric effector. The most prominent qualitative feature of the control characteristics is the algebraic _sign_ of the slopes, being designated as activational if plus, since the rate increases for an increase of effector; and inhibitory if the slope is negative, corresponding to a decreasing rate for an increasing effector; these features are collectively referred to as the "control character" of the interaction. For many enzymes, the control character is independent of the concentrations, while for more complex enzymes in which a substance may be both a substrate (hence activating) and an allosteric inhibitor, the character of the interaction will depend on the concentration level. The second feature of the control characteristics is the magnitude of the slopes (as evidenced by the value of the partial derivatives or logarithmic derivatives) which are referred to as the control strength and which generally depend on the concentrations. However, the qualitative features of the _system_ dynamics generally depends only on the relative values (i.e. greater or less than) of different control strengths and not on the exact magnitudes. For example, the substrate for a simple enzyme system (irreversible) has a region of logarithmic control strength of one (i.e. a given percentage change in substrate causes the same percentage change in rate) when the substrate is much less than the Michaelis Constant; and a region of "no control" when the substrate is much greater than the K_m and changes in the substrate level do not effect the flux. The basic tenet of the control theoretic approach can be restated as the expectation that the entire set of qualitative features which can be exhibited by the system dynamics as a function of the parameters is entirely exhausted by the study of the small number of states in which the control strengths achieve all possible arrangements relative to one another. The difficulty at our current state of knowledge is that we are not always aware _a priori_ of the possible domains of arrangement for the relative strengths which can depend on inequalities among the sums and products of the individual reaction control strengths.

As just described, the control theoretic approach provides a level of description for the kinetic properties of enzymes which is intermediate between the crude expression of an enzyme as catalyzing a particular reaction and a highly detailed description which would provide the complete control characteristics and perhaps even the underlying elementary chemical reactions. In principle, if the detailed control characteristics and their parameters (V_{max}'s, K_m's, activation constants, enzyme concentrations, etc.) were known, it would be possible to determine (with computers) which, if any, of a number of specific mechanisms afforded a realistic basis for the experimentally observed dynamics. Unfortunately, there are both practical and theoretical reasons which dictate against any direct a priori application of such information.

First, the detailed control characteristics of individual enzymes, as determined in vitro are usually and often necessarily limited in scope. Thus for simple enzymes (e.g. one substrate and product), the available experimental data is often limited in range of pH, temperature and even substrate concentrations. For complex enzymes (such as phosphofructokinase) which have complex substrate and product control characteristics as well as other known activators and inhibitors, even a reasonably complete elucidation of the control space would require several million experimentally determined values, an impractical task at this time. Second, even if it were practical to obtain complete data for individual enzymes, there would be (and are) potential theoretical objections to the a priori application of such results to the behavior of the enzymes in vivo or in complex multienzyme systems such as cytoplasmic extracts. The in vitro control characteristics could be readily modified quantitatively (and even qualitatively) by enzyme-enzyme interactions, membrane binding effects and the presence of unrecognized modifiers such as other ions or substrates.

Certainly there can be no question that a knowledge of the in vitro properties of the individual enzymes provides one starting point for the analysis of in vivo or extract properties. But it should be anticipated that the quantitative values of the control parameters will be modified in many cases; while the broad control properties will be effected in far fewer cases and finally, but only in rare cases, will there be an entire loss of activity. As a result, the analysis of in vivo and extract data must be approached as a test of the applicability of the in vitro data with the expectation of modifications and as a guide to further in vitro studies.

Conversely, the direct analysis of in vivo and extract kinetic experiments through the control theoretic approach, can lead to the determination of the complete control characteristics for individual enzymes. Of course, there are limitations and assumptions involved which require support from in vitro experiments. Nevertheless, the method has the advantage of providing a procedure and method for the interpretation and utilization of in vivo (or extract) data independent of in vitro experiments. In terms of these three levels of describing enzyme kinetics properties, it may be noted that most of the direct in vivo studies have concentrated on the existence of enzyme activities as deduced from the occurrence of various metabolic and catabolic reactions. More recently, the analysis of in vivo kinetics has utilized the cross over theorem (15,16) which provides a direct interpretation and analysis of the stationary state perturbation data at the level of the broad control properties.

There are several reasons for emphasizing the analysis of the data at the level of broad control. First, as already noted, the broad control properties are relatively invariant to changes in the parameters. Hence they are more likely to persist between in vivo and in vitro experiments even though the detailed control characteristics are modified. Second, they provide a description of biochemical systems at the level of regulation and control which is considerably more detailed than the level of activity alone. Third, as we have already noted, there is a growing body of theoretical and computer studies which suggest that the qualitative features of the dynamics for chemical systems are primarily sensitive to the broad control properties.

III. THEORETICAL ASPECTS OF BIOCHEMICAL OSCILLATORS

A. TWO COMPONENT OSCILLATORS

Under some experimental circumstances, many of the

components of a complex system may lack relevant or significant control or may be maintained essentially constant and the system may be describable in terms of just two significant chemical components, say X and Y. The analysis of two component systems has been presented elsewhere (3,7) and the necessary conditions for oscillatory behavior can be summarized in terms of a net flux diagram:*

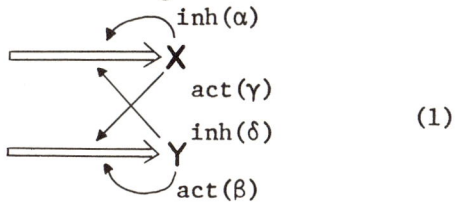

(1)

where the large arrows indicate the <u>net</u> rate of production for X and Y; that is, the sum of all reaction rates which can produce the component minus the sum of all reaction rates which can remove that component. The small arrows indicate the net effect of each component on its own net production (designated self coupling) and the net effect on the production of the other component (designated cross coupling). The labels (act. or inh.) on the small arrows indicate whether the net effect of an <u>increase</u> in the component is to activate or inhibit the net production; these terms correspond to the <u>sign</u> of the partial derivatives (evaluated at the singularity) when the rate laws are formulated as:

$$\dot{x} = \overline{X}(x,y) \qquad \alpha = \left.\frac{\partial \overline{X}}{\partial x}\right)^* \qquad \beta = \left.\frac{\partial \overline{Y}}{\partial y}\right)^*$$

and

$$\dot{y} = \overline{Y}(x,y) \qquad \gamma = \left.\frac{\partial \overline{X}}{\partial y}\right)^* \qquad \delta = \left.\frac{\partial \overline{Y}}{\partial x}\right)^* \quad (2a)$$

where the dot (°) indicates the time derivative and the star indicates evaluation at the singularity. The signs of the partial derivatives are interpreted as net activational (if plus) and net inhibitory (if minus) and are referred to as the control character of the interaction; these signs can usually be determined by inspection of the associated mechanism without any need for detailed analysis.

The necessary conditions for oscillations can then be

---

* Professor U.F. Franck has brought to my attention that a similar diagram was developed by Bonhoeffer (17) in his earlier studies of system dynamics.

stated as:

1. (a) The two self coupling terms must be of opposite character and
   (b) The two cross coupling terms must be of opposite character when evaluated at the singularity. (2b)
2. (a) The magnitude of the product of the cross coupling terms must be greater than the magnitude of the produce of the self coupling terms; and
   (b) The sum of the self coupling terms must be net positive.

Together these two conditions simply represent a restatement in control theoretic terms of the standard stability analysis of a singularity. To the first condition there corresponds four distinct net flux diagrams, determined by the location of the act and inh effects, and to each distinct diagram there corresponds an unique X,Y phase plane trajectory in terms of the direction of rotation and major axis (7).

The simplicity of applying these conditions is easily realized by some examples from glycolysis. Consider a reduced representation of glycolysis as:

$$[\text{GLU}] \xrightarrow{v_{in}} \text{F6P} \xrightarrow{v_p} \overset{\text{act}}{\frown} \text{FDP} \xrightarrow{v_o} \quad (3)$$

where v_{in} is the constant input flux for F6P as determined by the level of glucose; v_p is the flux through the phosphofructokinase reaction (the adenine nucleotides are considered essentially constant or out of control) and it is assumed that F6P exerts normal substrate control (activation) and the product, FDP, allosterically activates the reaction; v_o represents the flux through the lower glycolytic pathway (the fact that the actual control step may be through the glyceraldehyde-phosphate dehydrogenase reaction makes no difference to this analysis).

The corresponding net flux diagram is easily realized to be:

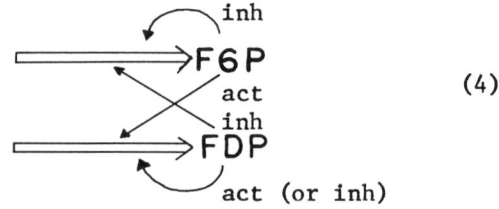

(4)

since: 1) an increase in F6P does not effect its input (v_{in}) but increases the rate of its output (v_p) thereby causing a decrease (inh) in its net production. At the same time the increase in v_p (the input for FDP) causes an increase (act) in the net rate of production of FDP. An increase in FDP increases the rate of removal of F6P (via the increased v_p) and hence represents an inhibition of the net rate of F6P production. The self coupling effect of FDP is somewhat more complicated: increasing FDP increases its input rate (v_p) but in general would also increase the rate of its output via activation (substrate control) of v_o. Thus, in order for FDP to exhibit a net self activation, the control strength for the FDP activation of v_p must be greater than the control strength for the output flux (v_o) as indicated by the dashed line in the reaction diagram. In this example, the conditions (2b) on the relative magnitudes of the control strengths can be easily fulfilled by choosing appropriate values of the parameters for typical enzyme reaction mechanisms.

Figures 1A and 1B show typical computer solutions for this mechanism for different parameter values corresponding to different control arrangements. Damped oscillations occur when the control strength of FDP on v_o exceeds its activation strength on v_p; and no oscillations occur when the FDP activation of v_p is essentially removed. This example should emphasize that the existence of the appropriate net flux diagram for oscillation (and the corresponding mechanistic controls) is a necessary condition for the oscillations to occur; the system will not oscillate when these controls are effectively removed by specific parameter values. However, if the necessary control is not present at all (in this example the FDP activation of pfk) then there are no parameter settings which could yield sustained oscillations, even though the individual enzyme reactions may involve rather complicated algebraic rate laws (given of course that they are consistent with the recognized control character).

In as much as the preceeding analysis has been largely analytic, it tends to obscure the actual basis of the oscillations and the power of the control theoretic approach as a basis for the interpretation of the qualitative aspects of the kinetics as well as the invariance of the results to the details of the individual reaction features. Figure 2A illustrates typical control characteristics which might be effective for the rate laws of v_p and v_o. Figures 2B and

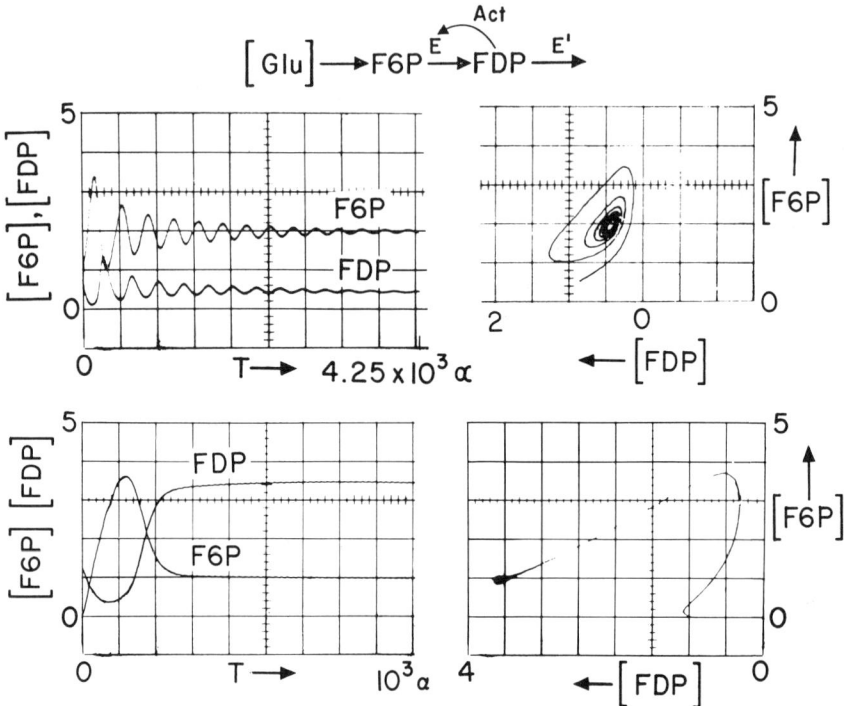

Figure 1A. Computer solutions and phase plane plots for the mechanism indicated. Upper graphs: for weak FDP activation of E. Lower graphs: for no FDP activation of E.

2C show typical (somewhat exaggerated) oscillatory solutions for this system and the corresponding individual reaction rates; the graph is divided into four regions according to the shifts of relevant control changes.

Note that the previous analysis with FDP = y and F6P = x demonstrated that for oscillations to occur the conditions as derived from Equation 2a for this system:

$$\left| \frac{\partial v_p}{\partial y} \right| > \left| \frac{\partial v_p}{\partial x} + \frac{\partial v_p}{\partial y} \right| \quad \text{and} \quad \left| \frac{\partial v_p}{\partial x} \right| > 0$$

must be satisfied for the control strengths and these are

BIOCHEMICAL OSCILLATORS

Back Activation Oscillator

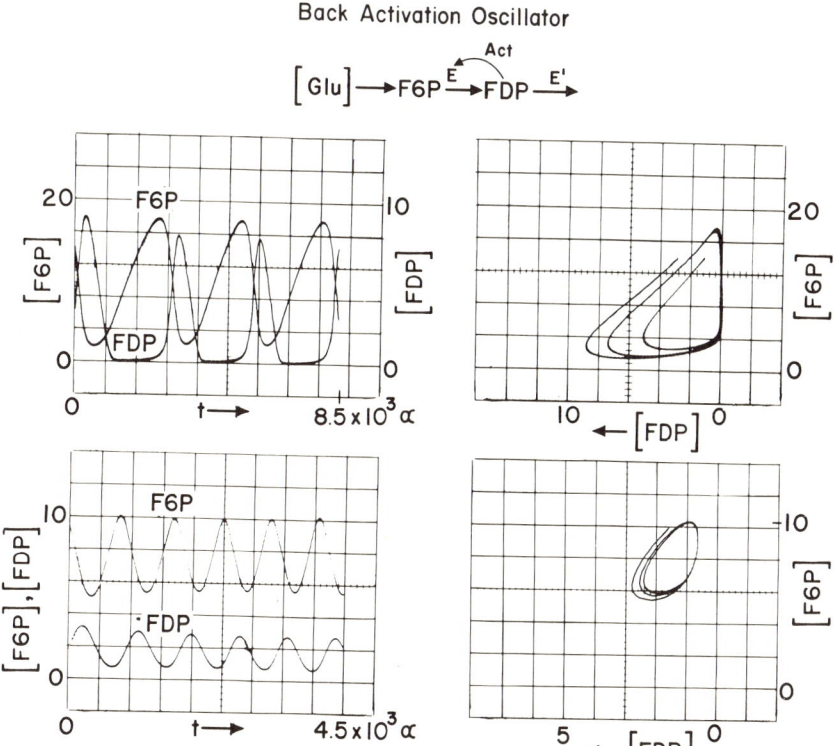

Figure 1B. Same as Figure 1A except upper graphs: strong FDP activation of E, giving relaxation limit cycle oscillations. Lower graph: same as upper graph except for different values of V_{maxes} giving rise to nearly sinusoidal limit cycle oscillations.

fulfilled by the graphs (Figure 2A). In regard to Figures 2B and 2C, the graphical analysis proceeds as follows: In region I F6P is falling and FDP is rising; the velocity of v_p continues to rise since FDP has the greater control strength. In the middle of this region the flux of v_p decreased due to the fact that the control by FDP has saturated (see Figure 2A) while the continued decrease of F6P brings it into dominant control. The flux v_p continues its fall and intersects the curve for v_o which has been relatively constant since FDP is well above the Km for v_o. At this intersection FDP has its maximum and begins to fall in

137

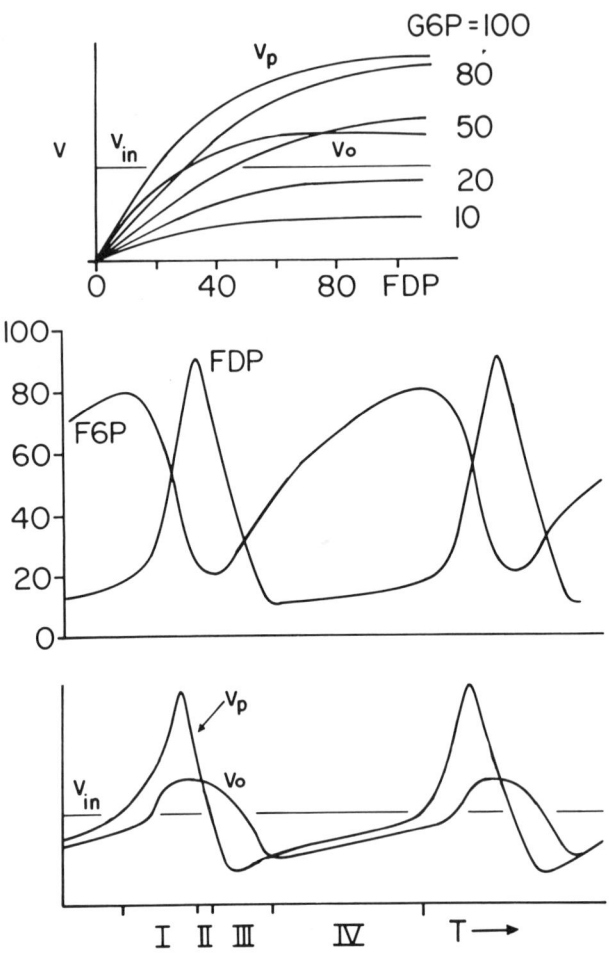

Figure 2. Graphical analysis of the back activation oscillator (see text Equation 3). Upper: typical control characteristics. Middle: kinetics of G6P and FDP. Lower: corresponding reaction rates versus time.

region II. The fall of FDP causes a further decrease in v_p as FDP reexerts its control and in turn the decrease in v_p causes FDP to fall even further; control of v_p is primarily due to FDP and the continued fall carries v_p below v_{in} giving rise to a minimum in F6P at the start of region III. The further fall of v_p causes F6P to rise more rapidly until it achieves control of v_p and reverses the fall of v_p. Note that the inflection of F6P occurs at the minimum v_p and necessarily before the minimum of FDP. The continued rise of F6P causes v_p to rise until it intercepts the v_o curve at the end of region III causing FDP to begin its increase. Shortly after the start of region IV, F6P control of v_p is very weak and the very slow increase in FDP is a result of its nearly equal control of v_p and v_o at these low levels. As FDP rises, its control on v_o is weakened relative to v_p causing the v_p and v_o curves to diverge faster with the resultant increase in the rate of FDP's rise. At the end of region IV, v_p intercepts the v_{in} line and F6P begins its fall; the cycle then repeats itself.

This discussion should help to clarify several basic results. First, no component (F6P or FDP) can reverse its own direction (rise or fall) due to its own control; that is, inhibitory self-coupling normally leads that component into a stable state while activation self coupling would normally lead that component into a continued rise (unstable singularity). Consequently, the minima and maxima in one of these components reflects the control of their rates by the other component. (This conclusion depends on the fact that the character of the control does not reverse itself in the region of interest which is typical of most enzyme reactions.) Second, it should be clear that the qualitative properties of the kinetics do <u>not</u> depend on the exact algebraic details of the rate laws. Any rate laws having the qualitative features illustrated in Figure 2 would yield similar results. Indeed even more major distortions can be made without affecting the existence of the limit cycle. For example, suppose the control characteristic (Figure 2A) for v_o exhibited an induction (e.g. quadratic) character at FDP = 0 and that its maximum control strength $(\partial v_o / \partial FDP)$ exceeded that for v_p at the singularity. Referring back to region III of Figure 2, we see that v_o would fall faster than v_p near the v_{in} level and then would fall (and rise) slower than v_p at the lower value of FDP occurring at the intersection of v_p and v_o (at the end of region III); thus the basic features of our graph would not be affected and the limit cycle would exist. However, the singularity would be stable and there would necessarily exist an anti-limit cycle. Essentially a limit cycle would occur so long as the effective <u>average</u> value of the control strength for v_o were less than that for v_p. We shall see additional examples of this phenomena in the next section.

Several other two variable oscillatory mechanisms can be derived for the glycolytic system. These differ in the choice of components and the requisite control features, but the method of analysis remains the same. We shall discuss these in a later section.

B. PHASE SHIFT OSCILLATORS

1. COMPUTER STUDIES

A second type of theoretical model for chemical oscillations has developed from studies of sequences involving end product inhibition as indicated by:

$$[S_0] \xrightarrow{E} S_1 \xrightarrow{k_1} S_2 \xrightarrow{k_2} S_3 \cdots S_n \xrightarrow{k_n} \qquad (6)$$

with inhibition (inh) from S_n on E; concentrations $x_0, x_1, x_2, x_3, \ldots x_n$.

where the x_i represents the concentration of the chemical species (S_i). The concentration x_0 is considered constant, as indicated by the brackets []. The first reaction is catalyzed by an enzyme which can be inhibited by the "end" product (S_n) according to the simplified rate law:

$$v_0 = \frac{V_0}{K_m + \alpha x_n^p} \qquad (7a)$$

where p is the effect Hill coefficient for the inhibition. Though the other reactions may also be enzymatically catalyzed they are taken here as irreversible pseudo-first order reactions, again for reasons of theoretical simplicity.

Sel'kov (6) was among the first to recognize that the system could have an unstable stationary state singularity for appropriate values of p, n and the various constants, a result which strongly suggests the possibility of limit cycle oscillations at least in the case of chemical systems. Goodwin (18) (in 1965) demonstrated such oscillations in computer studies for $n = 3$, $p = 1$, while Morales and McKay (19) (in 1967) carried out independent analog computer studies of the above system and demonstrated oscillatory computer solutions for the case of $n = 4$, $p = 4$ with different rate constants (see Figure 3); they also carried out computer studies to determine the dependence of the amplitude and

Goodwin (1965)

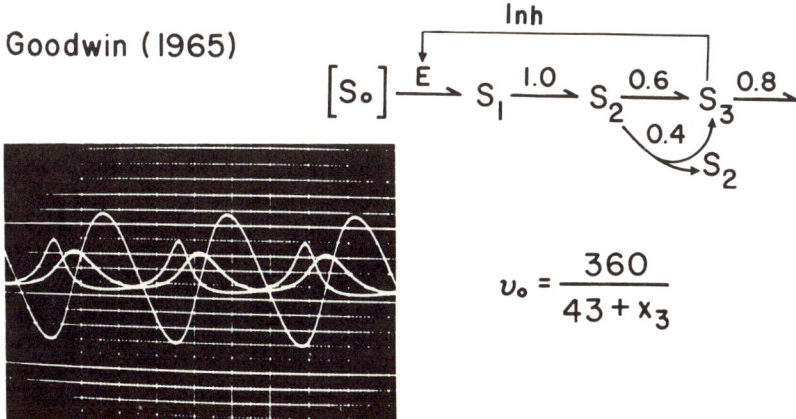

$$v_0 = \frac{360}{43 + x_3}$$

Morales – McKay (1967)

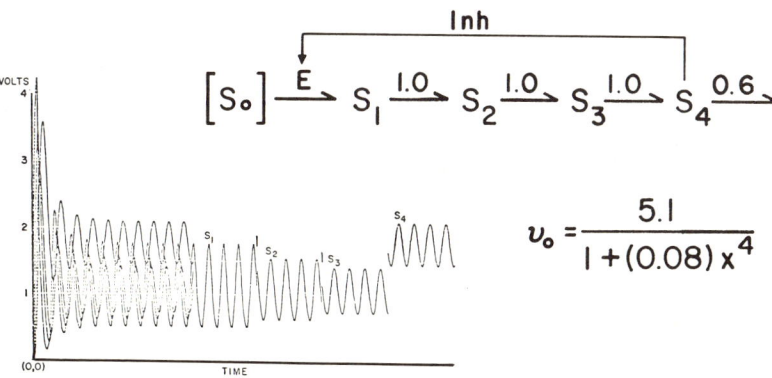

$$v_0 = \frac{5.1}{1 + (0.08) x^4}$$

Figure 3. Oscillatory computer solutions and the corresponding mechanisms obtained by Goodwin (1965) and Morales-McKay (1966).

frequency of the oscillations on the various rate constants.

We have carried out an analytical analysis to determine the conditions for the stability of the singularity. The analysis is presented below and the results can be summarized as follows:

The singularity must be stable unless the Hill

coefficient (p) is sufficiently large for a given value of n (the number of chemical species in the sequence according to the formula:

$$p > \frac{1}{[\cos \frac{\pi}{n}]^n} \quad (7b)$$

Thus a sequence of three species (n = 3) will be stable for p < 8 regardless of the values of the other parameters; for p > 8 the singularity may be made unstable by appropriate choice of the constants. For n = 4, we must have p > 4 in order to achieve stable oscillations.

It should be noted that these results are not directly compatible with the computer results obtained by Goodwin or by Morales and McKay. Their computer studies correspond to cases of a stable singularity. Consequently, either their solutions are computer artifacts or the system of equations must admit at least one antilimit cycle surrounding the singularity. Although this latter explanation seemed unlikely for the system or equations under study, we investigated the possibility on our analog computer. Somewhat to our surprise the computer solutions exhibited limit cycle oscillations and antilimit cycles as shown in Figure 4. Indeed, it was not difficult to obtain nested limit cycles. However, we remained suspicious of computer artifacts since the solutions were sensitive to changes in time scale and computer components. Consequently, the same system was studied with a digital computer. Here again the solutions could be made to give sustained (limit cycle) oscillations, under conditions on p, n which should give a stable singularity, these oscillations disappeared when the time step (Δt) was reduced in value and precision increased; the behavior of the system was then found to be entirely consistent with the theoretical expectations.

In summary, the system (Equation 6) has a globally stable singularity unless the conditions on p and n (given by Equation 7b) are fulfilled; in that case for appropriate values of the constants, the singularity can be made unstable and the system will possess one (and only one) limit cycle reflecting sustained oscillations in the time domain. The discussion has been presented in its historical sense of development to clarify the literature as well as the

motivation.* Though often stimulating, the precarious value
of computer solutions (whether analog or digital) is evident,
particularly if they are untested for computer artifacts or
unsupported by theoretical analysis. This specific system
would appear to have limited biochemical significance since
the large values of the Hill coefficient required for relatively short sequences are uncommon as are relatively long
sequences (to the "end product" causing inhibition) which
would allow more reasonable Hill coefficients. However, it
should be expected that minor modifications of the system,
such as additional back inhibitions to other reactions, could
readily make the oscillatory requirements more practical.

2. ANALYTICAL ANALYSIS OF THE PHASE SHIFT OSCILLATOR

As usual the stability of the system at the singularity
is determined by an examination of the roots of the characteristic equation for the linearized system. In general these
roots are complex numbers and the condition for the singularity to be unstable is that at least one root have a positive real part. For the n-component phase shift oscillator
the characteristic equation is an nth degree polynomial whose
coefficients are functions of the various rate constants and
the nature of the roots is diffucult to determine. Considerable simplification occurs if we assume that the greatest
liklihood for oscillations will be obtained when all the first
order rate constants are equal. This assumption seems reasonable on grounds of symmetry; in addition, for the 3-compon-
ent system, the roots can be solved as functions of the
different rate constants and it is easily demonstrated by
calculus that the maximum value of the real part occurs when
all the first order rate constants are equal. Consequently
we shall maintain this assumption and consider that it can
be justified.

For the n-component system with all first order rate

* Note: the senior author admits that this study was
carried out primarily because the results of Morales and
McKay were inconsistent with his own speculations (see
ref. 7) concerning the conditions for oscillations in
systems of more than two variables. Needless to say,
the final results demonstrate that speculation to be
wrong.

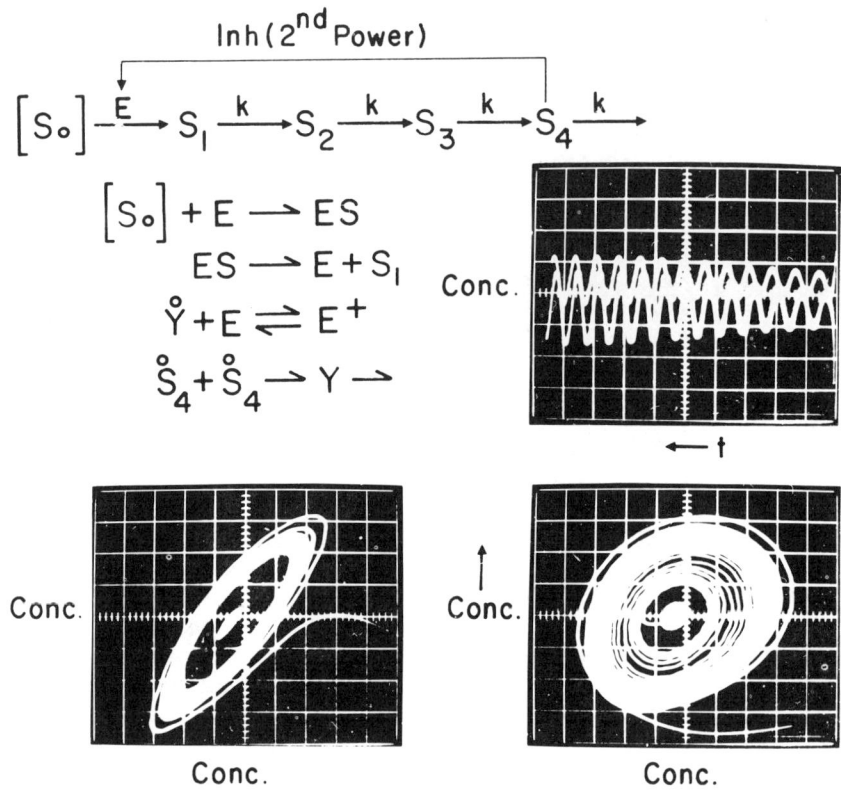

Figure 4. Oscillatory computer solutions obtained for the mechanism indicated. Note phase plane plots which demonstrate the existence of antilimit cycles and limit cycle oscillations. The detailed chemical mechanism (required by this computer) for the first enzymatic reaction is shown. The dots (°) above the chemicals (e.g. $\overset{\circ}{S}_4$) indicate that the substance has rate control but is not consumed by the reaction; as such, it is simply a computer artifice for establishing the desired rate law. Appropriate chemistry is easily developed.

constants equal (to k), that constant can be set equal to one by a change of time scale and the differential equations appear as:

$$\overset{\circ}{x}_1 = v(x_n) - x_1 \tag{8}$$

$$\overset{\circ}{x}_\ell = x_{\ell-1} - x_\ell \quad \text{for} \quad \ell = 2, \ldots, n \qquad (9)$$

where
$$v(x_n) = \frac{a}{1 + b(x_n)^p} \qquad (10)$$

To linearize these equations it is only necessary to replace $v(x_n)$ in the first equation by αx_n where α is the partial derivative $\partial v / \partial x_n$ evaluated at the singularity. The singularity is determined by setting all $\overset{\circ}{x} = 0$ and we find that at the singularity $x_i^* = x_j^* = x$ for all i and j and

$$v(x) = x \qquad (11)$$

where x indicates the common value of the x_i^*. Using Equation 10, Equation 11 can be rewritten as:

$$bx^{p+1} = a - x \qquad (12)$$

which can be graphically solved for x. The value of x lies between 0 and a and is smaller the larger the value of b. Taking the partial derivative of equation 10 and using equations 10 and 12 to simplify, we find:

$$\alpha = \frac{\partial v}{\partial x} = -p(1 - \frac{x}{a}) \qquad (13)$$

and we note that the magnitude of α is larger the smaller the values of x. Setting $\overset{\circ}{x}_i = A_i e^{\lambda t}$ in the linearized equations yields the following characteristic equation:

$$(1 + \lambda)^n = \alpha \qquad (14)$$

whose roots are:

$$\lambda_k = -1 + |\alpha|^{\frac{1}{n}} \cos\left(\frac{2k+1}{n}\right)\pi + i|\alpha|^{\frac{1}{n}} \sin\left(\frac{2k+1}{n}\right)\pi \qquad (15)$$

for $k = 0, 1, \ldots, n-1$

The real part will achieve its most positive value (least negative) when $|\alpha|$ is greatest (i.e. the smaller x) and when $k = 0$ (which gives the largest positive value for cos). The condition for positive real part, hence an unstable singularity, is easily realized to be:

$$p > \frac{1}{[\cos \frac{\pi}{n}]^n} \qquad (16)$$

and the specific values discussed in the previous section

are easily obtained.*

3. POSTSCRIPT: BIOCHEMICAL SYSTEMS WITH ANTILIMIT CYCLES.

Although this section was developed after the meeting, it seems appropriate to include it here for completeness. The computer solutions exhibiting antilimit cycles provide an interesting dynamical behavior whose potential biological interpretations (e.g. to circadian rhythms) should not be overlooked if it could be found to reside in reasonable chemistry rather than computer artifacts. There are two basic types of computer artifacts. The first corresponds to the finite value of the time step (Δt) which can be interpreted as a phase shift and occurs in both analog and digital computers. For the system (Equation 6) the attractive weakness of the singularity means that the oscillations will be slowly damped; the small additional phase shift introduced by the computers would correspond to larger values of n and could readily yield sustained oscillations. Such artifacts disappear (as they did in the digital computer study) with decreasing phase shift (or Δt). But in general such computer problems do not seem likely (at least to this author) to introduce multiple limit cycles which depend on the non-linearities of the differential equations. This aspect -- namely, misrepresentation of the differential equations -- is prevalent in analog computers which can only approximate the non-linear features of the required differential equations; the problem is largely absent in digital computers (which have high precision).

However, the analysis of the system (Equation 6) is greatly facilitated by its strict analogy to an equivalent electronic device, namely, the phase shift oscillator whose schematic design is indicated as:

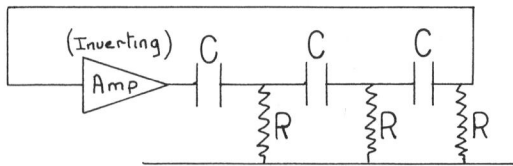

* Since the analysis given here was carried out, a later paper by Sel'kov (20) has come to our attention and presents a similar analysis of closely related mechanisms.

The amplifier section (which normally inverts and hence provides negative feedback as well as gain) plays the equivalent role of the enzymatic inhibition reaction step. As shown elsewhere (7) each elementary irreversible first order reaction step introduces a phase shift and a concentration amplitude reduction equivalent to that of the R-C circuits in the schematic diagram above. In either system (chemical or electronic) the analysis proceeds as follows: 1) Calculate the frequency which provides a total of 180° phase shift through the RC (or first order reaction) steps; 2) Compute the corresponding amplitude reduction (gain loss) through the same steps; in the case of three steps, the gain loss for the electronic oscillator is 1/29. If the amplifier is linear and has a gain exactly equal to the loss, semi-stable oscillations will ensue, the amplitude being determined by the initial conditions as well as noise variations. If the amplifier gain is less, the system will exhibit damped oscillations and a return to the stationary state; if the gain is greater, the amplitude of the oscillations will increase indefinitely (or until the amplifier saturates). Normally the amplifier is designed to have a gain which is slightly greater than required (e.g. > ∿29 for three steps) and a sharp cutoff. The ensuing oscillations are then slightly non-linear (i.e. nonsinusoidal) but exhibit a stable amplitude; the phase plane exhibits a limit cycle. In a crude sense the amplitude is determined such that the gain has the required "average" value* as indicated in Figure 5.

Multiple limit cycles are easily realized from nonlinear control characteristics which have several regions in which the requisite "average" value can be realized as indicated in Figure 5. In the case of the analog computer simulation, the additional non-linear artifacts introduced by the computer components can readily yield such control characteristics as noted above.

There is no *a priori* reason not to expect similar dynamical phenomena in realizable biological and biochemical systems. The necessary non-linear control characteristics can be realized in single or multiple enzymatic reactions, per-

---

*A careful theoretical analysis and interpretation of "average" can be carried out by the methods of Krylov--Bogoliubov (see Minorsky, N., <u>Nonlinear Oscillations</u>, Princeton, N.J.: D. van Nostrand Co., 1962).

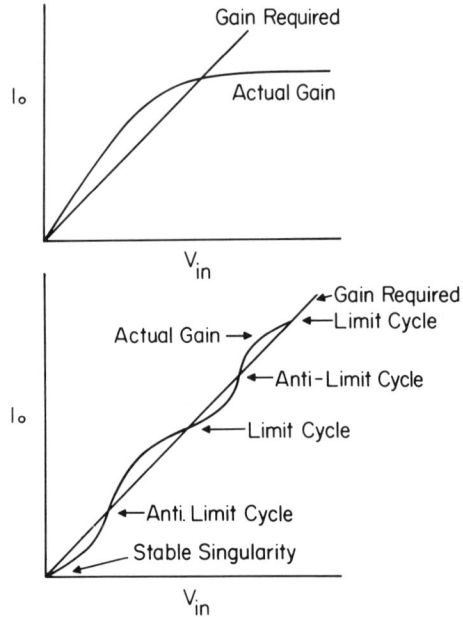

Figure 5. Typical control characteristics for the phase shift amplifier (Sec. 11B3) showing output current (i) versus input voltage (v). Top: control characteristic for single limit cycle oscillations. Bottom: control characteristic for nested limit cycle oscillations. Corresponding enzymatic control characteristics for flux vs. conc. are similarly constructed.

haps within systems or organs at the physiological level
and even in intracellular interactions. The biological systems could typically manifest a stable singularity surrounded by limit cycles and the system could be triggered by environmental perturbations from stable (temporal states) to oscillatory states (of different amplitudes and frequencies), as with any system involving multiple "stationary" states (5). Such dynamical behavior may provide an explanation of some aspects of physiological and circadian rhythms.

IV. STUDY OF GLYCOLYTIC OSCILLATIONS

As realized in the previous section, the occurrence of oscillatory kinetics requires only the existence of simple feedback pathways and appropriate activational and inhibitory control characters, properties which are common to a variety of known biochemical pathways. The case of most current interest is the glycolytic system for which there has been considerable experimental study of the oscillatory kinetics in both whole (yeast) cells and cell extracts (21,22,23).

A. THEORETICAL CONSIDERATIONS

Based on the known control character of the individual glycolytic enzymes, there are many distinct control mechanisms which would cause the glycolytic pathway to exhibit oscillatory kinetics. Particularly important in these mechanisms are the potential control properties of phosphofructokinase where there is evidence for FDP activation, ATP inhibition, ADP and AMP activation, and F6P activation through substrate control (although other effectors are known they need not concern us here). These controls may be separately or simultaneously operative depending on the concentration levels, pH, temperature, etc. of the experimental situation. The complete glycolytic sequence with known control features which might be expected for the cell extracts appears as:

(17)

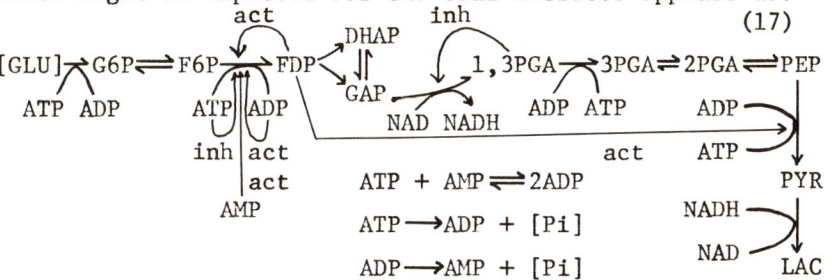

where we have indicated some reactions which usually are
fast and therefore expected to be near equilibrium.* In our
discussion we shall often ignore these reactions, treating
the components as making a pool (e.g. G6P and F6P form a
"G6P" pool). In this formulation of glycolysis we have also
ignored the influence of other effectors (e.g., Pi) which we
shall presume are in large concentration, buffered or simply
unchanging in the experimental situations of interest here.
Further, we shall assume that ATP and ADP do not have control
of hexokinase so that the first reaction may be considered
to have an essentially constant rate depending only on the
fixed glucose concentration. We can only emphasize that
these assumptions and further assumptions as to specific control situations will depend on the particular experimental
situation and can be ultimately justified only by a careful
study of the experimental results. We shall list in abbreviated fashion some of the specific control situations which
can give rise to sustained oscillations.

We have previously (Section IIA) discussed the mechanism:

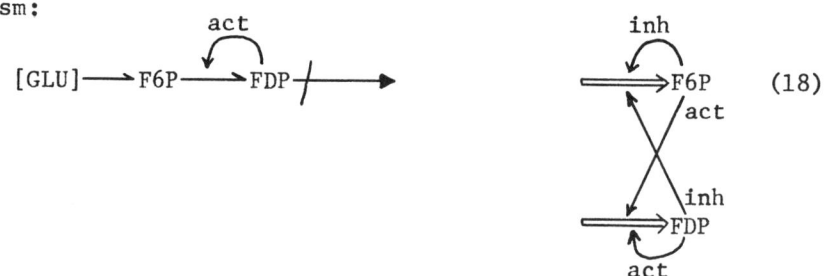

(18)

which requires F6P substrate activation and FDP allosteric
activation of pfk. Control of the lower glycolytic path

* Note the following abbreviations:
GLU = Glucose; G6P = Glucose-6-phosphate; F6P = Fructose-6-phosphate; FDP = Fructose-1,6-diphosphate; DHAP = Dihydroxyacetone phosphate; GAP = Glyceraldehyde-3-phosphate; 1,3PGA= 1,3 diphosphoglyceric acid; 3PGA = 3 phosphoglyceric acid;
2PGA = 2 phosphoglyceric acid; PEP = Phosphoenolpyruvate;
PYR = Pyruvate; LAC = Lactate; NAD = Nicotinamide-adenine dinucleotide; NADH = Reduced NAD; ATP = Adenine triphosphate;
ADP = Adenine diphosphate; AMP = Adenine monophosphate;
Lgp = Lower glycolytic pathway (The sequence from GAP to
Lactate); pfk = Phosphofructokinase; gapdh = Glyceraldehyde
phosphate dehydrogenase.

(henceforth abbreviated Lgp) by FDP must be relatively weak compared to FDP activation of pfk. This mechanism assumes little or no control of pfk by the adenine nucleotides although of course these substances would still be consumed or produced in their appropriate reactions.

A similar mechanism can be based on the F6P, ADP activation of pfk as:

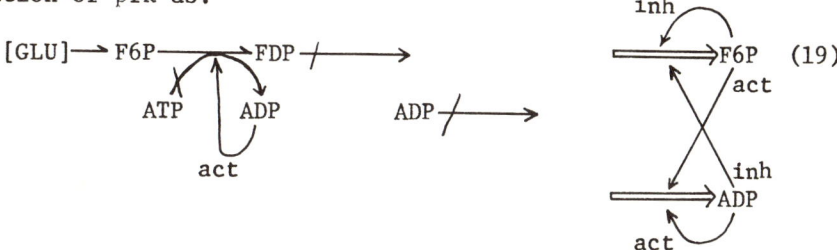
(19)

This mechanism is dynamically identical to mechanism (18) save that ADP replaces FDP. But this mechanism will operate in the absence of any effect of FDP on pfk and requires relatively weak control of ADP on its removal reaction (via the Lgp or other kinases). Given sufficient control strength of ADP on pfk the mechanism will function in the absence of any other adenine nucleotide (ATP, AMP) control on pfk.

(20)

A similar mechanism is given by:

which is based on ATP inhibition and F6P activation of pfk. Again the control of ATP on its removal reactions must be relatively weak compared to its inhibitory control of pfk.

In this same class, we may recognize a similar mechanism based on the concerted effects of the adenine nucleotides on pfk as:Equation 21. This mechanism recognized that the conservation of the total adenine nucleotides requires that ATP <u>decreases</u> as ADP (and AMP) increase. Thus an increase in ADP

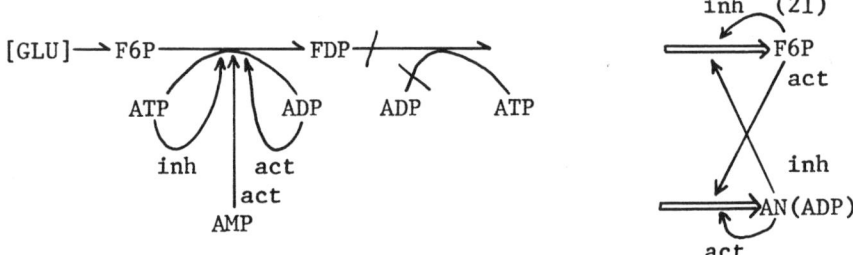

will directly activate pfk while the accompanying decrease in ATP will release the ATP inhibition of pfk and act as an effective activation of pfk. This concerted influence of the adenine nucleotides on pfk effectively enhances the activational control strength of ADP.

All four of these mechanisms are in essentially the same class, being based on substrate activation and product activation of pfk. The nature and shape of the oscillations which can be obtained would be virtually identical to those discussed in Section IIIA. As such the phase plane behavior would be identical and there is no way a priori to distinguish the likelihood of these mechanisms solely on the basis that the oscillations exist. Such distinctions must be based on direct observation (and experimentation) as to which of these variables can be in control. Thus if some are not changing significantly or if it can be demonstrated that additions of certain ones does not effect the state of the oscillations, those can be discarded as significant in the control factors which determine the oscillations. But in some cases all the components may be oscillating and may exert indirect control on the state of the other variables so that it becomes difficult to determine the control factors directly responsible for the oscillations. As we shall see later, such cases require more detailed methods.

A second class of mechanisms can be realized as:

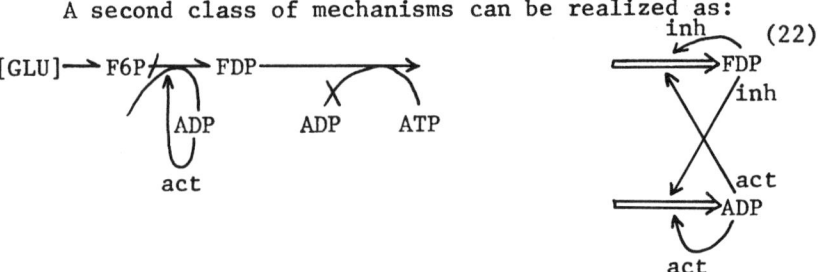

Here the pertinent coupling derives from ADP activation of pfk and the effect of FDP on the removal of ADP via the Lgp. Here we do **not** require F6P control of pfk but we do require that FDP exert control on the Lgp. Again, the ADP control of its removal reactions must be weaker than its control of pfk. Analog computer solutions for this mechanism are shown in Figure 6. Another variation in this class can be realized as:

$$[GLU] \longrightarrow F6P \longrightarrow FDP \xrightarrow{\text{act}} PEP \longrightarrow PYR \qquad (23)$$

with ADP (act) and ADP couplings as shown.

where the effect of FDP is not directly through the Lgp but due to its activation of pyruvate kinase. Other variations of this class can be obtained by shifting to ATP control or concerted adenine nucleotide control of pfk as in the first class. This class of mechanisms would differ from the first class in its direction of rotation of the relevant phase plane plot.

Still another class of mechanisms can be based on the phase shift oscillator as in:

$$[GLU] \longrightarrow G6P \rightleftarrows FDP \longrightarrow GAP \longrightarrow 1,3PGA \longrightarrow 3PGA \longrightarrow PYR \longrightarrow \qquad (24)$$

with ATP (inh), ATP, ATP, and ADP couplings as shown.

where we recognize end product (ATP) inhibition. Although we think this mechanism somewhat unlikely in view of the required delays and strength of inhibitory control, it does not seem that it should be ruled out entirely, since the concerted effects of the adenine nucleotides could provide sufficient strength and some of the intervening reactions may not be as fast as often believed (e.g. aldolase).

We may expect still other mechanisms based on combinations of the above classes. In some cases mixing of these classes will provide more likely conditions for the occurrence of oscillations while in other cases mixing will have the opposite effect and make the singularity more stable (12). Each case must be examined in more detail. In brief,

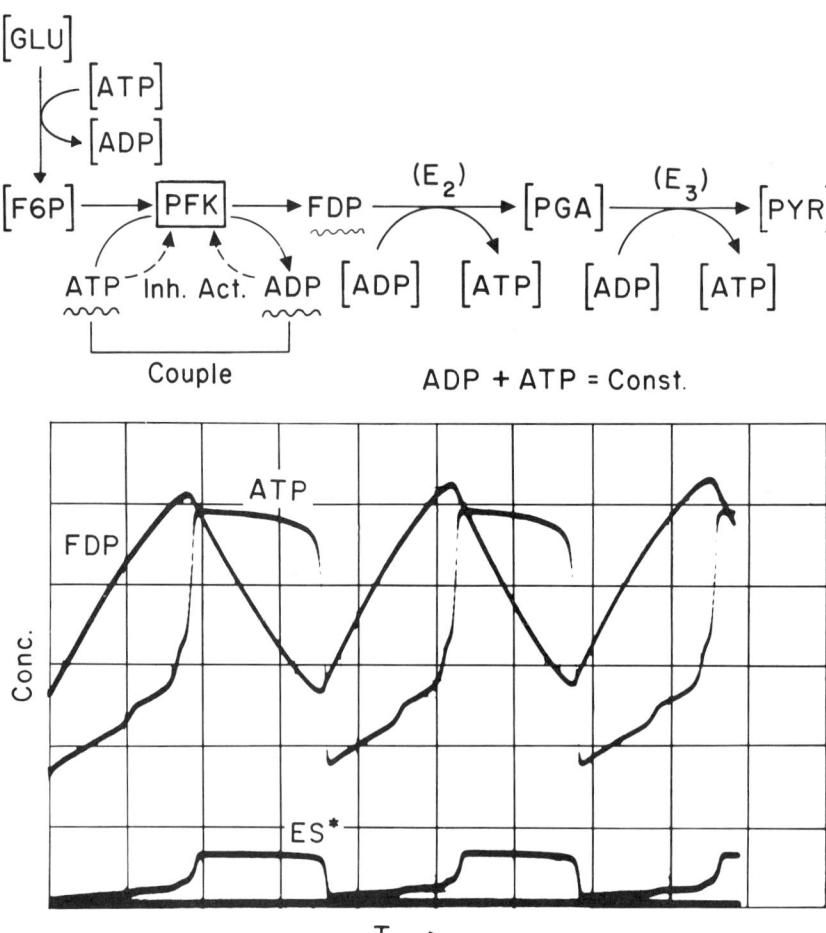

Figure 6. Oscillatory computer solutions based on FDP control of the lower glycolytic pathway and ADP (adenine nucleotide) activation of pfk as indicated in the mechanism. Brackets indicate substances which are consumed by a reaction but do <u>not</u> have any control on the reaction rate. As shown above, the first and last reactions simply control the position of the singularity.

the essence of these theoretical considerations is that there are many distinct control situations for the glycolytic system which can give rise to oscillatory kinetics. Although these specific mechanisms can provide a starting point for analysis of the experimental data, it is clear that the central problem is one of determining the specific control situation under the circumstances of the particular experiment. As such, the problem and approaches are just as viable whether or not the system oscillates, and as we have seen by the plethora of potential mechanisms, it is hardly, if at all, ameliorated by the observation of the oscillatory state per se.

B. REPRESENTATION OF THE EXPERIMENTAL DATA

We have seen that the occurrence of oscillatory kinetics places very minor constraints (i.e. feedback) on the mechanism. Consequently, the determination of the specific control properties and enzymes responsible for the oscillations is in no way different from the analysis of any dynamical response of the system (whether or not oscillatory).

However, there are certain experimental advantages to the analysis of the system when it is oscillating. When the assay techniques require temporary discrete measurements, as in the case of the glycolytic intermediates (except NADH), the repetition of the basic kinetic behavior over several cycles provides an internal verification of the existence of various features. In contrast to a complete repetition of the experiment which is subject to changes in initial conditions or preparations, the cycles provide automatic repetition under relatively consistent conditions, subject only to systemic errors (e.g. in the assays).

In addition, if the metabolic concentrations are of large amplitude, a greater range of control states will be covered and will typically be reflected as a more varied display of qualitative kinetic features. Indeed, if the individual metabolites for a particular enzyme were to oscillate over a large amplitude and cover random relative phases, the entire control characteristic could be determined directly for the in vivo or extract system. Unfortunately, such an event seldom occurs and the usual data provides only a partial description, leaving questions as to the specific control states which must be confirmed by further studies.

Experimental kinetic data has been obtained by one of us for the complete set of glycolytic intermediates for the oscillating heart muscle extract (23). The data are shown

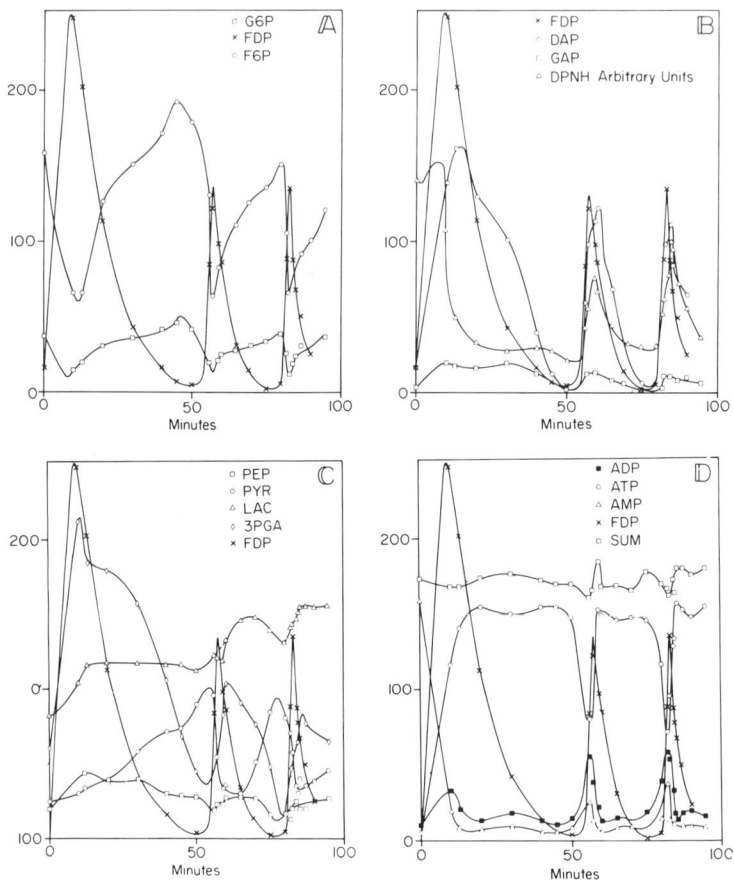

Figure 7. Experimental data (points) obtained for oscillating heart muscle extract. Lines have been drawn to reinforce features common to more than one cycle. Concentrations are in μmolar units and sum represents the sum of the adenine nucleotides. The DPNH (NADH) data were actually continuous and obtained by fluorescence monitoring.

as points in Figure 7, while the smooth curves have been drawn such that qualitative features evident in one cycle were confirmed (or at least not disproved) by the data points of another cycle. In addition to the basic oscillation with an apparent increasing frequency within the cycles observed, there are many particular points to be noted: (1) The inflection and sudden rise of G6P (or F6P) in the latter portion of each cycle (where the start of each cycle is taken as the peak of FDP). This behavior of G6P is reflected (and confirmed) by F6P (where we anticipate a near equilibrium with G6P). (2) The spike-like character of FDP showing a sharp rise, a relatively slow decay, and a minimum just before the sharp rise; (3) the adenine nucleotides show similar spikes at the ends of the cycles; there is a secondary minimum (or maximum) in the middle of the cycle which is confirmed by its presence in each cycle and by the expected correspondence between the ATP (a minimum) and ADP, AMP (maximum), thereby maintaining the essential constancy of their sums. It seems clear that these substances have a frequency component approximately twice the fundamental frequency; (4) although the data for DHAP show that it lags behind FDP in its rise and fall, there is clear evidence of a distinct inflection in its rise and fall which is not evident in FDP. Within the experimental accuracy, the data for GAP appear to reflect the qualitative behavior of the DHAP. (5) The behavior of 3PGA and PEP (within experimental error) are qualitatively similar as would be expected if there is a rapid equilibration between 3PGA, 2PGA (not measured) and PEP. While these components lag behind FDP in their rise and fall, the fall, as seen in 3PGA, is delayed well beyond that of FDP, after a short rapid drop. (6) While the kinetics of PYR, LAC and NADH reflect the fundamental oscillation, certain aspects of their behavior should be particularly noted. First, the rise of pyruvate begins and continues after the fall of PEP and ADP while its fall continues well after the rise of PEP and ADP. Second, there is an evident disappearance and minimum in lactate towards the end of each cycle, a result which is incompatible with the formation of lactate as an irreversible end produce of glycolysis. Furthermore, the minima in lactate appear to be coincident with the maxima in pyruvate. The rise and fall of NADH lags behind the rise and fall of FDP. It should be noted that the NADH was measured by fluorescence techniques; the curve is continuous although the units are arbitrary (no standard was taken) and the zero point of NADH is not known.

C. COMPUTER STUDIES

As already noted, one approach to the understanding of glycolytic dynamics is through the formulation of chemical models (mathematical) and the use of computers for their analysis. In view of the previous discussion concerning the existing knowledge (or lack thereof) of the detailed in vitro control characteristics for the individual glycolytic enzymes as well as the questions concerning their applicability to in vivo and extract systems, we have made no attempt to formulate models providing a quantitative fit. Rather our interest has centered on simple formulations containing the essential control properties -- in terms of activation and inhibition. Our aim has been to establish a qualitative computer fit to the experimental data and to determine which specific control properties are responsible for various qualitative features by comparative studies of models with modified control properties.

Although the use of simplified enzyme models produces a minimal number of parameters the available parameter (control) space must be thoroughly explored in order to verify that specific qualitative features are a consequence of the control properties and not the details of the particular models. Theoretical considerations (noted in Section II) can further reduce the ranges of parameter values which are of interest, but in general there still remains a large number of computer solutions (for specific parameter values) which must be examined as a result of our current theoretical uncertainties as to the domains of arrangement for the relative control strengths.

The studies illustrated below were carried out on the Johnson Foundation Electronic Analog Computer (JFEAC), Mark II (25). Although the capacity of this computer (about 20 mass action reactions maximum) was somewhat limited even for this control theoretic approach, the speed and low cost of operation more than compensated. The chemical equations for the basic mechanism studied are given below. Solutions, as illustrated in Figure 8, were obtained in a real time of 0.5 to 2.0 seconds for direct visual inspection and parameter settings could be changed in less than 0.1 seconds. The results presented below represent several hundred hours of computer studies (over 500,000 specific solutions) for various parameter values and control situations.

The basic chemical reactions studied with the computer are:

BIOCHEMICAL OSCILLATORS

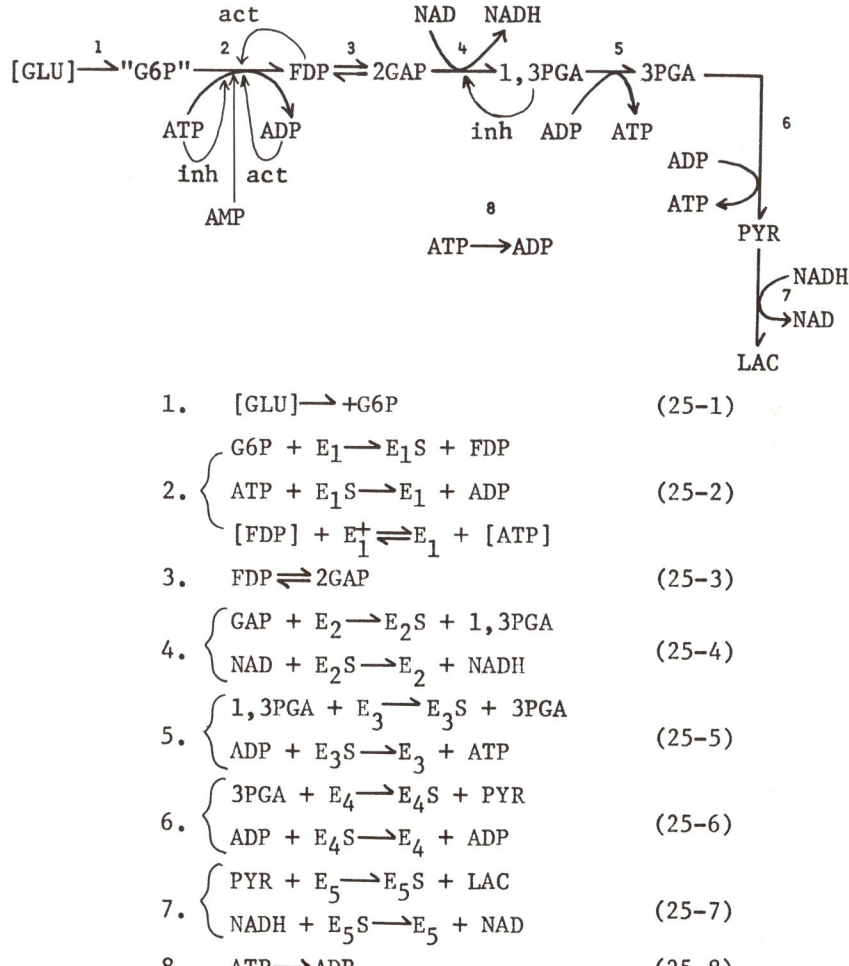

1. $[GLU] \longrightarrow +G6P$ (25-1)

2. $\begin{cases} G6P + E_1 \longrightarrow E_1S + FDP \\ ATP + E_1S \longrightarrow E_1 + ADP \\ [FDP] + E_1^+ \rightleftharpoons E_1 + [ATP] \end{cases}$ (25-2)

3. $FDP \rightleftharpoons 2GAP$ (25-3)

4. $\begin{cases} GAP + E_2 \longrightarrow E_2S + 1,3PGA \\ NAD + E_2S \longrightarrow E_2 + NADH \end{cases}$ (25-4)

5. $\begin{cases} 1,3PGA + E_3 \longrightarrow E_3S + 3PGA \\ ADP + E_3S \longrightarrow E_3 + ATP \end{cases}$ (25-5)

6. $\begin{cases} 3PGA + E_4 \longrightarrow E_4S + PYR \\ ADP + E_4S \longrightarrow E_4 + ADP \end{cases}$ (25-6)

7. $\begin{cases} PYR + E_5 \longrightarrow E_5S + LAC \\ NADH + E_5S \longrightarrow E_5 + NAD \end{cases}$ (25-7)

8. $ATP \longrightarrow ADP$ (25-8)

and represent an abbreviated form of glycolysis with only the most fundamental control properties present. Although the computer design required that enzymes reactions be represented in terms of mass action chemistry, the rate constants were such that enzymatic intermediates were operating under essentially steady state conditions. The design of the computer allowed a ready conversion between substrate control and "no" control states for the reaction indicated. For example, in Equation 25-6, ADP could be removed from

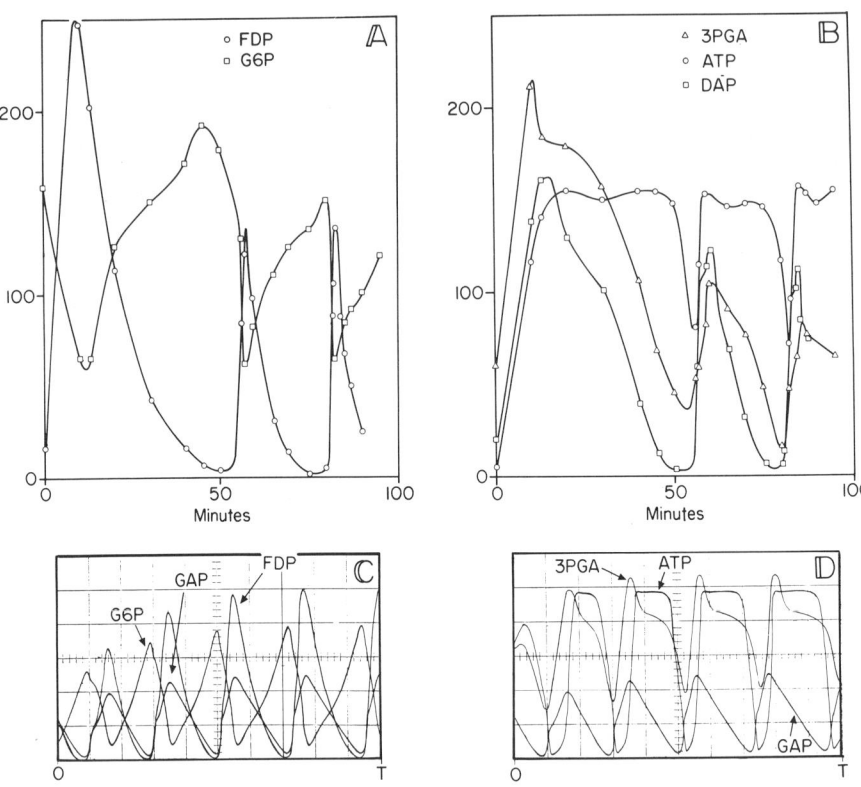

Figure 8A. Comparison of experimental data and analog computer solutions obtained for the simplified control theoretic model of the glycolytic pathway (Eq.25). Note the general agreement of qualitative features, particularly in the later computer cycles, for the metabolite shown.

control so that the rate of the second reaction in Equation 25-6 was given by $v = k[E_4S]$ rather than could otherwise be realized by the limited computer capacity. The abbreviation of the glycolytic intermediates is based on the previously noted essential equilibria (i.e. G6P to F6P and 3PGA to PEP). Although there are relevant questions concerning the relative rate of the time isomerase reaction, DHAP and GAP were assumed to be in essential equilibrium. Although AMP is indicated in the general mechanism, it has not been included

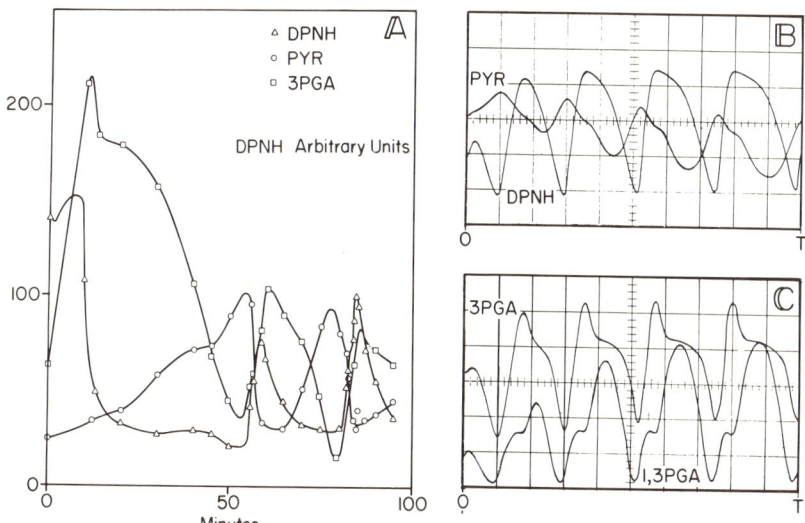

Figure 8B. Comparison of experimental data and analog computer solutions obtained for the simplified control theoretic model of the glycolytic pathway (Eq.25). Note the general lack of qualitative agreement for DPNH (NADH) and PYR.

in the computer study since its qualitative experimental behavior is equivalent to ADP (after the initial transient) as is its dynamical (activation) effect on phosphofructokinase; the experimental proportionality of ADP and AMP is expected due to the presence of adenylate kinase and ATPases in the experimental extract. The hexokinase reaction was compensated for by adjustment of the ATPase reaction (Eq. 25-8). Although the potential inhibition of glyceraldehyde phosphate dehydrogenase by 1,3PGA is indicated in the table, it was not utilized in the particular studies discussed below.

Figure 8 shows the best qualitative computer fit to the experimental data for the system studied. Attention should be focused on the later cycles of the computer solutions since no attempt was made to fit the initial transient. Excluding NADH, PYR and LAC (not shown) which are discussed below, it may be noted that the computer solutions demonstrate every qualitative feature of the experimental data although the features may be more or less exaggerated. Note in particular, the inflection during the rise of "G6P", the occurrence of the FDP minimum just before its rise, the secondary minimum of ATP just after its rise (though not shown, ADP imaged the behavior of ATP as in the experimental data). Note also the shoulder in the rise of GAP and the hint (at least) of a shoulder in its fall as well as the relatively faithful reproduction of 3PGA.

With the exceptions noted above (NADH, PYR, LAC) we may conclude that the simplified glycolytic scheme, based on known control properties of the individual enzymes provides a potential model capable of reproducing all the qualitative features of the experimental data. This result has merit in demonstrating that additional control properties need *not* be introduced to explain the basic features of the data and that we might anticipate a quantitative fit with relatively minor modifications. Yet, the result *per se*, does not afford us any better theoretical understanding of the specific control properties responsible for the oscillations or the various shoulders and inflections in the particular glycolytic components. Such understanding can only be achieved through the combined use of theory and comparative studies of the system under different control situations. While such comparative studies would be difficult and laborious, if not impossible, to achieve for real systems, they are relatively simple and fast to obtain from the computer model. Two examples follow:

1. It is well known that doubly periodic oscillations can be generated by the interaction of periodic function with itself after being phase shifted (7). Such an occurrence is possible in glycolysis. If the oscillations are generated through interactions localized around phosphofructokinase, then these oscillations will be generated directly in ATP ADP and FDP. The oscillations in FDP will be projected down the glycolytic pathway, and would appear in 1,3PGA and PEP, with the same fundamental frequency but phase shifted with respect to FDP. These oscillations would then interact with the unshifted oscillations of ATP and ADP at phosphoglycerate kinase and pyruvate kinase and could induce beat frequencies of twice the fundamental period. The expectations are confirmed by the computer solutions shown in Figure 9. The upper left figure shows typical solutions for ATP and GAP for the complete system (Eq.25) already discussed and the secondary minimum in ATP is easily recognized. This result

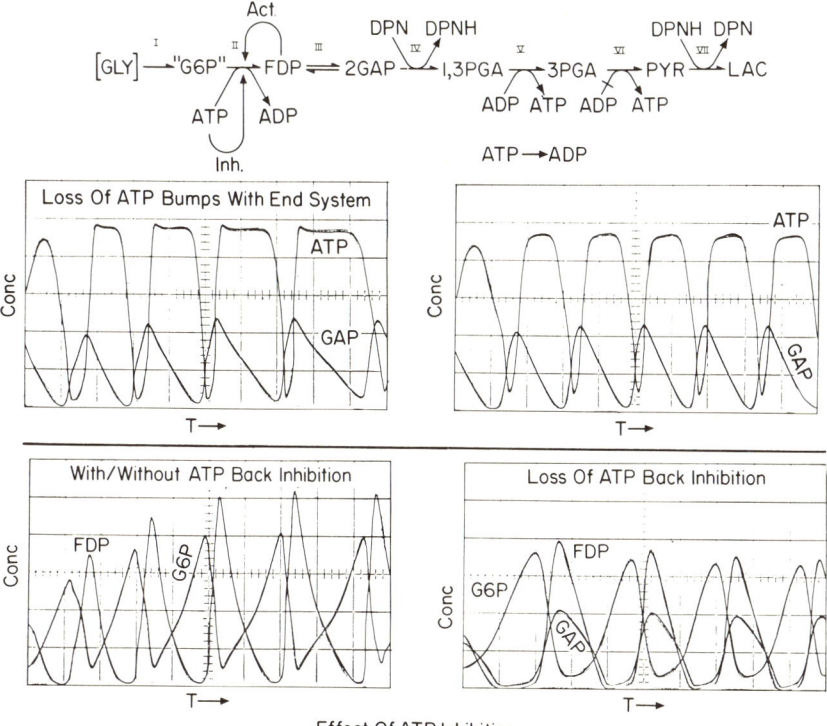

Figure 9. Comparative computer solutions for different glycolytic control states. Graphs on left-hand side show solutions for the complete system (Eq.25) and are the same as for Figure 8. Upper right: solutions when ADP control is removed from the lower glycolytic pathway. Lower right: solutions when adenine nucleotide control is removed from pfk (reaction II).

should be compared to the figure on the upper right which shows typical computer solutions when ADP (and ATP) have no control in the Lgp (i.e. reactions 25-5 and 25-6) which eliminates the possibility for the interaction of the oscillations. As can be seen, the secondary minimum in ATP is gone. Consequently, the computer studies provide substantiation that the secondary minimum in ATP (ADP and AMP) is due to interactions in the Lgp.

2. As noted in Section IVA of this paper there are many different control situations which could give rise to the existence per se of the experimentally observed oscillations. In view of the activation strength required of a phase shift oscillator, the most likely mechanisms are based on either FDP or ADP (or both) activation of pfk (or their equivalent in terms of adenine nucleotides), and are difficult to distinguish on a priori theoretical grounds since they would lead to basically similar experimental results as to the broad features of the oscillations although differences in detailed qualitative features might well be expected. Figure 9 shows typical computer solutions resulting from a comparative study of these two mechanisms. The graph on the lower left essentially repeats the results of the mechanism already discussed (Eq. 25) while the solution on the lower right shows the best qualitative fit achieved when the ATP inh (ADP act) of pfk was removed from the mechanism. The latter solutions are similar to those obtained in Section IIA (Figure 1B). In particular it should be noted that the minimum of FDP occurs after its rapid fall, and long before its sharp rise; in addition, the upward inflection previously noted in the middle of the "G6P" rise is not gone. The qualitative behavior of G6P now resembles its previous (Figure 1B) behavior in the latter portion of its rising cycle (after its inflection). The conclusions just noted for the case without ATP inhibition persisted over the entire range of parameter values; we were unable to obtain the minimum of FDP just before its sharp rise and the upward inflection in G6P never reappeared. Likewise, an attempt to explain these particular oscillating experimental kinetics based solely on the F6P-ADP (activation) couple at pfk (Eq. 19) would yield results similar to the G6P-FDP (activation) couple which is dynamically equivalent. Consequently, it seems necessary to anticipate that both the inhibition of pfk by ATP (or ADP activation or both) and FDP activation play a role in the understanding of this particular data, in particular with respect to the inflection in the G6P (F6P) kinetics.

D. ANALYSIS OF THE DATA

The preceding theoretical and computer studies indicate that these data cannot be explained on the basis of a two component model. The model which appears to provide a satisfactory explanation is based on the following system:

$$[\text{GLU}] \xrightarrow{v_{in}} \text{"G6P"} \xrightarrow[\text{(ATP)}]{v_p} \overset{\text{act}}{\curvearrowright} \text{"FDP"} \xrightarrow{v_o^{FDP}} 1,3\text{PGA} \xrightarrow{v_o^{ADP}} \quad (26)$$

(with ATP/ADP inh/act on v_p; ADP (ATP) on removal step)

As indicated in this model the rate for the removal of FDP need not be identical to that for ADP because of the intervening glycolytic intermediates as well as the non-glycolytic kinases. Although a detailed analysis (similar to that of section III a) is given elsewhere (24), we shall briefly summarize the major events. Concentrating on the first cycle beginning at the FDP maximum (t ∿ 10 minutes), the system behaves like an oscillator based on the F6P and adenine nucleotide couple at pfk (see Eq. 21) with no control by FDP. The high ATP, low ADP places pfk in a highly inhibited state causing a rise in F6P (G6P) and a fall in FDP, until t ∿ 20 minutes. At this time, the rising F6P causes an increase in pfk flux as evidenced by the decreasing net rate of G6P production, the rise in ADP and the fall in ATP. Up to this point the detailed behavior is identical to that previously presented (Section III a) for the back activation oscillator save that the activator is ADP. Yet we interpret the subsequent events as follows. The rise of F6P at t = 30 minutes is due to the fall of FDP into its activation control region for pfk and that fall of FDP gives rise to an effective inhibition of pfk, causing F6P and ATP to rise while ADP falls. Next (at t ∿ 40 minutes) the continued fall of FDP, or its delayed control effect (via GAP and 1,3PGA) causes a decrease in the flux through the Lgp thereby causing ATP to fall and ADP to rise. These latter changes cause an increase in pfk flux and a subsequent rise in FDP (since its production flux exceeds its removal flux through the Lgp). All these changes (FDP, ADP increase and ATP decrease) cause a further rise in pfk flux and a further enhancement of their own changes (i.e. the self-acceleration effect). Shortly thereafter (t ∿ 52 minutes) FDP passes through the control range for the Lgp and for pfk. The rise in ADP and fall in ATP are slowed in part by the rise in flux through the Lgp and subsequently by the continued fall of F6P and its control of pfk flux.

Although the basic features of this interpretation are consistent with the other cycles, certain aspects such as the location (in time or concentration) of the relative position of FDP control of pfk or the Lgp may be subject to modification. Further, there is evidence of ADP control in the intermediates of the Lgp. The analysis of such details depends strongly on the exact location of the various maxima relative to each other and will require further experiments for confirmation.

When a system can exhibit multiple control factors and shifts in the control regions as in the case above, there may be several explanations consistent with the observed dynamics. Any particular interpretation generally requires confirmation through other experiments. For example, the interpretation we have given above suggests that the K_a of pfk for FDP is on the order of 20 µM since there is no evidence of significant FDP control for values of FDP much greater. That the Km for F6P must be relatively high (greater than 30) since F6P must exert control in that region and that ATP always appears to be in its inhibitory region (as opposed to substrate activation region) in regard to pfk, suggesting that the ATP activation region must be for values of ATP less than 70 µM.

E. CONFIRMATORY STUDIES

One confirmation of the interpretation and results could be found by carrying out additional experiments on the extract under different initial values of the glycolytic intermediates. The different dynamical behavior would allow the variables to pass through both different and similar control regions thereby verifying and further clarifying the conclusions reached from this one experiment. Such studies provide information directly applicable to the reactions in the extract, without the need for extrapolating between in vitro and extract data.

A second point of confirmation derives from a comparison of the individual enzyme properties in vitro under the conditions of the extract. If such properties agree with the detailed results of the extract data then there is strong support for the interpretation given; however, if they do not agree, then it can only be concluded that either the interpretation is wrong or the enzyme behaves differently in the extract due to unrecognized factors. A study of purified beef heart pfk was carried out by one of us (26) for the purpose of comparison. The reaction rate was measured under

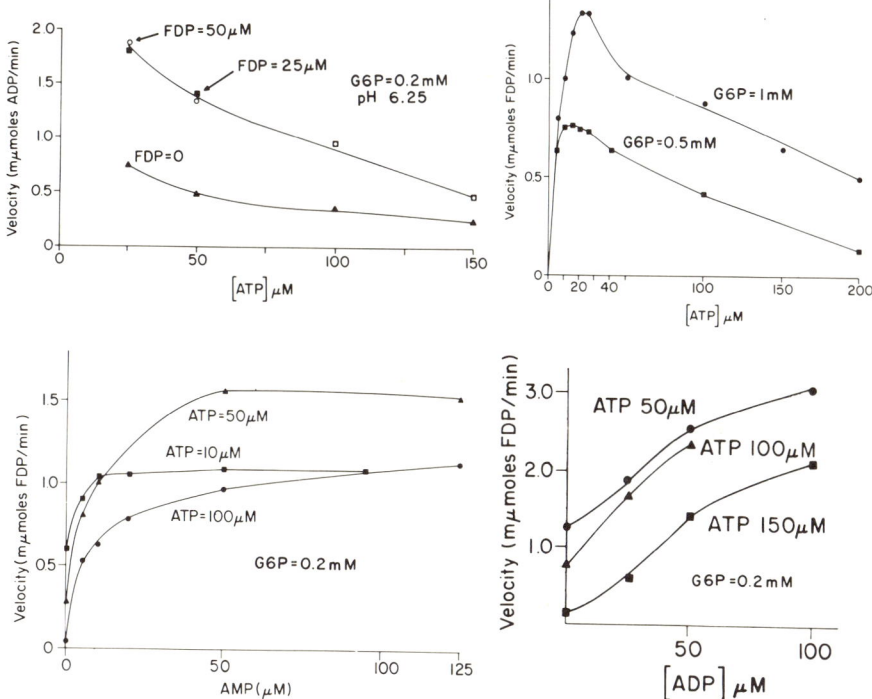

Figure 10. In vitro control characteristics obtained for isolated heart muscle phosphofructokinase under concentration and pH conditions observed in the oscillating extract.

the pH condition of the extract (pH 6.3) and for appropriate (constant) levels of the various effectors. Thus, the effect of FDP on the rate was measured at the high and low levels of ATP, ADP and AMP as determined from the extract kinetics (e.g. ATP = 150 and 50). The results are shown in Figure 10, and it can be readily seen that the various Michaelis, inhibition and activation constants are in agreement with those directly deduced from the interpretation of the extract dynamics. We feel confident in our interpretation of the kinetics of the pfk reaction; the behavior of the Lgp still requires further confirmation if we are to fully accept the interpretation given to the observed oscillations.

V. DISCUSSION

We have seen that the basic oscillatory behavior of this

particular heart muscle extract is complex and cannot be reduced to a simple two variable system. Further, the analysis demonstrated the existence of various control properties such as could allow any one of several oscillatory mechanisms to operate under different conditions. This appears to be the case. In the yeast cell extracts, Chance and Pye (21) found that FDP had no effect on the oscillatory waveforms observed by NADH fluorescence while the adenine nucleotides were always effective. On occasion Frenkel (27) found similar results with the heart muscle extracts. Such results confirm the existence of oscillations in the absence of any FDP control (on pfk or the Lgp) and support mechanisms based on adenine nucleotide control, although they do not provide any detailed confirmation. The studies presented here suggest that oscillations based solely on the F6P-adenine nucleotide couple could be easily achieved (due to their strong concerted control strength on pfk); while a mechanism based solely on the F6P-FDP activation of pfk would require more delicate adjustment since the relative effective control strengths of FDP on pfk and the Lgp are comparable and occur in the same concentration range.

Although the detailed analysis is given elsewhere, the preceding theoretical and computer studies have provided a reasonable and consistent qualitative and semi-qualitative explanation for the observed dynamics of the adenine nucleotides and all the glycolytic intermediates except for NADH, PYR and LAC. For these three components, we were unable to provide a qualitative fit through computer studies based solely on the glycolytic pathway. Elementary quantitative considerations confirm that result; namely, though the only recognized explanation for the periodic decay of lactate would be the reversibility (28) of lactic dehydrogenase, the amount (millimolar) which disappears is not accounted for by the increase in pyruvate (around 20 to 60 µM). In addition, the breakdown of the conservation condition for NAD oxidizing equivalents requires either other pathways for the turnover of NAD, or for the removal of at least one of the glycolytic intermediates involved in the conservation condition (i.e. from 1,3PGA to PYR) or both. Although further experiments are required we suspect that some alternate pyruvate pathway, probably coupled to the turnover of NADH is functional and that the correlated disappearance of lactate is via the reversal of lactic dehydrogenase and the alternate pyruvate pathway.

VI. CONCLUSIONS AND SUMMARY

In this study we have attempted to present the elements of the control theoretic approach as it applies to the analysis of multienzyme systems, in particular to the theory of oscillating reactions and the analysis of the experimentally observed oscillating heart muscle extract.

The control theoretic description is based on the qualitative properties of the individual enzymatic rate laws, the topology and feedback properties of the multienzyme pathway and a semi-quantitative breakdown of the dynamics into regions of relative dominance between the various activational and inhibitory control elements. Substantially, this approach presents a level of analysis and knowledge which is intermediate between the crude expression of an enzyme as catalysing some particular reaction and the highly detailed knowledge required for an exact quantitative description of the enzymatic rate laws.

The control approach to the theoretical understanding of two component oscillating reactions leads to simple necessary conditions in terms of either forward inhibition or back activation of the pathways. The conditions are readily expressed in terms of a net flux diagram which can be deduced by inspection of the control properties of the pathway. Additional conditions on the relative strengths of the controls prove to be sufficient (for all the enzymic systems studied) to lead to limit cycle oscillations for the kinetics. Equally simple conditions are derived for the n-component phase shift oscillators (n > 2) which can yield limit cycle oscillations with pathways involving forward activation or back inhibition. In both cases, the analysis can be carried out in graphical terms and the conclusions depend only on the qualitative properties of the control characteristics, not on the exact quantitative details. As such, the results are easily extended to illustrate mechanisms which can exhibit antilimit cycles and nested limit cycle oscillations.

In terms of the known control properties of the individual enzymes of the glycolytic pathway, application of the theoretical results leads to a multitude of distinct glycolytic control states which could yield oscillatory kinetics. And the determination of that particular control state responsible for the observed dynamics in a given situation requires confirmation internal to the given experiment as well as other (external) experiments. A careful study of the qualitative details of a kinetic response can delineate between various control states. That approach is greatly aided by comparative computer studies of the qualitative kinetic features derived from different states; and the computer models can be simplified to the extent that they will allow a representation of the various control states. While this provides some distinction between various control states, the resolution is generally not unique, depending on the extensiveness of the pathway and the variety of control states available to it. Further delineation of the specific control state must be found in related <u>in vivo</u> experiments and associated <u>in vitro</u> studies of the control factors which cannot otherwise be distinguished.

Such studies were carried out on the oscillating heart muscle extract. The kinetics of these oscillations exhibited a variety of complex qualitative features and their analysis suggested the presence of F6P, FDP and ADP (or all the adenine nucleotides) activational control of pfk as well as FDP and ADP control of the lower glycolytic pathway. Comparative computer studies of the glycolytic pathway demonstrated that such controls could account for the essential qualitative features of the experimental kinetics. <u>In vitro</u> studies of pfk confirmed the presence of all these control features under the conditions of the extract but the control strengths were such that the occurrence of the oscillations was dependent primarily on the concerted action of the adenine nucleotides (and F6P) in this particular case. The latter result is further confirmed by other experiments on the extract which demonstrated the presence of oscillations in the absence of any FDP control. Yet, it should be appreciated that the system demonstrates all the requisite control features to exhibit oscillations based on other control states (i.e. not dependent on the adenine nucleotides) although such states appear more difficult to obtain. An unexpected side result of the extract study was found in the kinetics of NADH, LAC and PYR. The qualitative features of these components were incompatible with the dynamics of the glycolytic pathway alone and strongly suggested that the lactic dehydrogenase reaction was oscillating about the equilibrium state and that pyruvate was being removed through some other pathway which was probably coupled to the turnover of NADH and subject to control by the other glycolytic components.

One obvious conclusion, both theoretically and experimentally, regards the potential dynamical complexity of control by glycolysis and correlated pathways; the system is capable of oscillatory states as well as multi-stable stationary states, and appears to be influenced by a number of metabolic factors. Prior treatments (29) which represent glycolysis as a single stationary state subject to control by just one or two factors are certainly suspect.

We believe that these theoretical, computer and experimental studies strongly support the usefulness of the control theoretic approach and our contention that the control theoretic properties are the prime determinant of the qualitative aspects of the kinetic behavior of the system. In many respects, this approach resembles the group theoretic approach in the prediction and interpretation of the qualitative features of atomic and molecular spectra. Given the symmetry properties of a specific system, group theory allows the prediction of the nature and number of line splittings which can be expected under certain circumstances. Conversely, group theory is a considerable aid in the direct interpretation of spectral data, though the interpretations are not unique and specific confirmation requires atomic or molecular compositions and even structures. Analogously, the control theoretic approach provides a basis for the prediction and interpretation of the qualitative features of the kinetics which might reasonably be termed the "temporal spectra". However, there is one major fault in the analogy at this time: while the group theoretic approach is mathematically well founded and can be readily related to the underlying physical equations, the control theoretical approach is only weakly founded and a great deal of future mathematical development will be required to establish the correct limitations, to delineate the relevant properties and to relate conclusions directly to the properties of the underlying nonlinear differential equations.

From the historical standpoint, the theoretical understanding and the experimental discovery (or acceptance?) (30) of oscillating biochemical reactions must certainly appear long overdue. The requisite features of enzyme control characteristics (e.g. activation and inhibition) have been fully recognized for well over thirty years. The theoretical methods, simple chemical mechanisms for oscillations and the role of non-linear phenomena in biology were fully appreciated by Lotka (31) in the early part of this century. For the past two decades, at least, computer studies of relevant models could have been and often were carried out, but those that were studied appear to have yielded only stable kinetics, artifactual oscillations, or (at best) little of relevant guidance. Experimental techniques for <u>in vivo</u> or extract metabolites assays and even for reconstituted (<u>in vitro</u>) studies of multienzyme systems* could have established or demonstrated (33) the existence of such phenomena long ago.

In part this historical situation can be attributed to a lack of communication between theoreticians, computerologists and experimentalists, but it must certainly be attributed as well to a general disbelief in the role of non-linear phenomena and an overemphasis, both theoretically and experimentally, on stable systems and linear (or linearized) equations. While the difficulties attendant to metabolic assays (for example) can be easily appreciated, it is clear that incomplete, sparsely distributed temporarily discrete data are valueless in the examination of the systems dynamics; it is not surprising that the experimental realization of oscillating biochemical reactions was based on the continuous monitoring of NADH fluorescence. While it is understandable that computer studies have been passing through a developmental stage, the evidence now appears strong that such studies are useless unless closely correlated with theoretical principles or at least with the guidance of logical concepts and well designed, careful methodology. Theoretically, we may recall the period at the end of the last century when unknown (and ad hoc) "vitalistic forces" were introduced to explain those aspects of biology which could not be obviously understood in terms of a rational (reversible) thermodynamic equilibrium state. Such ad hoc forces gave way to rational (yet stable) stationary state considerations, irreversible (but still linear) thermodynamics (34), and linearized studies of non-linear systems (35, 36) which culminated in the early 1950's. Yet such approaches failed to provide any understanding of the most obvious biological phenomena such as life and death, cell differentiation, memory and physiological rhythms, phenomena so readily

* The <u>in vitro</u> demonstration of multistable states was recently accomplished by H. Degn (32).

encompassed with the realm of nonlinear differential equations.

Along the way there have been many suggestions, hints and even theoretical models as to the relevance of nonlinear phenomena to cellular biology. Yet it seems likely that the discovery of biochemical oscillations represents the first clear cut confluence, at least at the cellular level, of experiment and theory, wherein the theory is well founded and the experimental variables are well defined and measurable. Nevertheless, such biochemical oscillations provide only one clear example of the role of nonlinear phenomena while the theory suggests that such phenomena can frequently occur. Consequently, we may question the extent to which such nonlinear effects may operate under normal physiological conditions or even whether non-linear differential equations provide the best basis for understanding the effects which are observed. Perhaps, as already suggested, the current lack of clear cut experimental examples is due to a lack of understanding and recognition, but it might also be that nature prefers to operate around a single stable stationary state under normal physiological conditions and that the observed nonlinear phenomena are a result of abnormal conditions.

The papers at this symposium already testify to our collective awareness of nonlinear phenomena and its <u>potential</u> theoretical and experimental consequences ranging from thermodynamics to circadian rhythms. The "lid" is off, but since virtually any known dynamical phenomena can be readily encompassed within the scope of nonlinear systems, let us hope it does not become a "Pandora's box" leading to a plethora of experimentally unrelated and unverifiable mathematical and computer models which can only demonstrate what we already know. As we have tried to indicate, the future theoretical problems require careful developments in the basic mathematics of nonlinear differential equations while the experimental studies will necessitate the continued close efforts of theoreticians and experimentalists as evidenced by this symposium.

ACKNOWLEDGMENTS

We wish to acknowledge the help and patience of Mrs. Carol Berger who typed this manuscript. This work was supported in part by PHS grants GM 12202-05 and 1-K04-GM-08313-01.

REFERENCES

1. Chance, B., Estabrook, R.W. and Ghosh, A. Proc. Natl. Acad. Sci. **51**, 1244 (1964).
2. Chance, B., Ghosh, A., Higgins, J. and Maitra, P.K. Ann. N.Y. Acad. Sci. **115**, 1010 (1964).
3. Higgins, J. Proc. Natl. Acad. Sci. **51** 989 (1964).
4. Higgins, J. In <u>Control of Energy Metabolism</u> (Eds. B. Chance, R.W. Estabrook and J.R. Williamson) Academic Press, New York, 1965. p.13.
5. Higgins, J. Royal Acad. of Medicine of Belgium (175 Anniversary Volume) pp. 235-268 (1966).
6. Sel'kov, E.E. <u>Scientific Thought</u>, Kiev, 1965.
7. Higgins, J. Ind. Eng. Chem. **59**, 18 (1967).
8. Minorsky, N. <u>Non-linear Oscillations</u>. Van Nostrand, Princeton, 1962.
9. Chance, Hughes, MacNichol, Sayre and Williams (editors) <u>Waveforms</u>. Radiation Laboratory Series, McGraw Hill, New York, 1948.
10. Cohen, A.R. <u>Outline of Electronic Circuit Analysis</u>. Regents, New York, 1964.
11. Brown, R.G. and Nilsson, J.W. <u>Introduction to Linear Systems Analysis</u>. Wiley, New York, 1962.
12. Higgins, J. In preparation.
13. Higgins, J. Ph.D. dissertation, Univ. of Pennsylvania, 1959.
14. Higgins, J. Ann. N.Y. Acad. Sci. **108**, 305 (1963).
15. Chance, B., Holmes, W. Higgins, J. and Connelly, C.M. Nature **182** , 1190 (1958).
16. Williamson, J.R. In <u>Control of Energy Metabolism</u> (eds. B. Chance, R.W. Estabrook, J.R. Williamson) Academic Press, New York, 1965. p.333.
17. Bonhoeffer, K.F. Naturwiss, **31**, 270 (1943).
18. Goodwin, Brian C. Adv. Enzyme Regul. **3**, 425 (1965).
19. Morales, M. and McKay, D. Biophys. J. **7**, 621 (1967).

20. Sel'kov, E.E. Proc. of Fed. Symp. on Unstable Processes in Biol. & Chem. Systems. Puschino on Oka, 1966.
21. Pye, E.K. Can. J. Botany. $\underline{47}$, 271 (1969).
22. Chance, B., Williamson, G., Lee, I.Y., Mela, L., DeVault, D., Ghosh, A. and Pye, E.K. This volume, p. 285
23. Frenkel, R., Achs, M.J. and Garfinkel, D. This volume, p. 187
241 Higgins, J. In preparation.
25. Higgins, J. In. Computers in Biomediacl Research (Eds. R. W. Stacy and B.D. Waxman) Academic Press, New York, $\underline{11}$, 101 (1965).
26. Hulme, E. In preparation.
27. Frenkel, R. Personal communication.
28. Cantarow, A. and Schepartz, B. Biochemistry (Fourth edition) Saunders, Philadelphia, 1967.
29. Chance, B. and Hess, B. Ann. N.Y. Acad. Sci. $\underline{63}$, 1008 (1956).
30. Dusens, L.N.M. and Amesz, J. Biochim. Biophys. Acta $\underline{24}$, 19 (1957).
31. Lotka, A.J. Elements of Mathematical Biology. Dover Publ. New York, 1956.
32. Degn, H. Nature $\underline{217}$, 1047 (1968).
33. Hess, B. and Boiteux, A. This volume, p. 229
34. Prigogine, I. Thermodynamics of Irreversible Processes. Wiley, New York, 1955.
35. Hearon, J.Z. Bull. Math. Biophys. $\underline{15}$, 121 (1953).
36. Bak. T. Fr. Baggs Kgl. Hofbogtrykkeri, København, 1959.

PROBLEMS ASSOCIATED WITH THE COMPUTER SIMULATION
OF OSCILLATING SYSTEMS

E. M. Chance

Department of Biochemistry, University College
London

Ever since the discovery of glycolytic oscillations in yeast cells (1) and the cooperative amplification of enzyme catalysis by allosteric mechanisms, much speculation has arisen about the nature of these as separate phenomena. The oscillatory tendency of feed-back systems has been well known in electronics, and on this basis the possibility of oscillators in biochemical systems has become accepted. Feed-back in such systems could be external or internal, depending on the relationship of the feed-back component to the oscillator. Higgins has produced an oscillating model of the heart supernatant system with a single enzyme oscillator of theoretical design. Selkov (2), however, has produced a model from the data of Betz for yeast phosphofructokinase (3).

Simulation of glycolytic enzymes has been of interest because of the allosteric nature of the key kinases. As is well known, two main schools of thought exist about the mechanism of these enzymes. The first is based on the Monod theory (4) that cooperativity is a consequence of a concerted process and the second theory, of which Koshland (5) is the main proponent, is that cooperativity is a consequence of a sequential process.

It was decided to attempt to see if either or both of these theories of allosteric mechanism could produce a continuous enzymic oscillator similar to the reconstructed yeast system of Hess and Boiteux(6). Furthermore, this problem led to the implementation of digital computer software to allow algebraic rate laws to be used in conjunction with mass law differential equations to solve the complex rate equations involved.

Method of Simulation

At University College, London, we do not have access to a large scale analog computer so are forced to make do with the facilities of a large scale digital computer (IBM 360/H65). Methods have been devised by which an enzyme mechanism may be translated automatically into mass law differential equations. Another approach is to use a state function for the binding of the enzyme in terms of its substrate and/or modifier concentration. Such a state function can be derived to fit the overall kinetic pattern of the enzyme. Certain differential equations are then eliminated, usually the ones which are difficult to integrate numerically because they involve the multiplication of a small variable by a large constant. Higgins (7) and Selkov (2) have adopted the use of rate laws in analog computer studies. Higgins, however, has not been able to solve the algebraic equations directly on an analog computer and has to represent these by differential equations which have the same mathematical properties as the algebraic equations. Furthermore, these differential equations are likely to have been scaled as it is very difficult to represent the product of a large constant times a large variable accurately on an analog computer. The general simulation program developed in this laboratory has been extended to include this capability. The program will permit the writing of mass-law differential equations from reaction images on IBM punched cards in a format shown in Figure 1. It will also allow flux statements to be over written at run time and one is able to include algebraic rate laws written in Fortran. By successive runs with the simulation program, one is able to compare the kinetics of an enzyme represented by a detailed chemical mechanism (as far as this might be known) and represent an algebraic statement of its overall kinetics in terms of substrate K_m and V_{max}.

Differential equation representation:

$$E + S \underset{k_2}{\overset{k_1}{\rightleftharpoons}} ES \overset{k_3}{\rightarrow} E + P \quad (1)$$

$$\dot{P} = k_3 [ES] \quad (2)$$

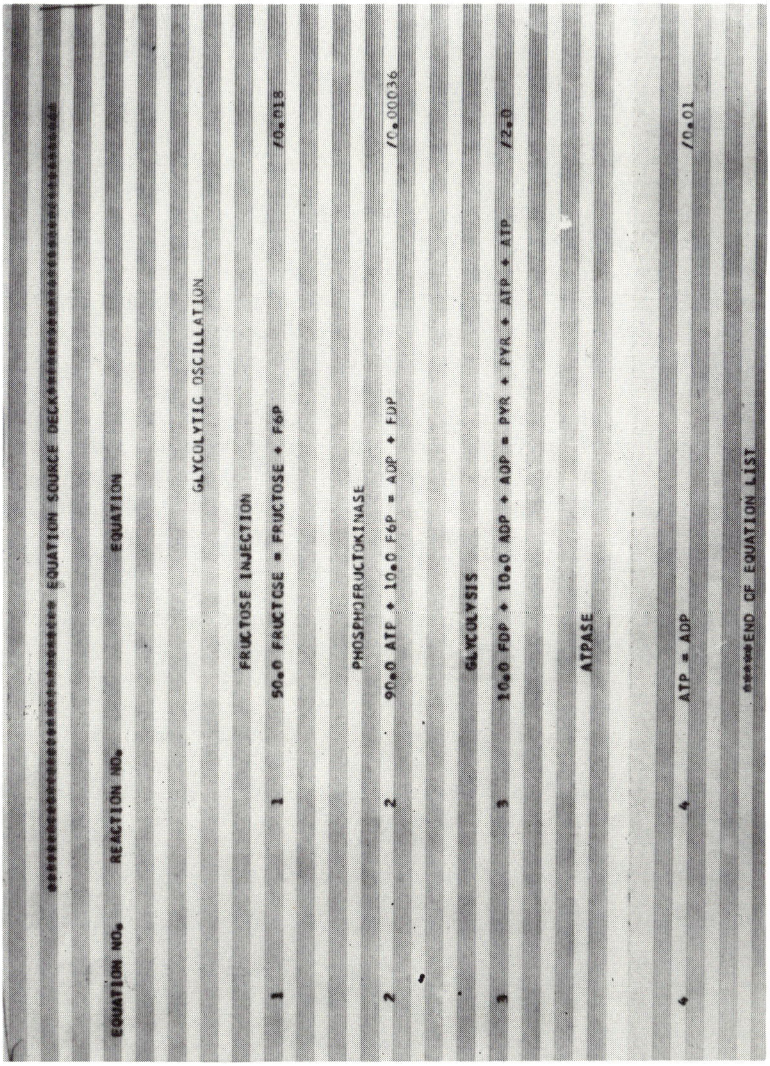

BIOCHEMICAL OSCILLATORS

EQUATION NO.	REACTION NO.	EQUATION
		GLYCOLYTIC OSCILLATION
		FRUCTOSE INJECTION
1	1	50.0 FRUCTOSE = FRUCTOSE + F6P /0.018
		PHOSPHOFRUCTOKINASE
2	2	90.0 ATP + 10.0 F6P = ADP + FDP /0.00036
		GLYCOLYSIS
3	3	10.0 FDP + 10.0 ADP + ADP = PYR + ATP + ATP /2.0
		ATPASE
4	4	ATP = ADP /0.01

*****END OF EQUATION LIST

Figure 1.

Rate law representation:

$$S \xrightarrow{V_{max}} P \quad (3)$$

$$\dot{P} = \frac{V_{max} S}{S+K_m} = \bar{y} V_{max} \quad (4)$$

This flexibility of the simulation program allows for easy examination of the situation without the necessity of rewiring, which would be incurred with an analog computer. One main disadvantage to digital computing has been the fact that the large scale digital computer is very seldom available in a 'hands on' manner, but due to the kindness of Professor B. Hess at the Max-Planck-Institute, Dortmund, an IBM 360/F44 has been made available, 'hands on'. A subroutine has been developed to allow conversational access to the computer so that one can vary constants during a solution. This system, with visual display of results, would have most of the attractive features of the analog computer, without the limitations of scale, size and accuracy.

The Model System

The model was based originally on the analog computer studies of Higgins and was chosen because the square wave nature of the solution was rather similar to the experimental square wave system observed in reconstituted systems (6). It must be stated, however, that the digital computer solution is not exactly identical to the analog solution as published, due probably to small fluctuations of kinetic constants as a result of drift.

The original model can be shown diagramatically as follows:

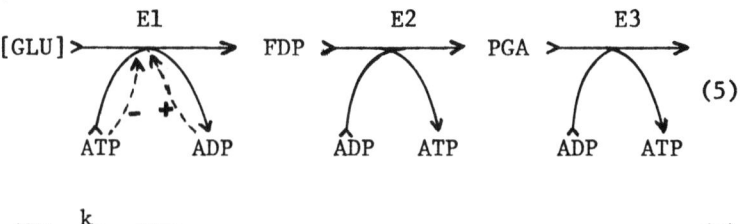

$$ATP \xrightarrow{k} ADP \quad (6)$$

One can notice by inspection that the above model consists of an input step, one non-Michaelis enzyme, two ordinary enzymes and a non-enzymic ATPase. This model may be further reduced, as shown below, to an input step and two enzymes, considering the enzymic nature of the conversion of FDP to PEP, and retaining the non-enzymic breakdown of ATP to ADP.

$$\text{FRUCTOSE} \xrightarrow{k1} \text{F6P} \quad (7)$$

$$[\text{F6P}] \xrightarrow[\text{ATP} \quad \text{ADP}]{E1} \text{FDP} \xrightarrow[\text{2ADP} \quad \text{2ATP}]{E2} 2\text{ETOH} + 2\text{CO}_2 \quad (8)$$

$$\text{ATP} \xrightarrow{K2} \text{ADP} \quad (9)$$

A rate law for glycolysis (E2) can be written:

$$v_{E2} = V_{max_{E2}} \cdot \left(\frac{\text{ADP}}{\text{ADP} + K_{M1}^{E2}\text{ADP}} \right) \cdot \left(\frac{\text{ADP}}{\text{ADP} + K_{M2}^{E2}\text{ADP}} \right) \cdot \left(\frac{\text{FDP}}{\text{FDP} + K_M^{E2}\text{FDP}} \right) \quad (10)$$

Constants for glycolysis which produced oscillations in the Higgins system have been retained, so the simulation problem is reduced to the question of what type of rate law and constants are required for the E1 step to produce oscillations. Two rate laws were tried – the allosteric rate law of Monod and a simple sequential rate law, which represents some of the ideas of Koshland. The allosteric rate law in a looped system is convenient because the α in the Monod-Wyman equations can be taken as ADP since a decrease of ADP in a looped system, that is an increase of ATP, will inhibit the enzyme. The glycolysis rate law of E2 is there merely to provide a sufficient sink, according to the ideas of Selkov (2). Therefore, in the allosteric model the state of activity of the enzyme is determined by the ADP concentration and the actual enzymic flux is the product of $V_{max} \cdot \text{ATP} \cdot \bar{Y}$, and the actual rate law is as follows:

$$\bar{Y} = \frac{L'C[ADP](1+C[ADP])^{n-1} + [ADP](1+[ADP])^{n-1}}{L'(1+C[ADP])^n + (1+[ADP])^n} \quad (11)$$

In this preliminary study the F6P and ATP interactions are non-enzymic, but in this system with a controlled input and sufficient carbon sink, these concentrations are rigorously controlled by the PFK activity.

The allosteric constants were chosen as n=2 (Hill exponent), $C = 5 \times 10^{-3}$ and $L = 1.0$ for damped oscillation. When $L = 2.5$, sawtooth sustained oscillations were produced as shown in Figure 2. The sequential model is built entirely around a Michaelis rate law and it is the application of this Michaelis rate law that produces the quadratic effect.

The rate law is as follows:

$$v_{E1} = V_{max_{E1}} \cdot \left(\frac{ADP}{ADP+K_{M1}^{E1}ADP}\right) \cdot \left(\frac{ADP}{ADP+K_{M2}^{E1}ADP}\right) \cdot [F6P] \cdot [ATP] \quad (12)$$

The state function is determined by the ADP concentration and also produces oscillations, but of slightly different sort because of the rational constants involved. For the sequential system there are two allosteric constants only: one for the first binding of ATP, and the other for the second. The Km for the first binding ($K_{M1}^{E1}ADP$) was in this curve set equal to the second ($K_{M2}^{E1}ADP$). This solution produced continuous square wave oscillations, similar to the reconstructed system.

Discussion

By introducing the algebraic rate laws instead of differential equations for this simulation, the time required to compute each separate solution was decreased by a factor of about 100. This is roughly comparable with analog computer speeds, if one neglects the delay caused by a mechanical read-out device on the digital computer. It was, therefore, possible to use the IBM 360/44 successfully for simulation and vary constants on line as required. In this case, due to the constraints of the problem, only a small number

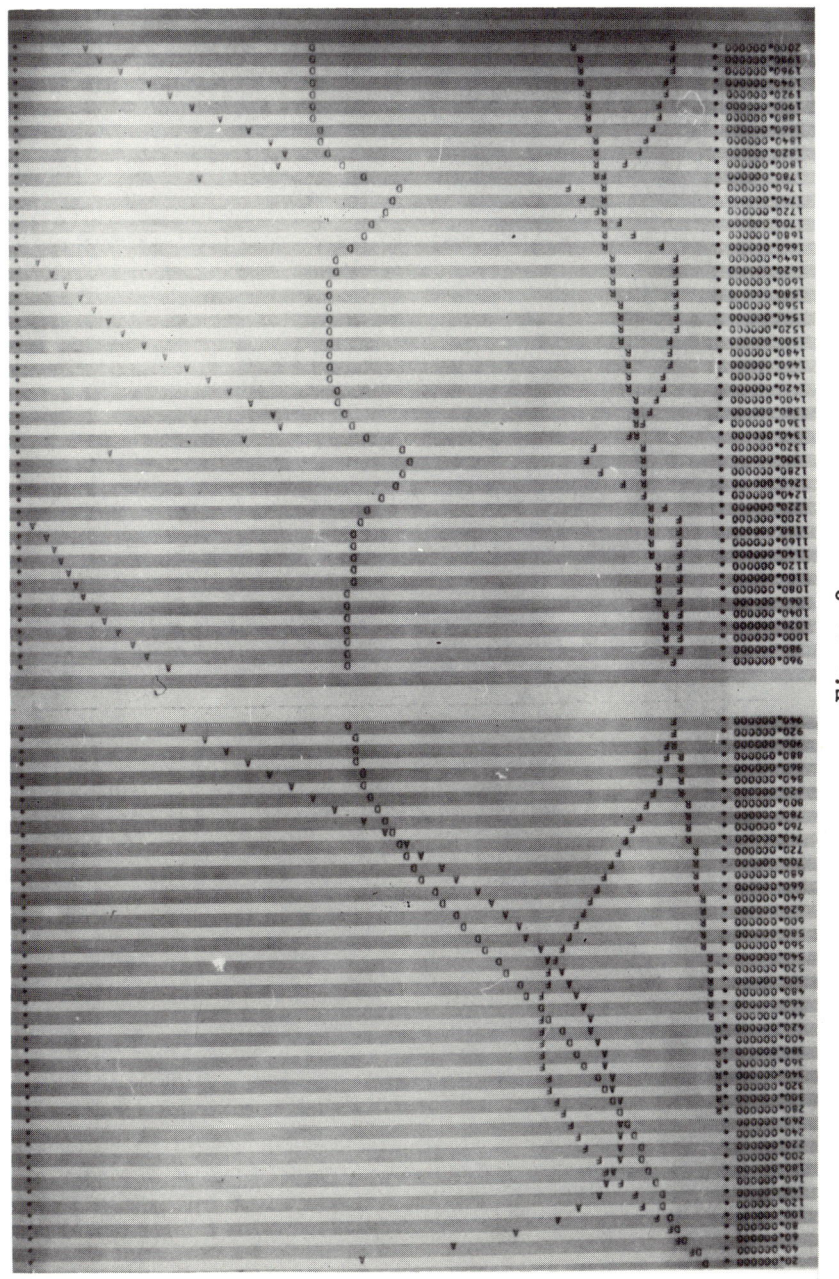

Figure 2.

of constants were varied, so that the actual 'fishing' was kept to a minimum. Input titrations were done on the faster IBM 360/65 because they require a systematic variation of a single parameter to assess oscillations over a range of inputs. The ideal system would be a visual terminal remote to a monolithic computing system such as the CDC 6600 installation proposed at London University.

One difficulty that has not been resolved is the definite differences between digital and analog simulations of supposedly the same problem. The causes of these discrepancies could result from errors due to analog drift, inaccuracies of digital integration techniques and minor differences of programming differential equations to substitute for equations in analog routines. Higgins (8) has suggested that these discrepancies are due to analog computer artifacts.

In this study, both the concerted and sequential mechanisms for the oscillating enzyme in glycolysis have been found to produce oscillations. In the concerted system, more varied forms of oscillation are possible because fewer components exist for the allosteric constants. The sequential system seems to produce square wave oscillations similar to those observed in the cell-free system. The constraints on the sequential system are, of course, greater. The K_m for each binding of the ADP activator was chosen so that the enzyme would be saturated at maximum ADP concentration. It would be difficult to define these K_ms in terms of the usual allosteric constraints, but the amplification is quadratic. In both cases V_{max} for all the enzymes is the same and the only parameters varied were the rate laws and associated constants for the two oscillators.

The Higgins system, as reproduced by our digital computer gave sinusoidal oscillations. Considerable activation of the enzyme was effected by the enzyme binding a significant amount of ATP. The possibility of this situation being relevant has come from immunological studies of pyruvate kinase by Hess and Krüger (9) which give the enzyme a rather high molarity. The rate law simulations as given in this communication would be invalid in this case, as the state function does not include a theory for removal of the modifier.

Further investigation is under way at present into the detailed nature of the forms of oscillation obtained with the concerted rate equation and its relation to the glycolytic input. It is required to establish a relationship for both mechanisms, between the FDP steady-state and the input, by varying the latter through a steady-state phase, an oscillatory stage and on to another steady-state phase. It can then be seen if this relationship is similar to that produced in the cell-free system.

Acknowledgments

The author wishes to thank Professor B. Hess for his advice and for the use of the Max-Planck-Institute's computer 'hands on'. The author further thanks Dr. J. Higgins for collaboration in implementing the rate law simulation, Mr. E. P. Sheperd for programming assistance, and the University College London Computer Centre for computing time.

This work was supported by U.S. Public Health Service Grant AM 10435.

References

1. Ghosh, A. and Chance, B. Biochem. Biophys. Res. Commun., 16, 174 (1964).
2. Selkov, E.E. and A. Betz. This volume, p.197.
3. Betz, A. This volume, p.221.
4. Monod, J., Wyman, J. and Changeux, J.P. J. Mol. Biol., 12, 88 (1965).
5. Koshland, D.E., Conway, A. and Kertley, M.E. In FEBS Symposium on Regulation of Enzyme Activity and Allosteric Interactions. (Ed. E. Kvamme and A. Pihl). Academic Press, New York, 1968. p. 131.
6. Hess, B. and Boiteux, A. In Abstracts 4th FEBS Meeting Oslo, 1968.
7. Higgins, J. Ind. Eng. Chem., 54, 19 (1967).
8. Higgins, J. Abstract No. 856, 5th FEBS Meeting, Prague, 1968.
9. Hess, B. and Kruger, J. Hoppe-Seyler's Z. Physiol. Chem. 349, 104 (1968).

THE EFFECT OF FRUCTOSE DIPHOSPHATE ACTIVATION
OF PYRUVATE KINASE ON GLYCOLYTIC OSCILLATIONS
IN BEEF HEART SUPERNATANT:
AN EXPERIMENTAL AND SIMULATION STUDY.*

Rene Frenkel,[+] Murray J. Achs, and David Garfinkel*

Johnson Research Foundation, University of Pennsylvania
Philadelphia, Pennsylvania 19104
and
Department of Biochemistry, University of Texas
Southwestern Medical School, Dallas, Texas 75235

As is evident from the other papers in this symposium, glycolytic oscillation has been primarily and most thoroughly studied in yeast and yeast extracts. However, this phenomenon also occurs in mammalian systems and may profitably be studied there. This communication presents an addition to a previously reported simulation of oscillating glycolysis in a beef heart supernatant fraction (1), and shows how simulation can be used to clarify control mechanisms. Specifically, the effect of activation of pyruvate kinase by fructose-1,6-diphosphate (FDP) was first simulated and then verified experimentally. This activation was found to be similar in both the glycolyzing system and the isolated enzyme.

DPNH oscillations have been observed to take place in a concentrated supernatant fraction prepared from beef heart at pH 6.25 and supplemented with a source of glucose-6-phosphate (glycogen and apyrase). At higher, more physiological pH, the same preparation shows active glycolysis but does not oscillate (2,3). Straightforward examination of

[+]Present address, University of Texas.
*Supported by grant FR-15 and Research Career Development Award GM-5469 from the National Institutes of Health.

the experimental data obtained, including concentrations as a function of time for all readily measurable glycolytic intermediates, indicates that oscillations are probably due to variations in activity of phosphofructokinase (PFK) under the influence of the adenine nucleotides and perhaps FDP (2,3); this is compatible with the properties of the PFK isolated from preparations of the type studied here (4). (A parallel study on the same experimental data with an analog computer is reported elsewhere in this symposium by Higgins (p.127), with similar representations of the enzymes, and with some different assumptions.

Digital computer simulations of this preparation have been carried out, both for pH 7.4 (5) and pH 6.25 (1). It is found, as would be desired, that substantially the same collection of enzyme models, suitably modified for the change in pH, behaves in the two quite different overall manners observed at these two pH's. It was necessary in both simulations to assume a number of presently unknown activations or inhibitions of the enzymes involved. At the lower pH it was possible to fit the oscillation data (3) to nearly within the experimental error (1); at the higher pH the fit to the kinetic data was not quite as good (5).

The principal basis of the oscillating model is that the oscillations depend on the behavior of PFK, whose activity is regulated almost exclusively by the adenine nucleotides. Sel'kov (6), in an analysis based primarily on yeast data, has since reached a similar conclusion. The basic control of the oscillation by this enzyme has been used to study some of its properties. Under these conditions several molecules each of ADP, AMP, and ATP are involved as PFK activators or inhibitors, the numbers involved varying from 2, 3, and 3 respectively to 4 molecules each. Perhaps two per cent of the total activation was contributed by FDP in the latter case (1). The exact wave-shape of the oscillation may be dependent on the detailed properties of the other enzymes, especially the control properties of the ATPase present, but this has been shown not to cause the oscillation itself (1).

This model (1) consists of 57 simultaneous differential equations representing 101 chemical reactions, most of which represent enzyme submodels carefully adjusted (5) to match available literature values. These equations were written and solved with a digital computer (PDP-6), with the aid of programs and methods described elsewhere (7,8). These programs accept as input biochemical equations, translate them to differential equations, solve them, and edit them into a form useful to the biochemist, without requiring much mathematical competence on his part. An example both of the type of experimental data used and of the fit of the model to it is shown in Figure 1.

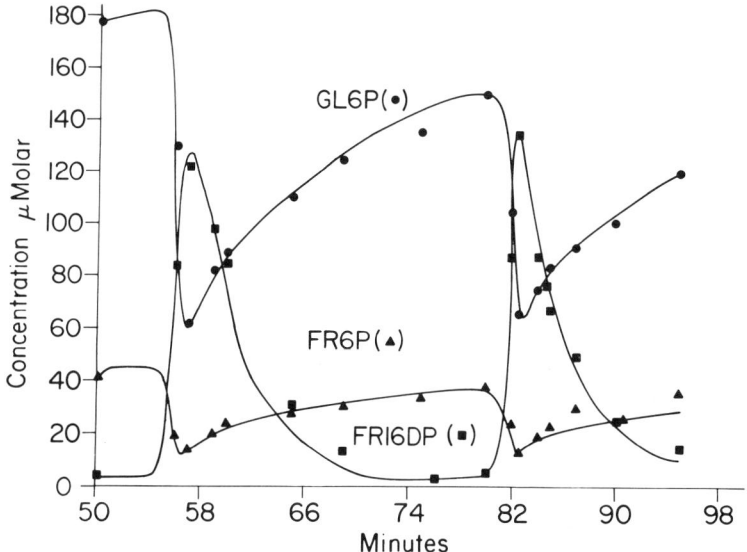

Figure 1. Fitting of hexose phosphates in the model to the corresponding experimental points. From (1), courtesy of Academic Press Inc.

Subsequent to the construction of this model, it was found that the pyruvate kinase (PYK) of beef heart is activated by FDP in a manner similar to that for yeast or liver (9). The omission of this activation from the original model leads to a small but systematic error for pyruvate and phosphoenolpyruvate concentrations.

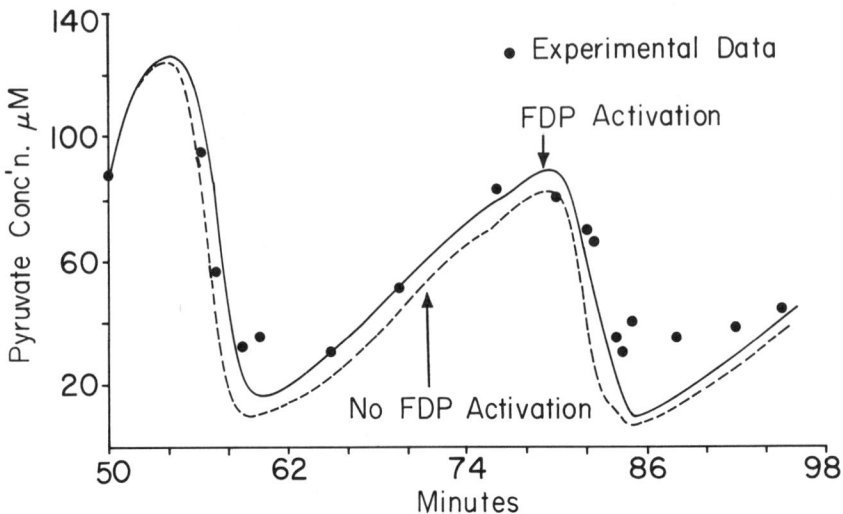

Figure 2. Pyruvate levels with and without activation of PYK by FDP. 'Best fit' method.

The effect of adding FDP activation of pyruvate kinase to the model is shown in Figure 2. Pyruvate concentrations were computed from the experimentally determined concentrations of all the contributing substrates (ADP, PEP, FDP; and DPNH for lactate dehydrogenase). Inversely, if pyruvate concentrations were set, PEP concentrations could be matched quite closely. The model for pyruvate kinase and its activation by FDP was assumed to be

$$\text{PYKASE} + \text{PEP} \underset{2}{\overset{1}{\rightleftarrows}} \text{PYKPEP} \qquad (1)$$

$$\text{PYKPEP} + \text{ADP} \underset{4}{\overset{3}{\rightleftarrows}} \text{PYKAPY} \qquad (2)$$

$$\text{PYKAPY} \overset{5}{\rightarrow} \text{PYKATP} + \text{PYR} \qquad (3)$$

$$\text{PYKATP} \overset{6}{\rightarrow} \text{PYKASE} + \text{ATP} \qquad (4)$$

PYKASE + PEP + (FDP) $\xrightarrow{11}$ PYKPEP (5)

The rate equation for the above is:

$$v = \frac{V(ADP)(PEP)}{(\frac{k_2}{k_1+k_{11}(FDP)})K_{mADP} + K_{mPEP}(ADP) + K_{mADP}(PEP) + (ADP)(PEP)} \quad (6)$$

This may be conveniently rearranged as:

$$v = \frac{V(ADP)(PEP)}{x \cdot K_{dissPYKPEP}(K_{mADP}) + x \cdot K_{mPEP}(ADP) + K_{mADP}(PEP) + (ADP)(PEP)}$$
(7)

where $x = \dfrac{k_1}{k_1 + k_{11}(FDP)}$ = non-activated fraction of total
PYKPEP formation flux, and K_{mPEP} is determined in the absence of FDP. The postulated (actually somewhat simplified) mechanism is irreversible and ordered; PEP adds first to form the PYKPEP complex (with and without FDP activation); PYR is released first.

Direct experiments using highly purified pyruvate kinase isolated from beef heart were then performed as a check on the proposed mechanism. Velocities (v), measured at 25° by following fluorimetrically the oxidation of DPNH by pyruvate and lactate dehydrogenase, are summarized in Table 1.

At pH 6.25, computer-optimized data fits were obtained with the mechanism outlined when the pyruvate kinase had the following kinetic parameters:

K_{mADP}	K_{mPEP}	V_{max}	K_{diss} for PYKPEP
48 μM	16.8 μM	10.2 mmoles/l/min	29.0 μM

Deviation of computed from observed velocities averaged about 5% which is well within the experimental error. (For equations relating kinetic constants to rate constants, see Cleland (10)).

TABLE 1

Experimental Measurements of Fructose Diphosphate Activation of Heart Muscle Pyruvate Kinase

ADP (mM)	PEP (mM)	Velocity (mmoles/l/min)	
		Control	+0.1mM FDP
	pH 6.25		
0.036	0.035	2.6	3.4
0.036	0.070	3.1	4.05
0.036	0.175	4.1	4.1
0.036	0.350	4.1	4.1
0.072	0.035	4.0	4.7
0.072	0.070	4.4	5.9
	pH 7.4		
0.072	0.0175	3.9	5.9
0.072	0.035	4.55	7.8
0.072	0.07	5.4	9.0
			+0.2mM FDP
0.36	0.35	6.7	12.7
0.36	0.07	11.2	17.7
0.36	0.14	14.6	25.0
0.36	0.175	19.6	28.5

Table 1 shows that when experimental concentrations are in the model range (ADP 36 μM, PEP 35 μM) the addition of 0.1 mM FDP causes the experimentally observed velocity to increase about 30%. Using the computer optimized properties of PYK obtained above, such an increase occurs when the FDP activated flux is 58% of the total for PYKPEP formation. In the oscillating beef heart model itself, pyruvate concentrations were matched most closely when the reaction velocity was increased 37% and the FDP activated flux was 62% of the total PYKPEP formation at 0.1 μM FDP. This causes the FDP-dependent K_{mPEP} for pyruvate kinase to vary by a factor of about 2.3 under the model conditions of FDP swings. The close match between computed and observed properties for the FDP activation of PYK indicate that the originally assumed mechanism of this activation is probably correct.

For the pH 7.4 data, however, no satisfactory fit could be obtained, unless the data for the two ADP concentrations were considered separately. Computer-optimized fits were obtained when the enzyme had the properties:

ADP	K_{mADP}	K_{mPEP}	V_{max}	K_{diss} for PYKPEP
27 μM	408 μM	29.6 μM	40.8	5.4 μM
360 μM	110 μM	198 μM	44.8	3.2 μM

This suggests that ADP may act as an allosteric regulator at the higher pH.

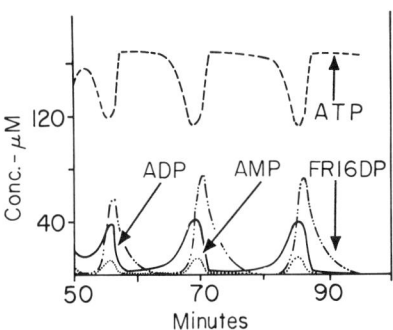

Figure 3. Adenine nucleotides and FDP. Activation of PYK by FDP at an activation strength causing 80% of total PYKPEP flux through activated branch at 0.1 mM FDP.

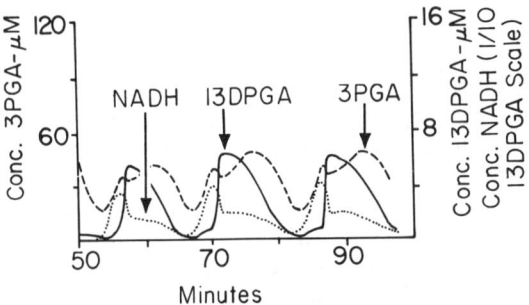

Figure 4. 3PGA, 13-DPGA and DPNH. Activation as in Fig.3.

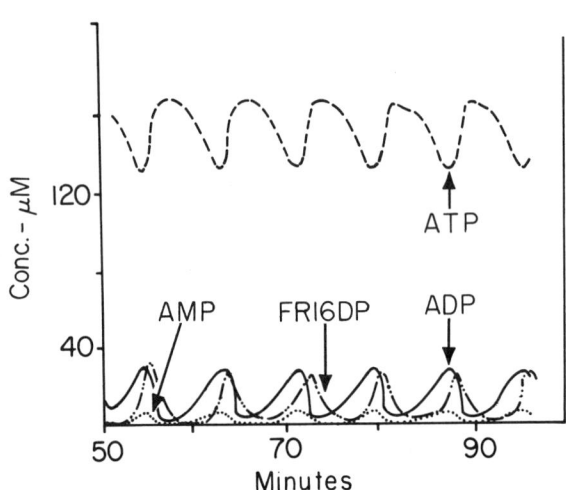

Figure 5. Adenine nucleotides and FDP. Activation strength causing 97.5% of total PYKPEP flux through the activated branch at 0.1 mM FDP.

BIOCHEMICAL OSCILLATORS

The effect of this PYK activation on the existence and shape of the oscillations was investigated. It has previously been noted by E.M. Chance (personal communication) in a preliminary exploration based on the "feed-forward" hypothesis of Hess (e.g.(11)) that the addition of PYK activation by FDP caused oscillation in a previously non-oscillating model of glycolysis (12) (as did cyclic pulsed injection of PEP). Since the oscillations were originally simulated (1) without considering this activation it cannot be responsible for the oscillations themselves or their general shape, but varying the strength of activation does vary the periodicity and detailed waveshapes as may be seen on comparing Figures 1, 3, and 5, and Figures 4 and 6. This activation could not completely damp out the oscillations (strengthening the FDP activation tends to lower FDP concentration). Damping is noticed with moderate activation (Figure 5) but not with the strongest activation (Figure 6). Figures 4 and 6 are quite different, although the FDP and adenine nucleotide profiles corresponding to Figure 6 are quite similar to those of Figure 3.

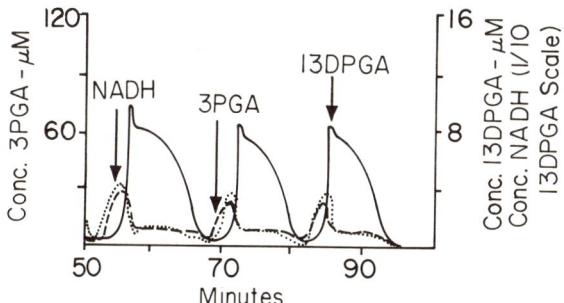

Figure 6. 3PGA, 13DPGA and DPNH. Activation strength causing 99.98% of total PYKPEP flux through the activated branch at 0.1 mM FDP.

In addition to the intrinsic value of studying these oscillations themselves, this method of study particularly permits obtaining the type of information regarding metabolic control of the glycolytic enzymes that is most easily seen during highly transient situations, and difficult to obtain from steady-state data. It is hoped future studies will continue to generate information of this type.

REFERENCES

1. Achs, M.J., and Garfinkel, D., Computers and Biomedical Research $\underline{2}$, 92 (1968).
2. Frenkel, R., Arch Biochem. Biophys. $\underline{125}$, 151 (1968)
3. Frenkel, R., Arch Biochem. Biophys. $\underline{125}$, 157 (1968).
4. Frenkel, R., Arch Biochem. Biophys. $\underline{125}$, 166 (1968).
5. Garfinkel, D., Frenkel, R., and Garfinkel, L., Computers and Biomedical Research $\underline{2}$, 68 (1968).
6. Sel'kov, E.E., European J. Biochem. $\underline{4}$, 79 (1968).
7. Garfinkel, D., Computers and Biomedical Research $\underline{2}$, 31 (1968).
8. Rhoads, D. G., Achs, M. J., Peterson, L., and Garfinkel, D., Computers and Biomedical Research $\underline{2}$, 45 (1968).
9. Haeckel, R., Hess, B., Lauterborn, W., and Wuster, K.H., Hoppe-Seyler's Z. physiol. Chemie $\underline{349}$, 699 (1968).
10. Cleland, W.W., Biochim. Biophys. Acta $\underline{67}$, 104, 173, 188 (1963).
11. Hess, B., Haeckel, R., and Brand, K., Biochem. Biophys. Res. Comm. $\underline{24}$, 824 (1966).
12. Garfinkel, D., and Hess, B., J. Biol. Chem. $\underline{239}$, 971 (1964).

ON THE MECHANISM OF SINGLE-FREQUENCY GLYCOLYTIC OSCILLATIONS

E.E. Sel'kov and A. Betz

Institute of Biophysics of the
U.S.S.R. Academy of Sciences,
Puschino, Moscow Region, U.S.S.R.

and

Institut für Molekulare Biologie
3301 Stöckheim b. Braunschweig, Germany

To explain the mechanism of glycolytic self-oscillations (1-9) a very simple model has been proposed by one of the authors (10-12) which represents a generalization of the Lotka mechanism (13):

$$\left. \begin{array}{l} \dot{x} = 1 - xy^r \\ \dot{y} = \alpha(xy^r - y) \end{array} \right\} \quad [1]$$

Initially (10,11) this model was intended to describe the phosphofructokinase reactions, where x and y are dimensionless concentrations of ATP and ADP, and r is the order of activation of phosphofructokinase (PFK) by ADP and AMP. Later, however, it was found (12) that this model describes the system of two coupled reactions catalyzed by phosphofructokinase and adenylate kinase (ADK). It has been found that in model [1] the condition $r \geqslant 2r'$ is satisfield where r' is the number of activating sites on PFK for AMP, x is the dimensionless concentration of fructose-6-phosphate (F6P), and y is the dimensionless concentration of ADP.

In order to describe the experimental data in greater detail, a more precise model has been constructed (14,15):

$$\left. \begin{array}{l} \dot{\sigma_1} = \vartheta_1 - \dfrac{\sigma_1(\vartheta_0 + \sigma_2^r)}{1 + \alpha_1\sigma_1 + \sigma_2^r(1 + \sigma_1)} \\ \dot{\sigma_2} = \alpha_2\left(\dfrac{\sigma_1(\vartheta_0 + \sigma_2^r)}{1 + \alpha_1\sigma_1 + \sigma_2^r(1 + \sigma_1)} - x_2\sigma_2\right) \end{array} \right\} \quad [2]$$

The designations in model [2] coincide with those used below, on page 205. This model, in contrast to model [1], allows for the fact that in the absence of AMP the relative PFK activity determined by the term ν_0 is not equal to zero. At $\nu_0 = 0$, model [2] can be reduced to model [1].

Model [2] is a further generalization of the Lotka mechanism (13) and has much in common with the model of a triode generator given by van der Pol (16) to which it can be reduced by a corresponding substitution of the variables. Model [2] gives a simple explanation for the mechanism of relaxation self-oscillations discovered recently by Hess and Boiteux (6-8) in the glycolytic system reconstituted from highly purified yeast enzymes.

When deriving models [1] and [2], it has been assumed that the PFK-ADK system is the single source of ADP in glycolysis. This means that models [1] and [2] describe the behavior of the glycolytic system under conditions of artificial injection of F6P, or other hexosemonophosphate.

This paper presents a modification of model [2], which describes an oscillatory behavior of the glycolytic system under more natural conditions. The model takes into account phosphorylation of the substrates in the hexokinase, fructokinase or glycogen phosphorylase reactions. The relationship between the levels of adeninucleotides and its effect on an oscillatory state of the PFK-ADK system has been analyzed in this work.

Relationship Between the Concentrations of AMP, ADP and ATP in the ADK Reaction

In order to substantiate a number of assertions given below, we have investigated the ADK reaction kinetics:

$$2 ADP \underset{}{\overset{ADK}{\rightleftharpoons}} ATP + AMP \quad [3]$$

Two assumptions have been made:
  a. that the activity of ADK is much higher than the activity of PFK. As shown earlier (12), it is equivalent to the assumption that equation [3] is in the equilibrium state.
  b. that the law of mass conservation holds in [3]:

$$[AMP] + [ADP] + [ATP] = A \qquad [4]$$

where A is constant.

In addition, dimensionless concentrations are introduced:

$$\beta_1 \equiv \frac{[AMP]}{A} \qquad \beta_2 \equiv \frac{[ADP]}{A} \qquad \beta_3 \equiv \frac{[ATP]}{A} \qquad [5]$$

From the first assumption it follows that

$$\frac{[AMP][ATP]}{[ADP]^2} \equiv \frac{\beta_1 \beta_3}{\beta_2^2} K \qquad [6]$$

where K is the equilibrium constant for the ADK reaction. Consequently, assumptions a. and b. may be represented by the following equation system:

$$\left. \begin{array}{c} \frac{\beta_1 \beta_3}{\beta_2^2} = K \\ \beta_1 + \beta_2 + \beta_3 = 1 \end{array} \right\} \qquad [7]$$

If β_2 is considered to be given, then the system has two solutions:

$$\left. \begin{array}{c} \beta_1' = 0.5\left(1 - \beta_2 - \sqrt{(1-\beta_2)^2 - 4K\beta_2^2}\right) \\ \beta_3' = 0.5\left(1 - \beta_2 + \sqrt{(1-\beta_2)^2 - 4K\beta_2^2}\right) \end{array} \right\} \qquad [8']$$

and

$$\left. \begin{array}{c} \beta_1'' = \beta_3' \\ \beta_3'' = \beta_1' \end{array} \right\} \qquad [8'']$$

which make sense at

$$\beta_2 \leq \beta_2^* \qquad [9]$$

where

$$\beta_2^* = \frac{1}{1 + 2\sqrt{K}} \qquad [10]$$

and is the value of β_2 for which solutions [8'] and [8''] coincide.

$$\beta_1^* = \beta_3^* = \frac{1 - \beta_2^*}{2} \qquad [11]$$

The plots of the functions $\beta_1 = \beta_1(\beta_2)$ and $\beta_3 = \beta_3(\beta_2)$

described by solutions [8] are quite identical with respect to form and represent a parabola, shown in Figure 1. It is

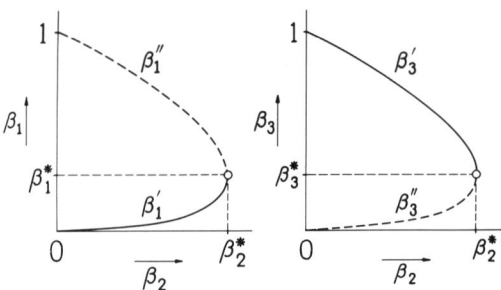

Figure 1. The relative AMP concentration, β_1, and the relative ATP concentration, β_3, as a function of the relative ADP concentration, β_2.

to be noted, however, that the values $\beta_1(\beta_2)$ and $\beta_3(\beta_2)$ lie on different branches of the parabola. These plots are projections of the parabola which lies on the plane $\beta_1+\beta_2+\beta_3=1$ determined by system [7] and shown in Figure 2. This curve determines a geometrical locus of the ADK reaction equilibrium states in its phase space.

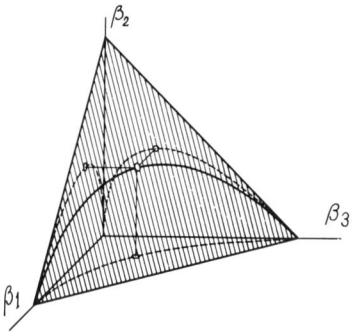

Figure 2. The phase space of the ADK reaction β_1, β_2 and β_3 are dimensionless concentrations of AMP, ADP and ATP respectively. The shaded place represents the conservation equation $\beta_1 + \beta_2 + \beta_3 = 1$. <u>Solid line</u>: a geometrical locus of the ADK reaction equilibrium states.

Since the functions $\beta_1(\beta_2)$ and $\beta_3(\beta_2)$ are double-valued, the question arises as to which of the two values of these functions is to be chosen in studies of oscillations. In order to answer this question, let us direct our attention to the experimental data presented in Figure 3.

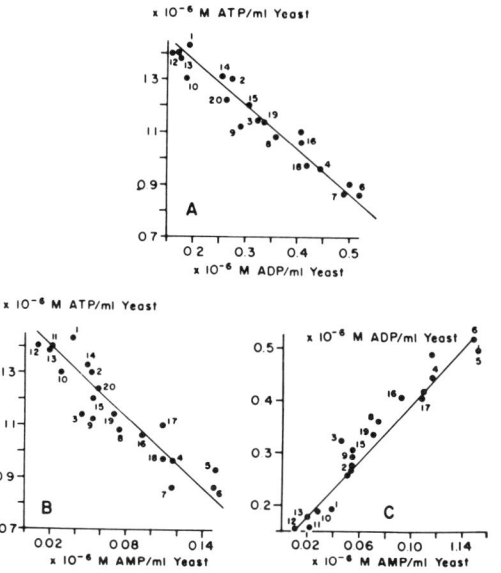

Figure 3. The relationship between the concentrations of adenylates in a suspension of yeast cells during two periods of oscillations. The numbered points indicate the order of consecutive samples. The data is taken from the work of Betz and Chance (17).

As seen from this figure, the relationship between adenylates in the glycolytic system corresponds to solution [8']. The branches of the curves described by solution [8'] are represented in Figure 1 by solid lines.

As seen from Figure 1, the change intervals of β_1, β_2 and β_3 are

$$0 < \beta_1 < \beta_1^*$$
$$0 < \beta_2 < \beta_2^*$$
$$\beta_1^* = \beta_3^* < \beta_3 < 1$$

[12]

Let us estimate the values of β_1^* and β_2^*. From the data of Betz (18), K = 0.48. Let us make the approximation K = 0.5. Then according to [10] and [11] $\beta_2^* = 0.4$, $\beta_1^* = \beta_3^* = 0.3$. Note that the lower limit of the working concentration for AMP and ADP is equal to zero, and for ATP it approaches a fairly large value. This gives a simple explanation for the fact that the relative amplitude of the oscillating ATP concentration is considerably smaller than that of AMP and ADP (6,17-19).

The choice of solution [8'] does not mean, of course, that solution [8"] has no physical meaning. This choice is based upon experimental data and has been made exclusively for the purpose of greatest possible simplicity. The model which allows for both solutions [8] will be described elsewhere (20).

Model of the PFK-ADK System

In order to avoid unnecessary complexity in the description of the PFK-ADK system, a series of simplifying assumptions may be made:

<u>a</u>. Since the total adenylate content in yeast extracts exhibiting oscillatory behavior normally exceeds 1mM, the minimum ATP concentration possible in the presence of ADK must be of the order

$$[ATP]_{min} = \frac{\beta_1^*}{A} \approx 0.3 \, mM$$

At [ATP]>0.3 mM the effect of ATP on PFK is practically absent (see Figure 4), provided that FDP, ADP, AMP and F-6-P are all in the normally observed concentration range.

<u>b</u>. In the presence of AMP and a high concentration of fructosediphosphate (FDP), ADP has no appreciable effect on PFK (see Figure 5).

<u>c</u>. At concentrations in the range of 1-10 mM, which are normally observed, FDP also does not substantially affect the PFK activity (see Figure 6).

<u>d</u>. AMP appears to be the only effective activator of PFK in the yeast extracts containing ADK (see Figure 7).

Assumption a allows the PFK reaction to be treated as a reaction with one substrate (F6P), and assumptions b and c make it possible to ignore the effect of ADP and FDP on the PFK reaction rate. This leads to a very simple kinetic

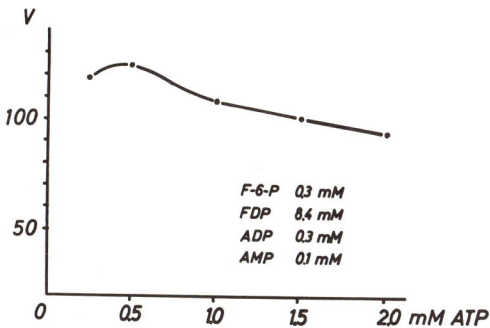

Figure 4. The phosphofructokinase reaction rate, V, as a function of the ATP concentration.

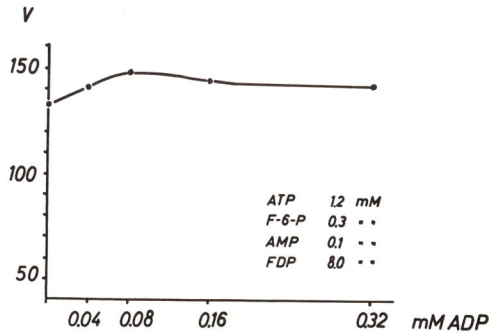

Figure 5. The effect of ADP concentration on the phosphofructokinase reaction rate, V, in the presence of AMP and a high concentration of FDP.

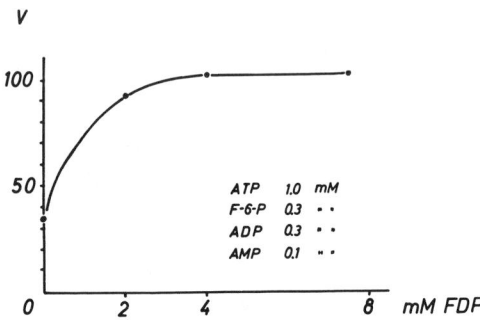

Figure 6. The effect of fructose-1.6-diphosphate (FDP) concentration on the phosphofructokinase reaction rate, V, in the presence of AMP and ADP.

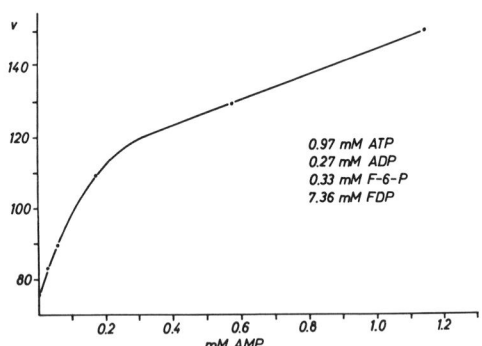

Figure 7. The activating effect of AMP on the phosphofructokinase reaction rate, V.

model of the PFK reaction:

$$\begin{aligned} S_1 + E &\rightleftharpoons S_1 E \\ S_1 + E S_3 &\rightleftharpoons S_1 E S_3 \\ S_1 E &\rightarrow E + S_2 \\ S_1 E S_3 &\rightarrow E S_3 + S_2 \\ S_3 + E &\rightleftharpoons E S_3 \\ S_3 + S_1 E &\rightleftharpoons S_1 E S_3 \end{aligned} \quad [13]$$

Here the following designations are used:

$$S_1 \equiv F6P, \quad S_2 \equiv ADP, \quad S_3 \equiv AMP, \quad E \equiv PFK$$

At $E \ll [S_1, S_2, S_3]$ the concentrations of S_1, S_2 and S_3 turn out to be slow variables as compared to the concentrations of different forms of the enzyme (E, S_1E, ES_3, S_1ES_3) (12). Therefore, the time required to establish equilibrium between the different forms of the enzyme can be considered as negligible. Thus, the rate of conversion $S_1 \rightarrow S_2$ in system [13] can be described by the steady state equation

$$v = V \frac{\sigma_1 (\lambda_0 + \sigma_3)}{1 + \alpha_1 \sigma_1 + \sigma_3 (1 + \sigma_1)} \quad [14]$$

where the following designations are used.

$$\sigma_1 = \frac{[S_1]}{K_1}, \quad \sigma_3 = \frac{[S_3]}{K_3}, \quad \alpha_1 = \frac{K_1'}{K_1}, \quad V = \max v$$

Here K_1 is the Michaelis constant for the unactivated enzyme, and K_1' is that for the activated enzyme. K_3 is the activation constant, σ_1 and σ_3 are dimensionless concentrations of F6P and AMP, α_1 is the parameter determining the effect of AMP on the affinity of F6P for PFK. At $\alpha_1 < 1$, AMP increases this affinity and at $\alpha_1 > 1$, it decreases it.

With the ADK activity much higher than the activity

of PFK

$$G_3 = \frac{\beta'_1 A}{K_3}$$

or with consideration for [8']

$$G_3 = \frac{A}{2K_3}\left(1 - \beta_2 - \sqrt{(1-\beta_2)^2 - 4K\beta_2^2}\right) \quad [15]$$

Substituting [15] into [14] we obtain

$$V = \frac{V G_1}{1 + G_1} \cdot \frac{\nu_0 + \delta(1 - \beta_2 - \sqrt{(1-\beta_2)^2 - 4K\beta_2^2})}{\frac{1+\alpha_1 G_1}{1+G_1} + \delta(1 - \beta_2 - \sqrt{(1-\beta_2)^2 - 4K\beta_2^2})} \quad [16]$$

where

$$\delta = \frac{A}{2K_3}$$

The factor

$$F = \frac{\nu_0 + \delta(1 - \beta_2 - \sqrt{(1-\beta_2)^2 - 4K\beta_2^2})}{\frac{1+\alpha_1 G_1}{1+G_1} + \delta(1 - \beta_2 - \sqrt{(1-\beta_2)^2 - 4K\beta_2^2})} \quad [17]$$

in equation [16] determines an apparent activating effect of ADP on PFK. (In the presence of excess ADK the concentration of the real effector, AMP, is a function of [ADP]).

The dependence of F on β_2 at different fixed values of the parameter δ is represented by the family of curves shown in Figure 8. Function [17] is too complex for investigation. At $\delta \gg 1$ and $\beta_2 < \beta_2^*$ this function can be represented by the simpler one given in Figure 9.

$$\widetilde{F} = \frac{\nu_0 + (\beta_2 k)^\gamma}{\frac{1+\alpha_1 G_1}{1+G_1} + (\beta_2 k)^\gamma} \quad [18]$$

The coefficients k and γ can be obtained from the formulae:

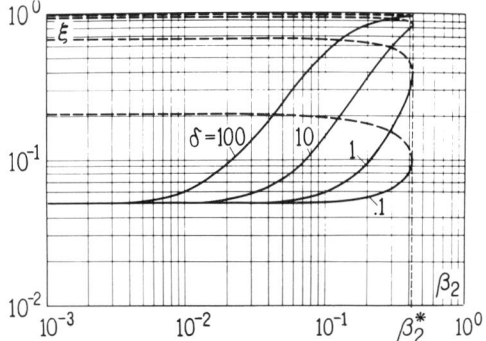

Figure 8. The effect of ADP concentration on the relative PFK activity in the presence of excess ADK (solid lines). β_2 is the relative ADP concentration. ξ is the relative PFK activity. δ is the relative affinity of AMP for PFK. The curves are calculated for the values $K = 0.5$, $\nu_0 = 0.05$, $\alpha_1 = 1$, by equation [17]. The dashed curves correspond to solution of [8"].

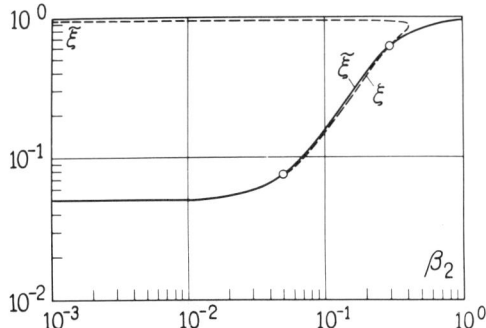

Figure 9. Approximation of the function ξ (dashed line) by the function $\tilde{\xi}$ (solid curve). The circles indicate the points which were used for calculation of the approximation function $\tilde{\xi}$. The parameters of the function ξ are: $K = 0.5$, $\nu_0 = 0.05$; $\alpha_1 = 1$. The parameters of the function $\tilde{\xi}$ are: $\gamma = 2.16$, $k = 3.78$.

$$\gamma = \frac{1}{\lg \frac{\beta_{21}}{\beta_{22}}} \lg \frac{(\mathcal{F}_1 x - \gamma_0)(1-\mathcal{F}_2)}{(1-\mathcal{F}_1)(\mathcal{F}_2 x - \gamma_0)} \qquad [19]$$

$$k = \sqrt{\frac{(x-\gamma_0)(\mathcal{F}_1 - \mathcal{F}_2)}{(1-\mathcal{F}_1)(1-\mathcal{F}_2)(\beta_{21}^r - \beta_{22}^r)}} \qquad [20]$$

Here

$$x = \frac{1+\alpha_1 \sigma_1}{1+\sigma_1} \qquad [21]$$

$\beta_{21} \neq \beta_{22}$ are the two values of β_2 which are chosen for reasons of the best approximation. $\mathcal{F}_1$ and $\mathcal{F}_2$ are the values of $\mathcal{F}$ at $\beta_2 = \beta_{21}$ and $\beta_2 = \beta_{22}$.

With consideration for approximation [18] the expression for the PFK reaction rate in the presence of excess ADK takes a simpler form:

$$v_1 = V \frac{\sigma_1(\gamma_0 + \sigma_2^r)}{1 + \alpha_1 \sigma_1 + \sigma_2^r(1+\sigma_1)} \qquad [22]$$

where

$$\sigma_2 = k\beta_2 = \frac{k[ADP]}{A} \qquad [23]$$

is a new dimensionless concentration of ADP.

Comparison between equations [22] and [14] shows that ADP in the presence of ADK produces an effect on PFK which is equivalent to the activating effect produced by γ molecules of AMP. From this it follows that we can consider a monosubstrate enzyme reaction

$$F6P \xrightarrow{E_e} ADP \quad (+) \curvearrowleft \gamma$$

as a simple kinetic model of the PFK-ADK system. Here the enzyme E_e is equivalent to the two enzymes, PFK and ADK, and is activated by γ molecules of ADP.

Model of the Glycolytic Self-Oscillator

A block-diagram of the glycolytic system is presented in Figure 10. The glycolytic system is conditionally divided into three blocks: block 1 - the source of F6P and ADP; block 2 - the PFK-ADK system; block 3 - the sink for ADP and FDP.

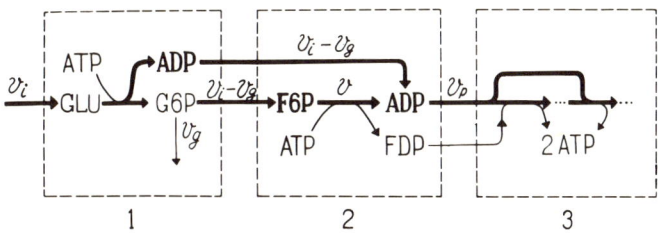

Figure 10. Block-diagram of the glycolytic system.
1 - the source of ADP and F6P; 2 - the PFK-ADK system; 3 - the sink of ADP and FDP. ν_i - the glucose injection rate, ν_g - the rate of leakage of G6P, ν - the PFK reaction rate, ν_p - the rate of phosphorylation of ADP in the lower part of the glycolytic system.

We can assume that the activities of all enzymes of the source and of the sink reactions are much higher than the activity of PFK. In this case the concentrations of all intermediate metabolites, with the exception of F6P and ADP, turn out to be fast variables and can be excluded from consideration by the corresponding mathematical procedure (21,22). The behavior of the slow variables, F6P and ADP, in time, t, can be described by the following equation system:

$$\frac{d[F6P]}{dt} = \nu_i - \nu_g - \nu \qquad [24]$$

$$\frac{d[ADP]}{dt} = \nu_i + \nu - \nu_p$$

where ν_i is the glucose injection rate, ν_g is the rate of leakage of hexosemonophosphates into collateral metabolic pathways, v is the PFK reaction rate determined by equation [22] and v_p is the rate of ADP phosphorylation.

Generally speaking, v_g and v_p are rather complex functions of F6P and ADP. In order to avoid unnecessary detail, we assume, as the simplest hypothesis, that

$$v_g = 0$$
$$v_p = k_p [ADP] \qquad [25]$$

where k_p is constant.

In this simplest case, system [24] can be reduced to the following form

$$\frac{d\sigma_1}{dt} = \nu_1 - \nu \qquad [26]$$

$$\frac{d\sigma_2}{dt} = \alpha_2 (\nu + \nu_1 - 2x_2 \sigma_2)$$

where

$$\nu \equiv \frac{\sigma_1 (\nu_0 + \sigma_2^\gamma)}{1 + \alpha_1 \sigma_1 + \sigma_2^\gamma (1 + \sigma_1)} \qquad [27]$$

The designations used here are

$$\tau = \frac{V}{K_1} t \;,\; \nu_1 = \frac{v_i}{V} \;,\; \nu = \frac{v}{V} \;,\; \alpha_2 = \frac{K_1'}{A} k \;,\; x_2 = \frac{k_p A}{k} \qquad [28]$$

The mathematical model [26] differs from model [2], proposed earlier, by the presence of the term ν_1 in the second equation. This term takes into consideration the rate of formation of ADP in the hexokinase reaction. Model [26] has the same form in the case of injection of any other non-phosphorylated carbohydrate (e.g. in the case of injection of hexosemonophosphates, the term ν_1 in [26] is zero, and model [26] is reduced to model [2] (15).

Investigation of the Model

System [26] has the only steady state

$$\bar{\sigma}_1 = \vartheta_1 \frac{1 + \bar{\sigma}_2^r}{\vartheta_0 + \bar{\sigma}_2^r - \vartheta_1(\alpha_1 + \bar{\sigma}_2^r)} \qquad [29]$$

$$\bar{\sigma}_2 = \frac{\vartheta_1}{x_2} \qquad [30]$$

As only $\bar{\sigma}_1 > 0$ has a physical meaning, the condition for existence of the steady state $\bar{\sigma}_1, \bar{\sigma}_2$ is

$$\vartheta_0 + \bar{\sigma}_2^r - \vartheta_1(\alpha_1 + \sigma_2^r) > 0$$

or

$$x_2 < \vartheta_1 \sqrt[r]{\frac{1 - \vartheta_1}{\vartheta_1 \alpha_1 - \vartheta_0}} \qquad [31]$$

The steady state is unstable if

$$2\alpha_2 x_2 > \frac{(\vartheta_0 + \bar{\sigma}_2^r - \vartheta_1(\alpha_1 + \bar{\sigma}_2^r))^2}{-\sigma_2^{2r} + (\frac{r}{2}(1-\vartheta_0 - \vartheta_1(1-\alpha_1)) - 1 - \vartheta_0)\sigma_2^r - \vartheta_0} \qquad [32]$$

This condition is satisfied if

$$-\sigma_2^{2r} + (\frac{r}{2}(1-\vartheta_0 - \vartheta_1(1-\alpha_1)) - 1 - \vartheta_0)\sigma_2^r - \vartheta_0 > 0$$

or

$$^1\bar{\sigma}_2 < \bar{\sigma}_2 < {}^2\bar{\sigma}_2 \qquad [33]$$

or if

$$^2 x_2 < x_2 < {}^1 x_2 \qquad [34]$$

where

$$^1\bar{\sigma}_2 = (a - \sqrt{a^2 - \vartheta_0})^{\frac{1}{\gamma}}$$

$$^2\bar{\sigma}_2 = (a + \sqrt{a^2 - \vartheta_0})^{\frac{1}{\gamma}}$$

$$^1x_2 = \frac{\vartheta_1}{(a - \sqrt{a^2 - \vartheta_0})^\gamma} \qquad [35]$$

$$^2x_2 = \frac{\vartheta_1}{(a + \sqrt{a^2 - \vartheta_0})^\gamma} \qquad [36]$$

$$a = \tfrac{1}{2}\left(\tfrac{\gamma}{2}(1-\vartheta_0-\vartheta_1(1-\alpha_1))-1-\vartheta_0\right) \qquad [37]$$

In turn, for the existence of $^1\bar{\sigma}_2$ and $^2\bar{\sigma}_2$ it is necessary that

$$a^2 - \vartheta_0 > 0$$

or

$$\tfrac{1}{4}\left(\tfrac{\gamma}{2}(1-\vartheta_0-\vartheta_1(1-\alpha_1))-1-\vartheta_0\right)^2 - \vartheta_0 > 0 \qquad [39]$$

From this it follows that at $\alpha_1 < 0$

$$\gamma > 2\frac{(1+\sqrt{\vartheta_0})^2}{1-\vartheta_0-\vartheta_1(1-\alpha_1)} \qquad [40]$$

i.e., the parameter γ must be greater than 2. Analysis of approximate function [18] shows that a good approximation of function [18] can be achieved with a large interval β_2 at $\gamma \approx 2$ (see, for example, Figure 9). Thus, model [26] comes in conflict with the experimental data, since at $\gamma = 2$ conditions [40] and [32] are not satisfied and, therefore, model [26] fails to yield oscillations.

This contradiction results mainly from the fact that in [25] we have ignored the nonlinearity of the dependence of v_p upon ADP concentration. A simple example demonstrates the effect of this nonlinearity.

Let

$$v_p = k_p [ADP]^{r_p} \quad [41]$$

In this case [26] takes the form

$$\dot{\sigma}_1 = \vartheta_1 - \vartheta$$
$$\dot{\sigma}_2 = \alpha_2(\vartheta + \vartheta_1 - 2x_2\sigma_2^{r_p})$$

By the substitution of

$$z = \sigma_2^{r_p}$$

this system can be reduced to the following

$$\left.\begin{array}{l}\dot{\sigma}_1 = \vartheta_1 - \vartheta \\ \dot{z} = \alpha_2 r_p z_2^{1-\frac{r}{r_p}}(\vartheta_1 + \vartheta - 2x_2 z_2)\end{array}\right\} [42]$$

where

$$\vartheta = \frac{\sigma_1(\vartheta_0 + z_2^{\frac{r}{r_p}})}{1 + \alpha_1\sigma_1 + \sigma_2^{\frac{r}{r_p}}(1+\sigma_1)}$$

For the existence of self-oscillations in [42] it is necessary that $\frac{r}{r_p} > 2$. This requirement means that $r_p < 1$ since $r \approx 2$. Consequently, a necessary condition for the appearance of self-oscillations in the glycolytic system fed by nonphosphorylated substrates is nonlinearity in the dependence of the rate of ADP phosphorylation on the ADP concentration as depicted in Figure 11.

This nonlinearity can be allowed for in model [26] by increasing the parameter r by a factor of $\frac{1}{r_p}$. As the behavior of model [26] far from the line $\sigma_2 = 0$ coincides qualitatively with the behavior of the more complex model [42], we use the first model for analysis of the case $r_p < 1$. In this case, model [26] has $r > 2$. If this condition is satisfied, condition [40] can be satisfied, as well as the condition of instability of the steady state [31]. As the steady state itself can exist only when inequality [31] is fulfilled, the region of unstable steady states must be determined by the system of inequalities

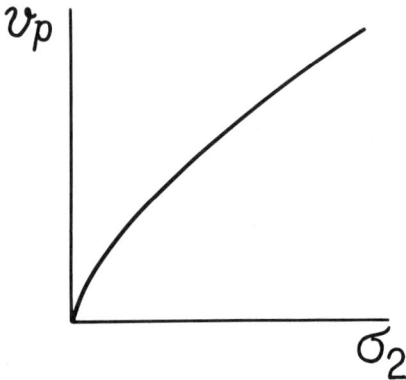

Figure 11. The rate of ADP phosphorylation, v_p, as a function of the relative ADP concentration, $\bar{\sigma}_2$. The figure shows the type of nonlinearity which is necessary for generation of self-oscillations.

$$\left. \begin{array}{c} x_2 < \vartheta_1 \sqrt[r]{\dfrac{1+\vartheta_1}{\vartheta_1 \alpha_1 - \vartheta_0}} \\[2ex] x_2 > \dfrac{(\vartheta_0 + \bar{\sigma}_2^r - \vartheta_1(\alpha_1 + \bar{\sigma}_2^r))^2}{2\alpha_2(-\bar{\sigma}_2^{2r} + 2a\bar{\sigma}_2^r - \vartheta_0)} \end{array} \right\} \quad [43]$$

At $\alpha_2 \gg 1$ the second inequality can be replaced with good accuracy by condition [34] which determines the limit region of instability which model [26] has at $\alpha_2 \to \infty$. Figure 12 shows how fast, with increase in α_2, the boundary of the instability region, determined by numerical solution of the equation

$$2\alpha_2 x_2(-\bar{\sigma}_2^{2r} + 2a\bar{\sigma}_2^r - \vartheta_0) - (\vartheta_0 + \bar{\sigma}_2^r - \vartheta_1(\alpha_1 + \bar{\sigma}_2^r))^2 = 0$$

reaches the boundary of the limit region constructed by equations [35] and [37].

The length of the limit region of self-oscillations is determined by the critical value of ϑ_1, which is equal to

$$\vartheta_1^* = \dfrac{1 - \vartheta_0 - \dfrac{2}{r}(1 + \sqrt{\vartheta_0})^2}{1 - \alpha_1} \quad [44]$$

for which

$$^1\mathcal{X}_2 = {}^2\mathcal{X}_2 = \mathcal{X}_2^* = \frac{1 - \nu_o - \frac{2}{\gamma}(1 + \sqrt{\nu_o})^2}{2\gamma\sqrt{\nu_o}\ (1 - \alpha_1)} \qquad [45]$$

As seen from equation [43], ν_1^* increases with an increase in γ, α_1 and with a decrease in ν_o.

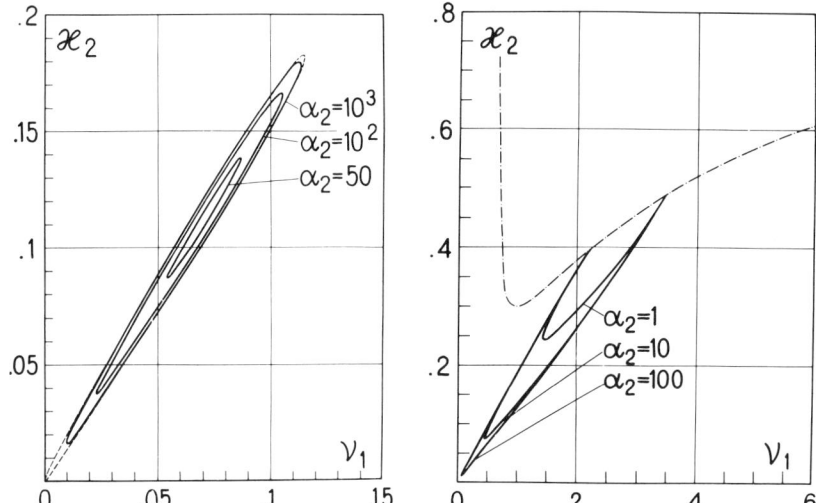

Figure 12. The boundaries of the region of existence of self-oscillations in model [26], in the plane of its parameters ν_1 and χ_2, for different values of α_2. $\gamma = 3.25$, on the left, and $\gamma = 3.5$ on the right. The boundaries (solid lines) were calculated from equation [46] for $\alpha_1 = 0.75$ and $\nu_o = 0.05$. The broken line on the left-hand diagram is the limit boundary calculated from equations [35] and [36]. The dashed-dotted line on the right-hand diagram is the boundary above which system [26] has no equilibrium states in the positive octant.

If condition [43] is satisfied, there occur self-oscillations in model [26]. Various waveforms of the oscillations are possible. Figure 13 shows some computer solutions for model [26].

From Figure 13 it is seen that at $\alpha_2 \gg 1$ self-oscillations are of the relaxation type. This can be explained, primarily, by the fact that the rate of change of the variable σ_2 outside the neighborhood of the quasi-steady state curve, $\dot{\sigma}_2 = 0$, is approximately α_2 times higher than that of σ_1 and, secondly, by a S-shaped character of the quasi-steady state dependence $\tilde{\sigma}_2 = \tilde{\sigma}_2(\sigma_1)$ determined by the equation $\dot{\sigma}_2 = 0$, or

$$\nu_1 + \frac{\sigma_1(\nu_0 + \sigma_2^r)}{1 + \alpha_1 \sigma_1 + \sigma_2^r(1 + \sigma_2^r)} - 2\chi_2 \sigma_2 = 0 \qquad [46]$$

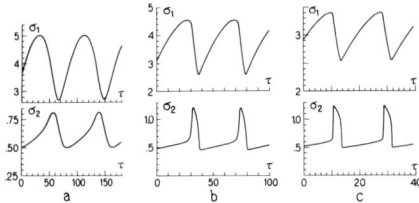

Figure 13. Computer solutions for model [26] at $\alpha_2 = 1$ (a), $\alpha_2 = 10$ (b) and $\alpha_2 = 100$ (c). The other parameters are: $\alpha_1 = 0.75$; $\gamma = 3.5$; $\chi_2 = 0.32$; $\nu_0 = 0.05$; $\nu_1 = 0.2$.

Figure 14 shows two families of quasi-steady state curves constructed from equation [44] for different fixed values of ν_1 and χ_2.

In Figure 15 is shown an evolution of the limit cycle with a change in α_2 from 1 to 100. As seen from this figure, with an increase in α_2, the limit cycle closely approaches the two branches of the quasi-steady state curve [44]. As a result, system [26] is almost all the time in a state of slow drift along the quasi-steady state curve [44]. Only during a short time (twice in a period) does system [26] take part in a fast motion when it jumps from one branch of the quasi-steady state curve to the other.

Since the length of the two portions of the quasi-steady state curve [44] near which system [26] drifts, increases with increasing χ_2, the increase in the period of relaxation self-oscillations with an increase in χ_2, or with a decrease in ν_1, can be easily understood.

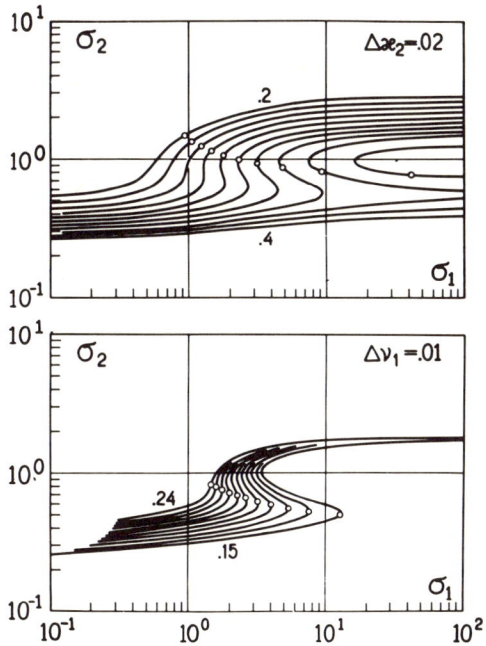

Figure 14. Two families of quasi-steady state curves in the phase plane of model [26], calculated from equation [46]. In the top diagram $0.2 \leq \chi_2 \leq 0.4$, changed by steps $\Delta\chi_2 = 0.02$ and in the bottom diagram, $0.24 \leq \nu_1 \leq 0.15$, changed by step $\Delta\nu_1 = 0.01$. Circles on each quasi-steady state curve indicate the equilibrium states of model [26].

The period of self-oscillations which approximate in their waveform to sinusoidal ones, can be calculated by the equation:

$$T = 2\pi \frac{\sqrt{(1+\bar{\sigma}_2^{\gamma})(\nu_o + \bar{\sigma}_2^{\gamma})(\bar{\sigma}_2^{\gamma} - {}^1\bar{\sigma}_2^{\gamma})({}^2\bar{\sigma}_2^{\gamma} - \bar{\sigma}_2^{\gamma})}}{(\nu_o + \bar{\sigma}_2^{\gamma} - \nu_1(\alpha_1 + \bar{\sigma}_2^{\gamma}))^2}$$

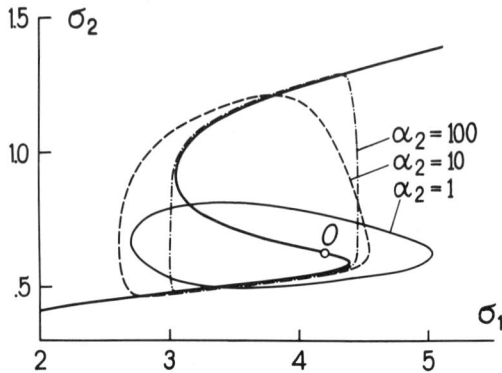

Figure 15. Limit cycles corresponding to the solutions presented in Figure 13. Solid line: quasi-steady state curve described by equation [46]. Point 0 is the equilibrium state.

Discussion

In this paper we have analysed some properties of the glycolytic generator under the conditions of non-phosphorylated substrate supply. The analysis supports the conclusion drawn earlier [12, 14, 15] that the PFK - ADK system is responsible for the self-oscillations in glycolysis. In general, the properties of the self-generator under conditions of hexosemonophosphate supply [14, 15] are mainly the same as in the case of non-phosphorylated substrate supply which is considered above. The basic distinction is that in the case of non-phosphorylated substrate supply a certain degree of inhibition of the lower part of the glycolytic system by ADP or AMP is necessary for the self-generation of oscillations.

As follows from the above analysis, the relationship between the adenylate concentrations appears to play an exclusively important role in the glycolytic system. Therefore, no matter which of the reactions is the key one (hexokinase, glycogen phosphorylase, fructokinase, PFK or pyruvate kinase, depending on the conditions), the ADK reaction is the main regulatory step in glycolysis. This reaction has two important functions. In the first instance, it is a highly sensitive primary element which determines by the level of ADP the intensity of the ATP consumption. Secondly, this reaction governs the activities of all of the key enzymes by changing the AMP concentration. From this point of view the role of adenylcyclase, converting AMP into cyclic 3',5' - AMP, becomes understandable.

Since 3',5'-AMP is an effector of the key enzymes in carbohydrate metabolism, (several times more active than AMP), the adenylcyclase reaction plays the part of an amplifier of signals transmitted from the regulator, the ADK reaction, to the executive mechanisms - the key reactions.

In this paper we do not take into account the effect of adenylcyclase on the oscillatory mechanism. It is quite clear, however, that an addition of adenylcyclase must be equivalent to an increase in the apparent effect of ADP on PFK and other key enzymes.

In terms of models [2] and [26] participation of adenylcyclase in glycolytic regulation is reflected in a corresponding increase of γ - an apparent affinity of ADP for PFK. According to condition [32], this must facilitate the generation of self-oscillations.

The above model, of course, cannot be expected to be the only true one. Under various limit conditions different parts of the glycolytic system can be a source of self-oscillations [12, 22, 23]. This necessitates the construction of quite dffferent models describing the behavior of the glycolytic system in various limit cases. These limit models may turn out to be quite uncomparable because they describe uncomparable limit situations.

Acknowledgements

The authors are thankful to Prof. A.M.Molcanov for valuable advice and Prof. B. Hess for the opportunity, kindly given, to study unpublished experiments.

References

1. Hess, B., Brand, K. and Pye, E.K. Biochem. Biophys. Res. Commun., 23, 102 (1966).

2. Hess, B. Studia Biophys., 1, 41 (1966).
3. Pye, E.K., and Chance, B. Proc. Natl. Acad. Sci., 55, 888 (1966).
4. Frenkel, R. Arch. Biochem. Biophys., 115, 112 (1966).
5. Pye, E.K. Studia Biophys. 1, 75 (1966).
6. Hess, B. and Boiteux, A. In **Regulatory Functions of Biological Membranes** Ed. J. Jarnefelt. Biochim. Biophys. Acta Library, 11, 148 (1968).
7. Boiteux, A. and Hess, B. This volume, p. 243
8. Hess, B. and Boiteux, A. This volume, p. 229
9. Pye, E.K. This volume, p.269
10. Sel'kov, E.E. In **Bionics. Simulation of Biological Systems** (in Russian), Naukova Dumka, Kiev, 1967, p.99.
11. Sel'kov, E.E. Europ. J. Biochem., 4, 79 (1968).
12. Sel'kov, E.E. Molekularnaja biologija (U.S.S.R.), 2, 252 (1968).
13. Lotka, A.J. J. Phys. Chem., 14, 271 (1910).
14. Sel'kov, E.E., Abstr. N 1052, FEBS Meeting, Prague, July 1968.
15. Sel'kov, E.E., In: **Mathematical Simulation of Biological Systems and Apparatus**, Nauka, Moscow 1969.
16. van der Pol, B. Philosophical Mag., 158, 240 (1922).
17. Betz, A. and Chance, B. Arch. Biochem. Biophys., 109, 585 (1965).
18. Betz, A. Physiol. Plant., 19, 1049 (1966).
19. Hess, B., personal communication.
20. Sel'kov, E.E., Betz, A. and Semashko, L.R., in preparation.
21. Tikhonov, A.N. Matem. Zbornik (U.S.S.R.) 22, 193 (1948).
22. Pontrjagin, L.S. Izv. Acad. Nauk S.S.S.R. ser matem., 21, 607 (1957).
23. Higgins, J.J. Ind. Eng. Chem., 59, 19 (1967).
24. Higgins, J.J. This volume, p.127

KINETICS OF YEAST PHOSPHOFRUCTOKINASE AND THE GLYCOLYTIC OSCILLATOR

A. Betz

Institut für Molekulare Biologie,
3301 Stöckheim b. Braunschweig,
Mascheroder Weg 1, Germany.

As we have seen from the papers of Sel'kov (1), Higgins (2) and Mayer (3), different models can be proposed to explain oscillatory patterns of changing metabolite concentrations in biochemical pathways. Common to all of them is the occurrence of autocatalytic activation at one enzymic step in a flux system, *i.e.* in a system which is displaced from equilibrium.

In yeast glycolysis, these minimum requirements fit well with the known kinetic data of phosphofructokinase (PFK). There is no doubt that this enzyme is a key step in the generation of oscillations in this metabolic sequence. But before going into any detail, I want to mention that in the living yeast cell PFK generates oscillations only in close connection with some other control site located further down the glycolytic sequence. In this respect I agree with Higgins (4) in the statement that PFK is not working independently. That is evident from phase relationships (5), from studies on the incorporation of ^{32}P into FDP and ATP and from very recent studies on phase shifts induced by the addition of pyruvate and acetaldehyde.

When we focus on PFK alone as the main oscillator it is still an open question as to which chemicals might be the real effectors in PFK regulation. Sigmoidal substrate/velocity curves suggest that yeast PFK might be an allosteric enzyme similar to that from E. coli (6). Using the Hill plot, with yeast PFK, coordination numbers up to 4 can be observed with F-6-P as an activating substrate, as shown in Figure 1A, while numbers of almost 2 can be demonstrated for AMP (Figure 1B).

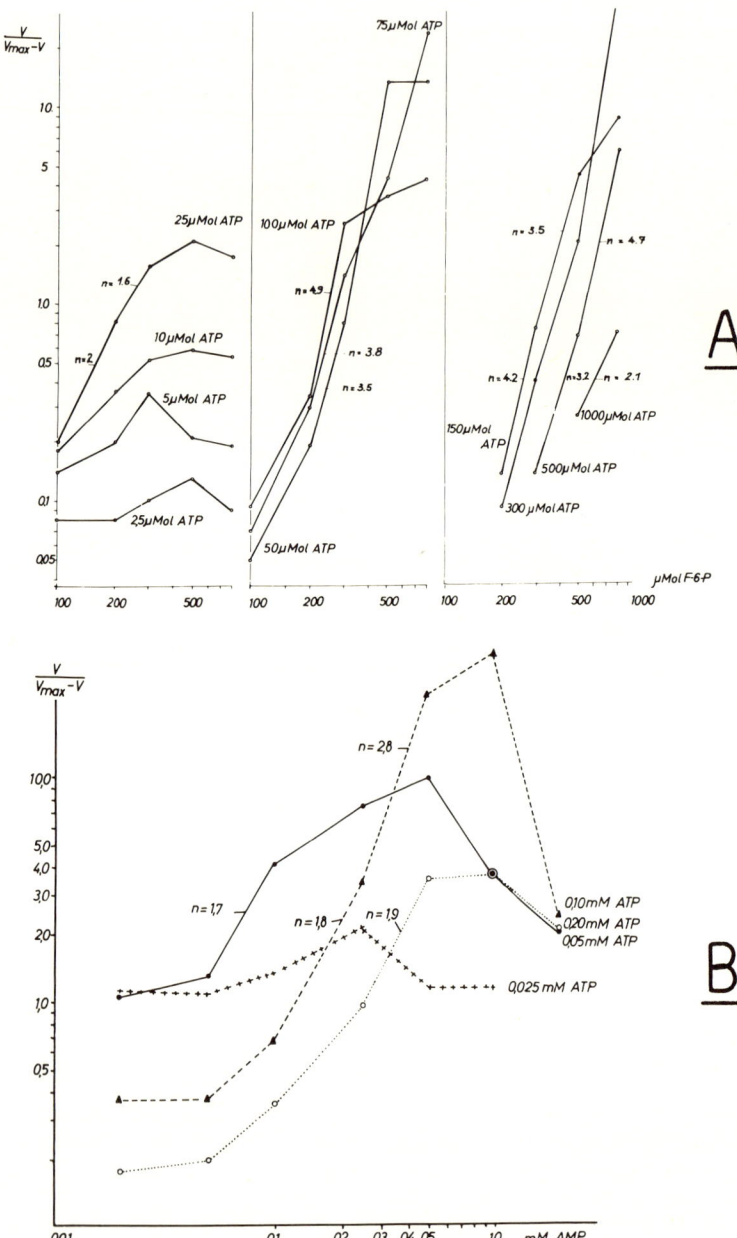

Figure 1. Hill plots for yeast phosphofructokinase. Conditions as shown. B, F-6-P = 0.3 mM.

Those for ATP are clearly elevated by the addition of either AMP or FDP, as seen in Figures 2A and 2B. ADP, on the other hand, does not raise the coordination numbers for ATP (Figure 2C).

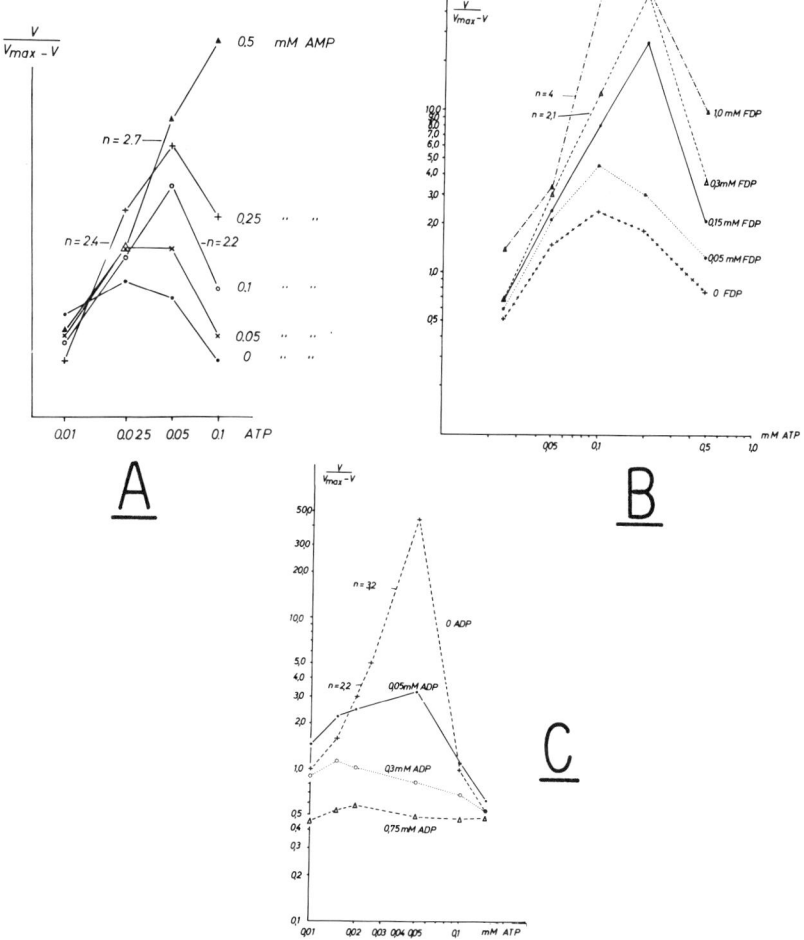

Figure 2. Hill plots for yeast phosphofructokinase. Conditions are as shown. A and B: F-6-P = 0.2 mM; C: F-6-P = 0.3 mM.

The coordination numbers are very similar with both stimulating and inhibitory concentrations of ATP. This is an important observation since the ATP concentration in cells and in the cell-free extract is always in the inhibitory range for PFK. It should be noted that PEP has very low inhibitory activity for the yeast enzyme, as shown in Figure 3, which is in contrast to the observation by Blangy et al. (6) on preparations from E. coli.

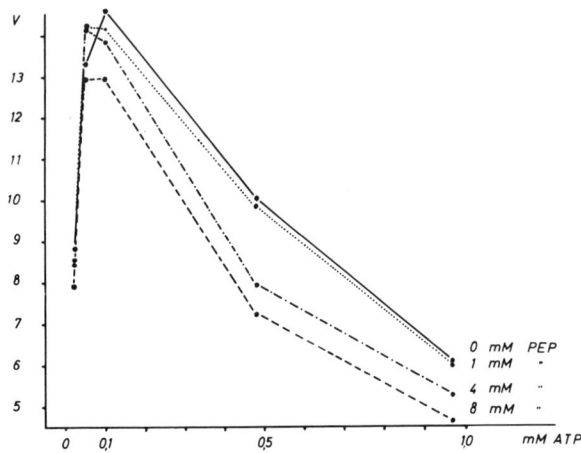

Figure 3. Effect of PEP on kinetics of yeast phosphofructokinase. Conditions as shown.

These data exclude ATP and PEP as effectors in the oscillatory control of PFK. But inhibition and activation by F-6-P, AMP and FDP must still be considered. With all these metabolites PFK shows coordination numbers indicative of cooperative processes that could qualify them for control interaction.

So far, the data might suggest that allosteric interactions are essential for oscillatory control in glycolysis. Thus we have followed the suggestions of Sel'kov (8) and rechecked the PFK kinetics under conditions similar to those in the cell-free extract, i.e., conditions under which oscillatory control actually occurs. Unless otherwise stated, we added approximately 1 mM ATP, 0.3 mM F-6-P, 0.3 mM ADP, 0.1 mM AMP and 8 mM FDP. Under these conditions the PFK kinetics become completely different. As is demonstrated in Figures 4A and 4B, with higher concentrations of FDP, yeast PFK changes its kinetic behavior and becomes rather ineffective

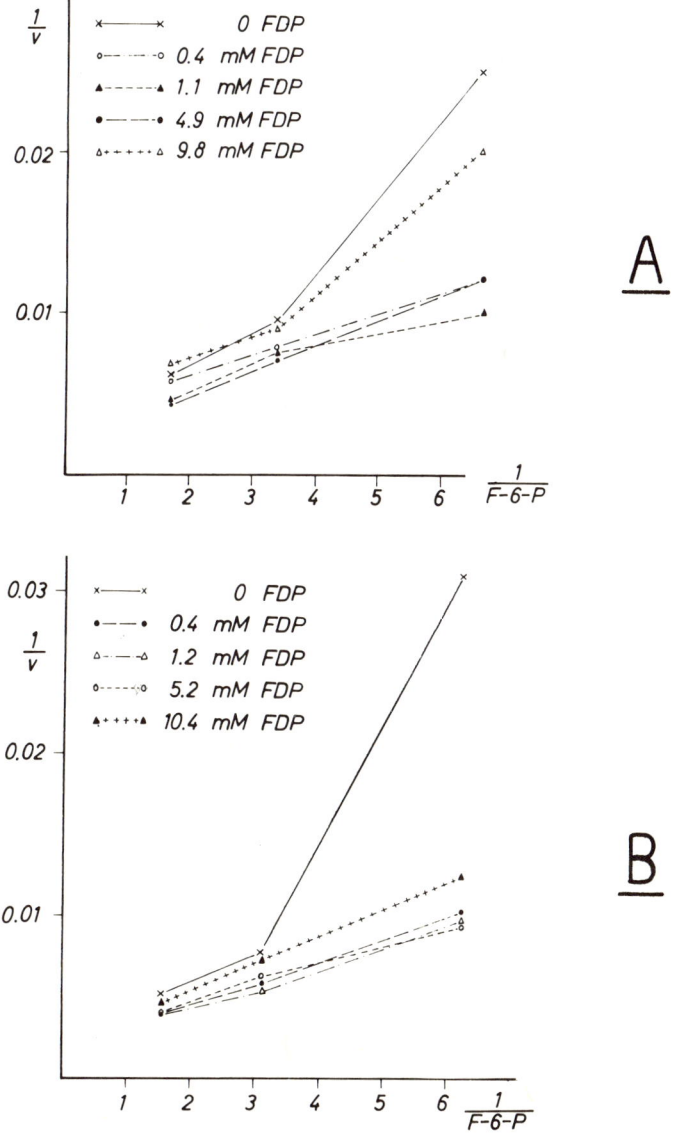

Figure 4. Effect of FDP on the kinetics of yeast phosphofructokinase. Conditions as shown. A: ATP = 0.4 mM, B: ATP = 0.8 mM.

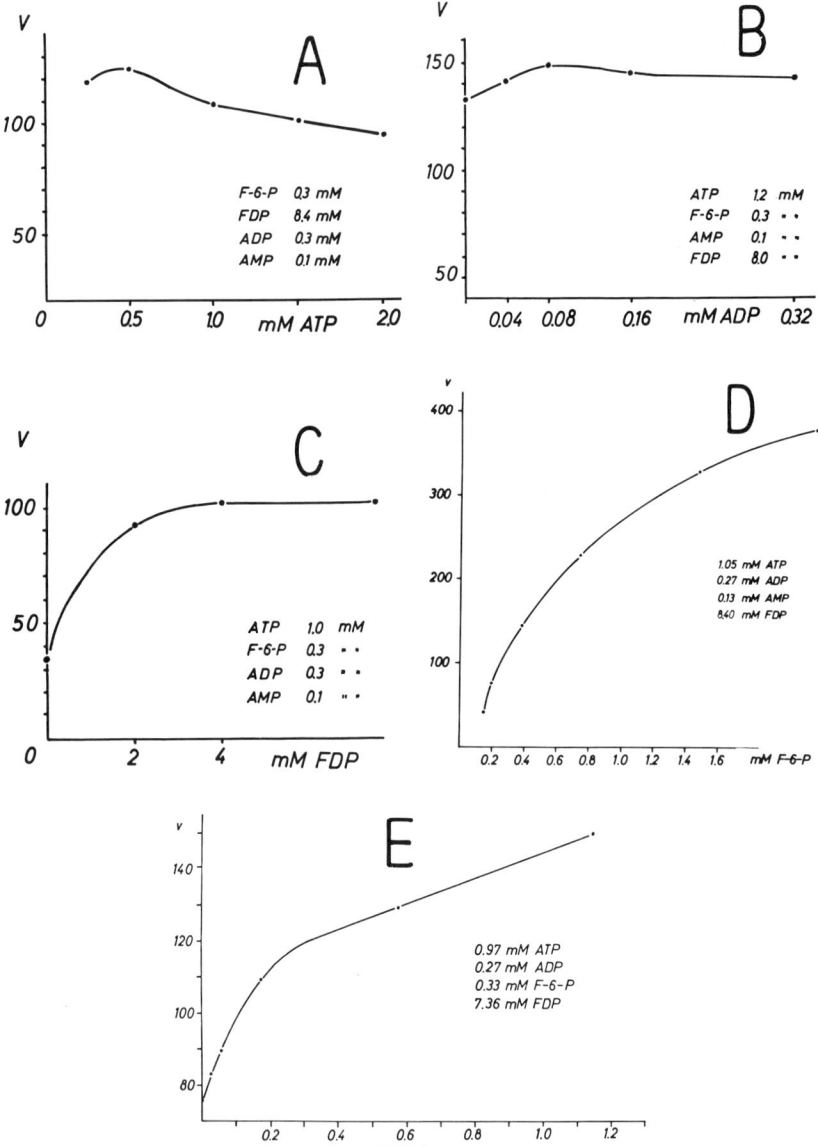

Figure 5. Effect of A: ATP, B: ADP, C: FDP, D: F-6-P and E: AMP on the kinetics of yeast phosphofructokinase. Conditions as shown.

as an inhibitor, as shown in Figure 5A;- from 0.5 to 2 mM ATP the velocity changes less than 30% - while ADP is still totally ineffective (Figure 5B). FDP below 2 mM enhances the activity but with higher concentrations of FDP there is no further change in velocity, as seen in Figure 5C.

In the cell-free extract of yeast, the concentration of FDP is very high (5-8 mM) and it changes only slightly. It cannot, therefore, be expected to have any control of PFK. In intact yeast cells the FDP concentration is lower and it oscillates with an amplitude of 2/3 of the total. Thus, FDP might act as a control chemical here, but even in this case its control characteristics are rather low. Furthermore, we cannot, at present, exclude the possibility that it could be unequally distributed in the living cell. If it were located predominantly in the cytoplasm, any control activity would be improbable even in the intact cell.

F-6-P is an efficient activator of PFK, even up to unphysiologically high concentrations, as shown in Figure 5D. From studies on the enzyme, it cannot be excluded as one of the most potent effectors of the PFK step. However, from the phase relationships, it seems improbable that F-6-P could trigger the PFK reaction, since F-6-P and FDP are shifted exactly 180° from each other. For this reason it is impossible that F-6-P should be the only control chemical; its role can be an accessory one at best.

The last chemical in this series, AMP, is the most potent activator of PFK, as seen in Figure 5E, even under the conditions mentioned previously. This is especially true in the concentration range at which AMP actually occurs in cell-free extracts of yeast.

In conclusion we can state that in generating glycolytic oscillations the real control chemical for PFK is AMP, which is produced in the adenylate kinase reaction and which is almost in phase with ADP. Thus far, our experimental results are in very good agreement with the model proposed by Sel'kov (7).

References

1. Sel'kov, E.E. and Betz, A. This volume, p.197.

2. Higgins, J. This volume, p.127.

3. Mayer, D. This volume, p.31.

4. Higgins, J. Proc. Natl. Acad. Sci. 51, 989 (1964).

5. Betz, A. and Chance, B. Arch. Biochem. Biophys. 109, 585 (1965).

6. Blangy, D., Buc, H. and Monod, H. J. Molecul. Biol., 31, 13 (1968).

7. Sel'kov, E.E. Europ. J. Biochem., 4, 79 (1968).

SUBSTRATE CONTROL OF GLYCOLYTIC OSCILLATIONS

Benno Hess and Arnold Boiteux

Max-Planck-Institut für Ernährungsphysiologie,
Dortmund, Germany.

Biochemical systems, like any other high order system, can, in principle, rise to a specific level of activity and break into continuous oscillations if proper parametric conditions are maintained. Recent investigations in a number of laboratories revealed that the classical pathway of energy metabolism, namely glycolysis, offers a beautiful example of a biochemical network involving multiple feedback interactions and capable of continuous oscillatory performance (for summary see references 1,2,3).

The relative completeness of our knowledge of the macroscopic features of the glycolytic enzymes, coenzymes and intermediates allow a reasonable analysis and interpretation of the conditions and mechanisms yielding oscillations of glycolysis. In this paper the relationship between the outer parameters of glycolysis, namely its substrates and the formal character of the glycolytic oscillations will be presented. This is followed by a subsequent paper in which the internal mechanism of the oscillation will be discussed (4).

In order to study the glycolytic oscillations the following experimental design was developed, as demonstrated in Figure 1. An injection technique was used to allow a continuous and controlled input of glycolytic substrates and intermediates for the induction of reproducible steady states of the pathway. Simultaneously, a continuous record of some indicator parameters of the system, namely pH, NADH and CO_2, was achieved by respective recording techniques. A pH electrode mainly reflects the dynamics of the ATP-system and directly indicates the activity state of the phosphofructokinase reaction, as will be discussed in the subsequent paper (4). The NADH concentration is measured fluorometrically

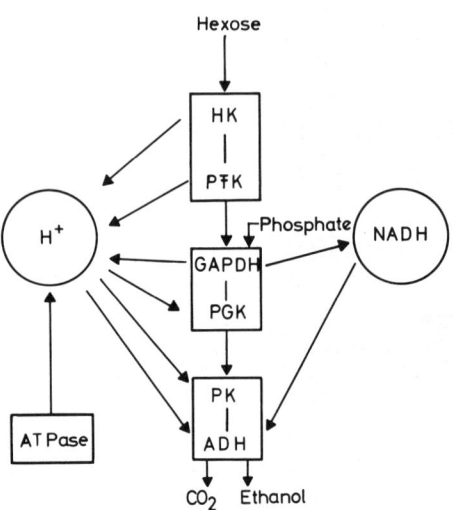

Figure 1. Control diagram of glycolysis.

and indicates the activity of the lower glycolytic pathway. A CO_2 electrode finally measures the CO_2 production. Thus, the input and one output parameter, as well as two internal parameters of the system are recorded. Furthermore, manual sampling allows the assay of glycolytic intermediates, usually by enzymic methods. The incubation equipment is mounted on a modified Eppendorf fluorometer. A special thermostated 4 ml cuvette is equipped with a pH electrode and CO_2 electrode, a thermoelement and a stirring rod driven by a motor. Stepwise and continuous addition of metabolite and substrate solution to the sample is achieved by a microdosimeter which is connected to the cuvette by a teflon tubing. The dosimeter is driven by a synchronous motor through a gear system allowing a tenstep linear speed variation between 0.02-0.2 rev/min. The rate of volume additions varies between 1 and 10 μl/min. The sensitivity of the pH and pCO_2 recording circuit is tested by addition of CO_2/HCl, as demonstrated in

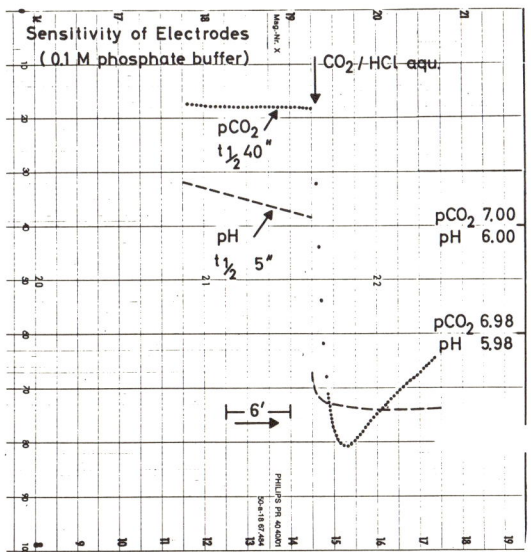

Figure 2. Sensitivity test of the pCO_2 and pH recording.

Figure 2. The half-time response is 5 sec for pH and 40 sec for pCO_2 recording within the indicated concentration range.

With this equipment, oscillations of glycolysis can be obtained (2) in a great variety of yeast strains grown under aerobic or anaerobic conditions and tested in the form of a cell suspension or a cell-free extract. Figure 3 demonstrates the time progression of pyridine nucleotide reduction and oxidation in a suspension of anaerobically grown yeast. Beginning the glucose injection (upper curve) at a rate of 150 mM/h leads to a rapid reduction of pyridine nucleotide and later, after about 30 sec, to a strong oxidation. The concentration of reduced pyrivine nucleotide shoots over to the oxidized state and settles into a small amplitude oscillation which can be recorded for several hours. This is illustrated in the lower trace which reports the state of oscillation 60 min later. The mean level of the NADH oscillation is somewhat more oxidized. The frequency is 1.5 per min and compares with the frequency observed in aerobically grown yeast. The frequency variation of all the cycles is 6.8 percent. Such a spontaneous oscillation in a cell suspension is remarkable and points to a synchronizing function

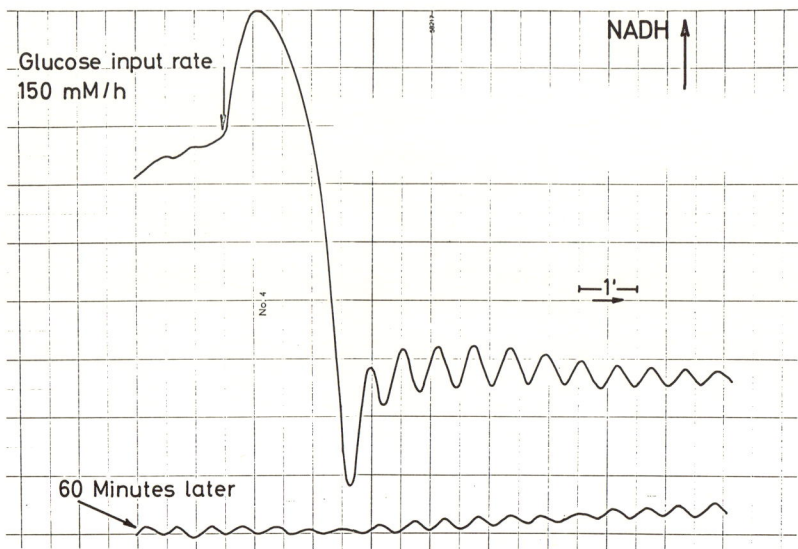

Figure 3. Continuous oscillations in a yeast cell suspension grown anaerobically. NADH is measured by fluorometry. Ordinate is in arbitrary units.

of glucose as the substrate of the oscillating metabolic circuit. A similar experiment is demonstrated in Figure 4. Here the rate of glucose input is increased to 220 mM/h. After an overshoot, the oscillation is displayed with a higher frequency of 2.5 per min compared to the frequency recorded in the experiment shown in Figure 3. Thus, in yeast cell suspensions a variety of frequencies can be demonstrated.

For an evaluation of the control properties it is more convenient to analyze a cell-free extract which prepresents an approximately 1:1-1:2 dilution of the cytosol of yeast cells and eliminates the control imposed by cell permeability on the glycolytic kinetics. Figure 5 demonstrates the response of glycolysis in a yeast extract to the addition of glucose at a rate of 36 mM/h. One can distinguish at the beginning a large amplitude of low frequency, with a period of approximately 15 min, which spontaneously changes to a small amplitude (approximately 1/3 of the initial amplitude) and a period of 6 min. Also, a high frequency modulation of the trace, especially in the later part, is obvious. This extrafrequency very much resembles the solutions of a Mathieu's

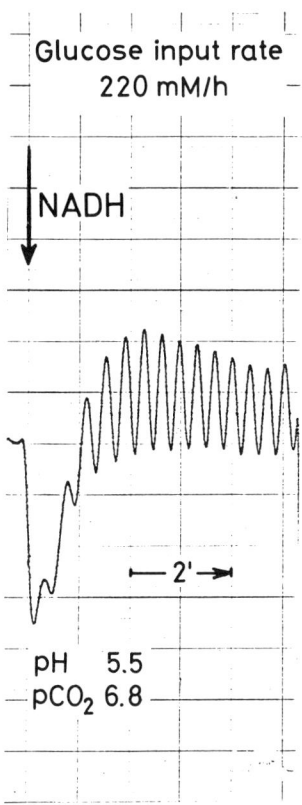

Figure 4. For explanation see Figure 3.

equation for nonlinear oscillators with periodic coefficients (5).

Similar amplitude and frequency modulations of the glycolytic kinetics are obtained by injection of fructose. This is demonstrated in Figure 6, where a spontaneous transition from the high to a low amplitude state of NADH concentration is recorded. The pH, as well as the pCO_2, is also given. The amplitude of the fluorescence trace falls spontaneously by a factor of 3.7 and the frequency rises from a period of 6.0 to 3.8 min. It can be seen that a coupled

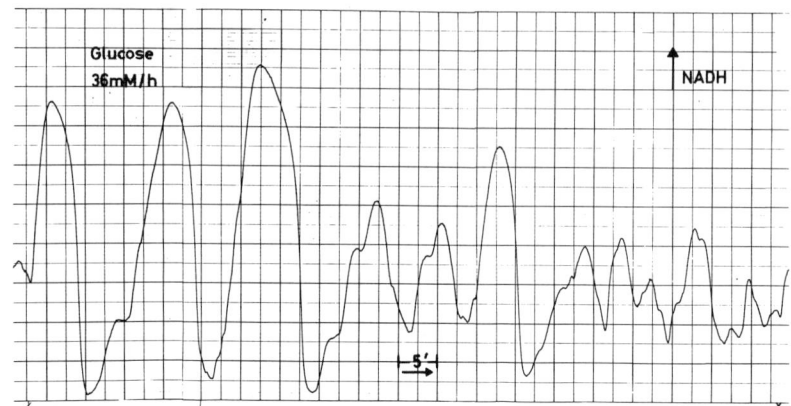

Figure 5. Oscillations in a cell free extract. Ordinate is in arbitrary units of fluorescence.

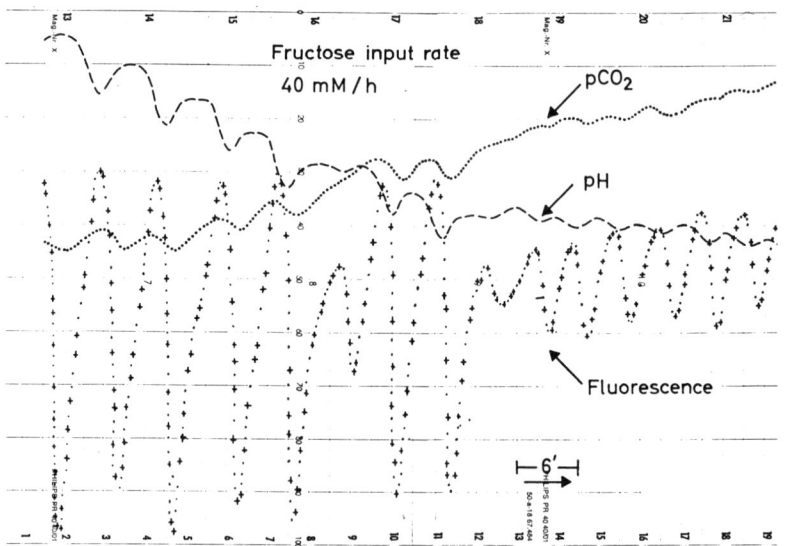

Figure 6. Oscillations of NADH, pCO_2 and pH in a cell free extract of yeast. For scale and conditions see text and previous legends.

change with a fixed phase angle occurs in pH and pCO_2 indicating that not only the NADH oscillation, but also the activity state of PFK and the CO_2 output is spontaneously changing under these conditions. Independent metabolite assays confirm this interpretation. From this experiment it can be concluded that glycolysis is able to oscillate within several - at least two - kinetic regions provided a certain critical input condition is maintained.

Indeed, double periodicities and non-sinusoidal wave forms (2) prevail at low input levels. In Figure 7 the addition of fructose, with an injection rate of 40 mM/h, drives the system into an oscillatory state with periods of 3.5 min and a non-sinusoidal waveform of NADH. A decrease in the injection rate to 20 mM/h leads directly to double periodicities. Increasing the rate from 40 mM/h to 80 mM/h yields almost a sinusoidal waveform of the NADH oscillation. This is shown in Figure 8. In this experiment, the phase angle shift of pH and NADH oscillation, in response to the increase of the input rate is indicated. The phase shift angle, $\Delta\alpha$, is increased from $87°$ to $120°$ with the increased glycolytic flux. Thus, the input rate of glycolytic oscillations clearly influences the phase angle between the two components. The mechanism of phase control will be discussed in the subsequent paper (4).

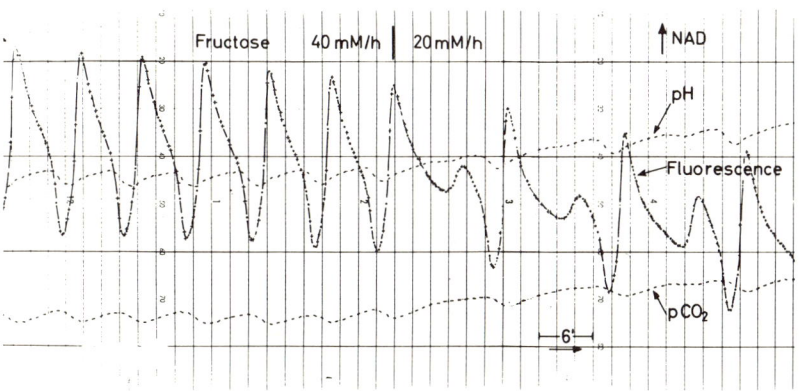

Figure 7. For explanation see Figure 6.

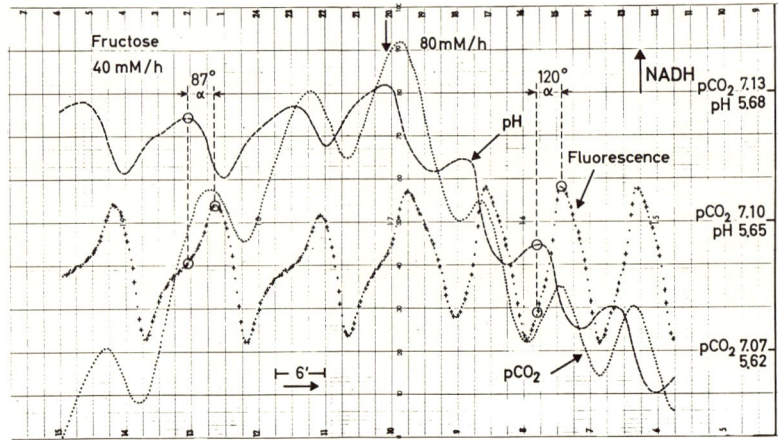

Figure 8. For explanation see Figure 6.

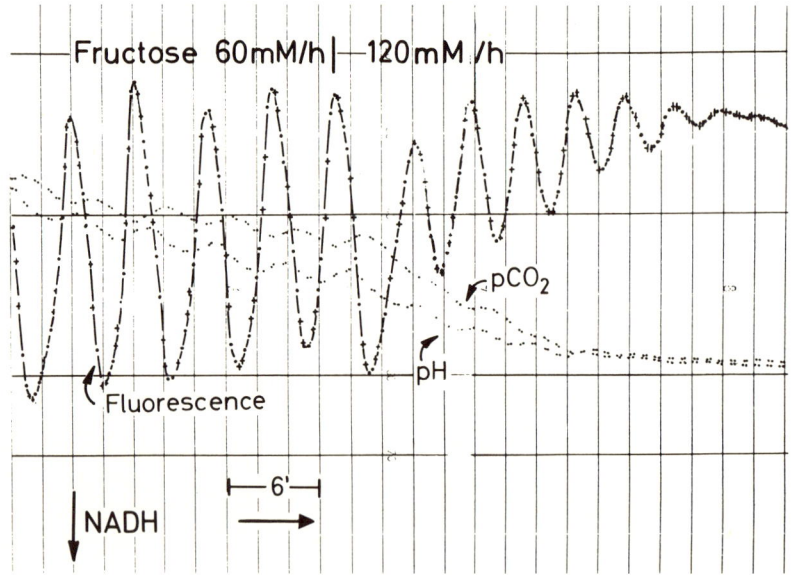

Figure 9. For explanation see Figure 6.

Clear sinusoidal oscillations are obtained on higher input rates of glucose, fructose, glucose-6-phosphate and fructose-6-phosphate (2,6). The experiment shown in Figure 9 tests the effect of higher input rates of fructose. An input of 60 mM/h leads to sinusoidal oscillations with an amplitude of 8.3 units and a period of 2 min. On doubling the rate the system is damped within 7 cycles and settles at the NAD-maximum. When the input is finally stopped, NAD is slowly reduced to a non-oscillatory state below the NADH-minimum of the fluorescence trace.

The mode of the responses of oscillating glycolysis to the rate of substrate additions is summarized in Figure 10.

Input rate[+] mM/h	period min	amplitude in mM NADH	Damping	waveform
< 20	—	steady high level of NADH	—	—
20	~ 8.6	0.2 – 0.4	—	double periodicities, nonsinusoidal
40	~ 6.5	0.6	—	nonsinus-sinus
60 – 80	~ 5.0	0.3	—	stabil sinus
120	~ 3.5	0.2	—	stabil sinus
> 160	—	steady low level of NADH	+++	

[+] fructose or glucose serve as substrates. cellfree extract of ~ 60 mgr/ml

Figure 10. Range of glycolytic oscillation in yeast extracts.

Generally it can be stated that three different oscillation ranges can be induced depending on the input rate of the source chemical. Each of the states can be reversibly obtained. They are not only displayed by the parameters demonstrated here, but also by the levels of glycolytic intermediates:

1. At a low input level (about 30 mM/h) a region with a low amplitude and frequency and double periodicities is observed. Below an input rate of 20 mM/h no oscillations can be recorded.

2. At an input level between 40 and 120 mM/h oscillations with high amplitude, high frequency and normal sinusoidal waveforms are found.

3. At high input levels, 120-160 mM/h, the system can maintain stable oscillations with higher amplitudes and frequencies. On further raising of the input rate the oscillations damp out.

The induction of glycolytic oscillation by mere setting of a steady input rate of the source chemical categorizes the glycolytic oscillator as a self-oscillatory system which is driven by a steady source of energy and not by a periodic forcing function.

The occurrence of double periodicities and pure sinusoidals in response to a variety of input rates clearly indicates the phase angle controlling function of the glycolytic input. Experiments carried out with this technique demonstrate that the minimum substrate to induce self-oscillation of glycolysis is fructose-6-phosphate (6). Thus, any other reactions "above" phosphofructokinase do not lead directly to self-excitation, but function only as additional input devices.

We reported earlier (2) that FDP is not able to produce oscillations under physiological conditions. However, it is important to note that we have observed that FDP does induce NADH oscillations, in a transient, under restricted conditions which can be obtained if phosphorylation and dephosphorylation rates are balanced by activation of the ATPase system with deoxyglucose.

Glycolytic oscillations cannot be induced only by hexoses but the second input metabolite of glycolysis, namely phosphate, is also needed if the system is phosphate-limited. Indeed, the Lynen-Johnson theory proves to be valid even under oscillating conditions. This is demonstrated in Figure 11 where the NADH fluorescence trace records initially some irregularities of the NADH level on injection of fructose at a rate of 60 mM/h. The traces of pH and pCO_2 indicate a steady rate of glycolysis, but no oscillation. Even on increasing the injection rate to 80 mM/h no change of the kinetics is observed. However, the addition of inorganic phosphate immediately induces oscillations, as recorded in

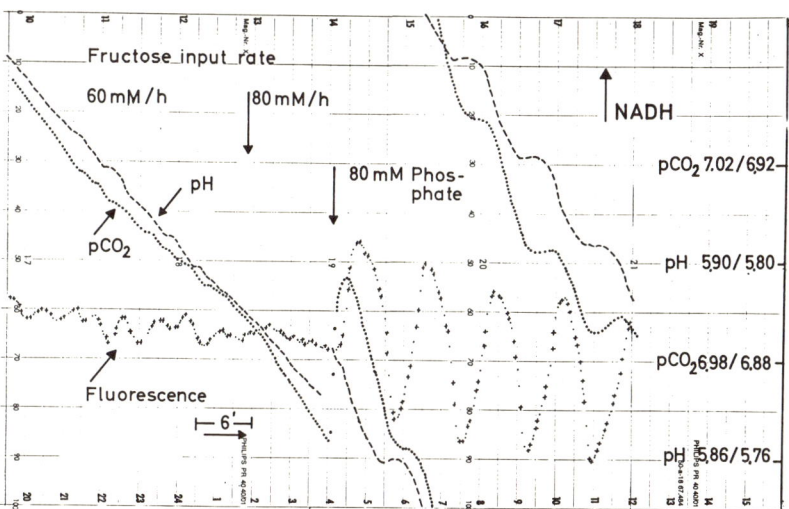

Figure 11. See Figure 6.

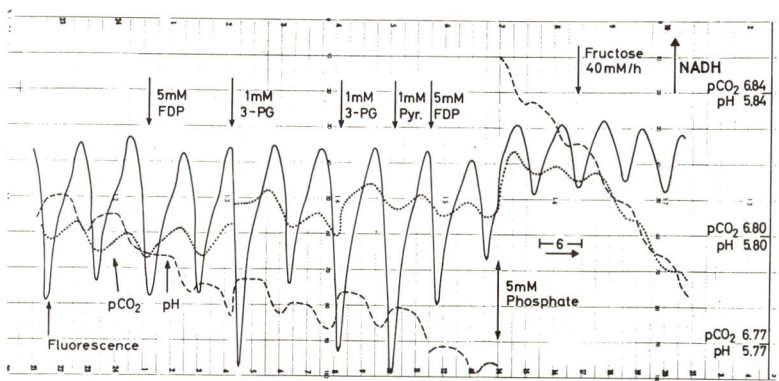

Figure 12. See Figure 6.

the traces of NADH, pH and pCO_2. The controlling function of phosphate can be strong enough to resist even a change in the type of glycolytic substrate. This is demonstrated in Figure 12 where the addition of phosphate first changes the character of the oscillation and, later, the injection of fructose to the phosphate-controlled oscillation increases only the rate of pH and pCO_2 changes but does not change the waveform of the NADH cycle. This experiment again demonstrates that the coupling between the pH and NADH oscillation is by no means fixed and even the waveform of the NADH oscillation can be constant on changing conditions.

The induction of self-oscillations of glycolysis can be achieved not only in cell-free extracts but also in reconstructed systems. We earlier reported the observation of sinusoidal and square wave motions of pyridine nucleotide in reconstructed systems (2,6). Here we would like to present the production of NADH pulses in reconstructed glycolysis as demonstrated in Figure 13. Here the rate of substrate injection again controls the waveform. In this experiment glycolysis is excited by the input of glucose at a rate of 18 mM/h. Within a time interval of almost exactly 1 min a spikelike production of NADH of 0.2 mM concentration is observed with a half rise-time less than ~1.5 sec. The

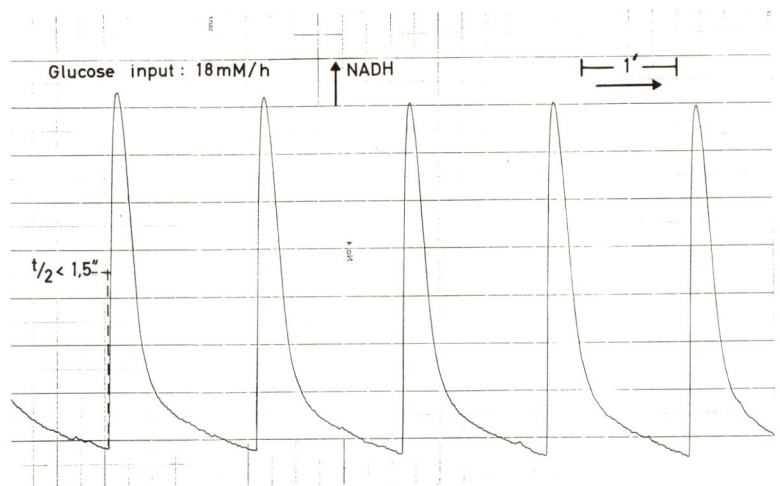

Figure 13. See Figure 6.

recording is instrument limited. It can be calculated that the true half rise-time should be in the order of 5-600 msec. The record is part of a series of 30 pulses. This experiment again demonstrates that the control of glycolytic oscillations by substrate injection is only a consequence of the property of self-excitation of glycolysis.

REFERENCES

1. Hess, B. In *Systems Theory and Biology* (Ed. D. Mesarovic), Springer Verlag Inc., New York, 1968, p.88.

2. Hess, B. and Boiteux, A. In *Regulatory Functions of Biological Membranes*, (Ed. J. Jarnefelt,) Biochim.Biophys. Acta Library, 11, 148 (1968).

3. Hess, B. and Brand, K. In *Control of Energy Metabolism*, (Ed. B. Chance, R.W. Estabrook and J.R. Williamson), Academic Press, New York, 1965, p.111.

4. Boiteux, A. and Hess, B. This volume, p. 243

5. Blaquiere, A. *Nonlinear System Analysis*, Academic Press, New York, 1966.

6. Boiteux, A. and Hess, B., Z. Physiol. Chem. 348, 1228, (1967).

References added in proof:

Hess, B. and Boiteux, A. Mechanism of Glycolytic Oscillation in Yeast (I). Hoppe-Seyler's Z. Physiol. Chem. 349, 1567 (1968).

Hess, B., Boituex, A. and Krueger, J. Cooperation of Glycolytic Enzymes, Advances in Enzyme Regulation 7, 149 (1969).

Hess, B. and Boiteux, A. Oscillatory Phenomena in Biochemistry. Annual Review of Biochemistry 40, 237 (1971).

CONTROL MECHANISM OF GLYCOLYTIC OSCILLATIONS

Arnold Boiteux and Benno Hess

Max-Planck Institut für Ernährungsphysiologie,
Dortmund, Germany

In the previous communication (1) we have discussed the input conditions necessary for the induction of glycolytic oscillations. These conditions, defined in terms of the steady state rate of substrate additions (hexoses and phosphate) to glycolysis, were found to be responsible for the frequency, amplitude, wave-form and damping constant of the oscillations and the turnover rates of the glycolytic intermediates. From the character of the input/turnover relationship it can be concluded that this metabolic oscillation is of the type which is commonly named a self-oscillatory system. Such a system is excited with the aid of a source reaction which has no periodicity and should clearly be distinguished from the operation of a periodic forcing function. Thus the mechanism, which generates glycolytic self-oscillation, is a property of the system itself. In this paper, a detailed analysis of glycolysis under oscillating conditions as well as the general mechanism of oscillation is presented.

The basis of a general mechanism of glycolytic oscillation is the establishment of a metabolite and energy balance of the pathway. In order to define the overall balance, glycolytic intermediates in a highly concentrated yeast extract (60 mg protein/ml) were analyzed over two oscillation periods of a total time interval of 14 min. The samples for the metabolic analysis were taken during the oscillation as recorded in Figure 1a, at the points A and B, marked with arrows. The concentrations of the compiled glycolytic intermediates and products were computed in relation to the substrate influx, as given by the rate of substrate injection. Figure 1a demonstrates an equilibrated balance, within the experimental errors of the metabolic assays. Eighty eight percent of the injected glucose is converted to ethanol and glycerol, the remainder is pooled in the form of glycolytic intermediates. It is obvious that there is no space for other glucose derivatives.

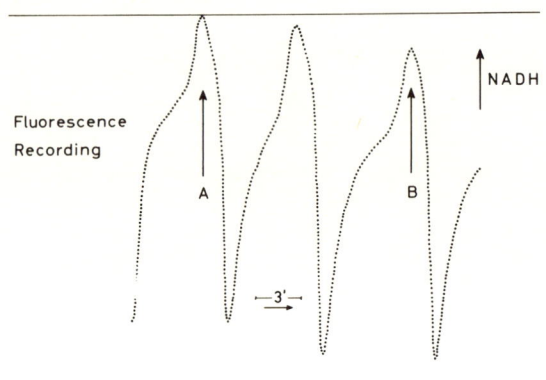

Substrate Balance
(calc. in μmoles glucose or glucose equivalents resp.)

	A	B	Δ_{B-A}
Ethanol	431.9	511.0	79.1
Glycerol	11.0	12.2	1.2
FDP	22.5	30.8	8.3
G-6-P	16.0	18.9	2.9
Σ other intermediates	19.0	19.0	0.0
Total	500.4	591.9	91.5
Glucose input			91.3

<u>Figure 1a</u>. Metabolic balance of glycolytic intermediates in an oscillating yeast extract.

The energy balance shown in Figure 1b is equally well equilibrated. Here the total amount of ATP synthesized is calculated on the basis of the ethanol formation. The ATP breakdown is calculated from the formation of hexose phosphates as well as from the measured activity of ATPase. Indeed, as we reported earlier, this Zn^{++}/ATP controlled ATPase equilibrates the overall ATP balance (2).

On the basis of this metabolite and energy balance, we would like to discuss the mechanism of the metabolite oscillation. As has been reported earlier by a number of authors and by us (for summary see ref. 2), the metabolite concentrations change during the oscillatory state with

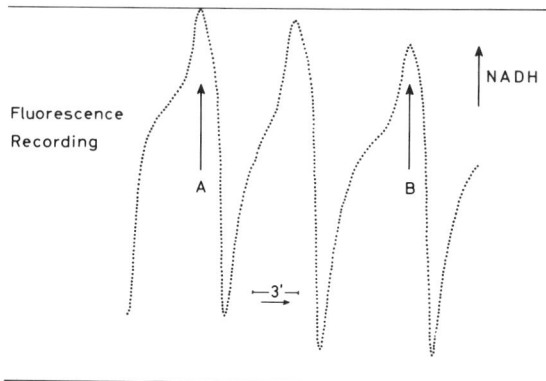

Figure 1b. Energy balance in an oscillating yeast extract.

equal frequency, but different phase angles relative to each other. This is demonstrated in Figure 2, where the directly recorded traces of NADH fluorescence and pH, as well as the concentration of the hexose-6-monophosphates are presented. The values of the hexose phosphates are obtained by enzymatic analysis of samples taken in a time series during the oscillation and drawn below the recorded traces.

The phase angle, $\Delta\alpha$, between the distinct maxima of hexose-6-phosphate and NADH concentrations indicates that the trigger function to initiate a new oscillation period is located in the first part of the glycolytic pathway. The steep slopes of the pH and hexose-6-phosphate traces coincide exactly. One free proton is produced per hexose-6-phosphate molecule converted to fructose diphosphate. Both metabolites show far larger concentration changes than any other glycolytic intermediate; they also display a relative

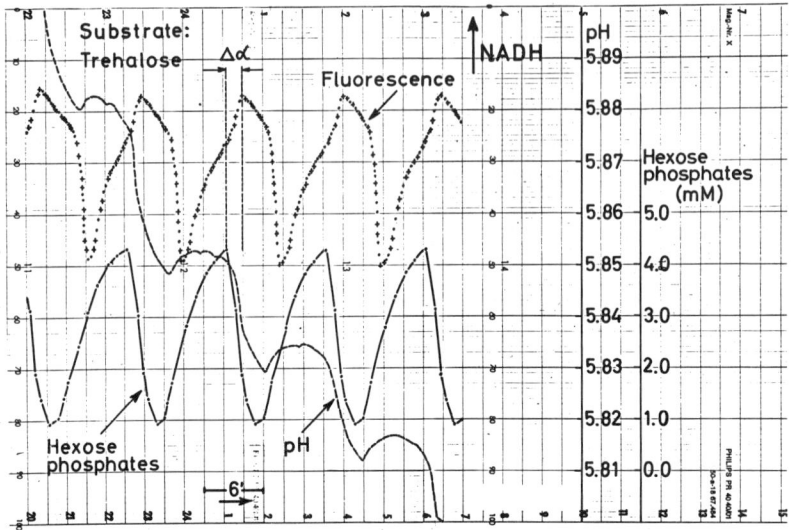

Figure 2. Trehalose induced oscillations of NADH, pH and hexose-6-monophosphates in a cell-free extract of yeast. For explanations, see text.

phase shift of 180° (see Figure 3). Thus the pH trace reflects a variable phosphofructokinase activity. Indeed, here we demonstrate direct evidence for the periodic activity change of this enzyme.

A detailed analysis of the phase relations of all the glycolytic metabolites under a great variety of input conditions yields an interesting property of glycolytic oscillation. The results of our studies are summarized in a schematic form in Figure 3. Each of the curves are obtained by at least 20 analytical points per period. It can be seen that according to their time dependent concentration change, the glycolytic metabolites can be summarized in two groups. In each of the groups the maxima and minima of the concentration changes coincide in time. The two groups differ by the phase angle $\Delta\alpha$. It is important to note that $\Delta\alpha$ can be varied depending on the experimental conditions (see ref.1). The phase angle is increased on increase of substrate injection rate, ATPase activity, or the adenosine phosphate pool. For diminution of the phase shift the reverse is true. This kinetic property can be predicted and explained as shown in our subsequent discussion.

BIOCHEMICAL OSCILLATORS

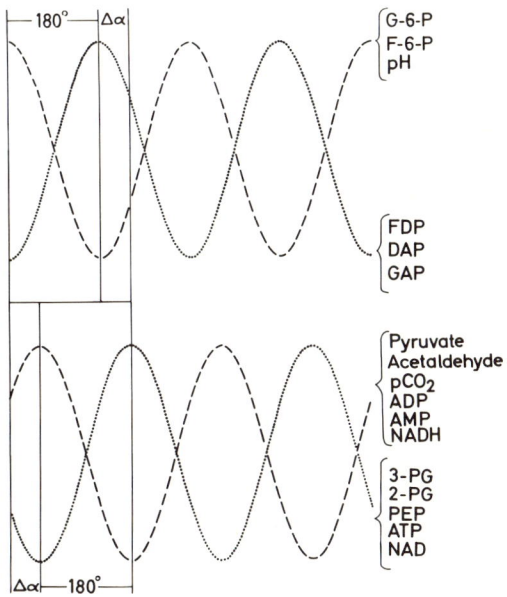

Figure 3. Phase shift of oscillating glycolytic intermediates. The amplitude of the oscillations is normalized.

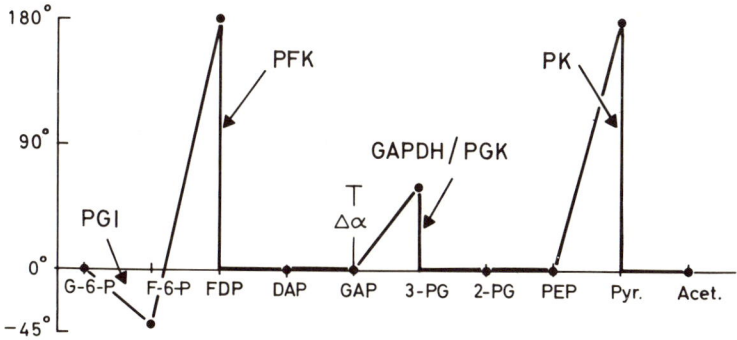

Figure 4. Phase shift of glycolytic intermediate concentrations during oscillations.

The location of the control points of the oscillation can be demonstrated with the help of a crossover diagram. In Figure 4 the phase shift in degrees between adjacent glycolytic intermediates is plotted as a function of their respective location in the glycolytic sequence.

It is obvious that the reaction step between glyceraldehyde-3-phosphate and 3-phosphoglycerate, catalyzed by the enzyme couple glyceraldehyde-3-phosphate dehydrogenase/phosphoglycerate kinase, corresponds to the variable phase angle $\Delta\alpha$. A phase shift by the 180° of the maximum versus the minimum is found between the metabolites fructose-6-phosphate and fructose-1,6-diphosphate, catalyzed by phosphofructokinase, as well as between phosphoenolpyruvate and pyruvate, catalyzed by pyruvate kinase. The two latter observations confirm earlier results of several laboratories (see references 2 and 3). Thus we find three crossover points corresponding to well-known control points of the glycolytic pathway. All the points are coupled to the adenosine phosphate system taking part either in ATP synthesis or ATP breakdown.

We would like to discuss the sequence of events during glycolytic oscillations in a cell-free extract with the help of the flow diagram presented in Figure 5 where the network of the various phosphokinase and dehydrogenase reactions of the pathway is summarized. The circle on the right represents the NADH pool which accepts, per mole of hexose turnover, 2 moles of NADH via the glyceraldehyde-3-phosphate dehydrogenase reaction and gives 2 moles of NADH to the alcohol dehydrogenase reaction. Any variation of the glycolytic flux which does not occur synchronously at both enzymes will produce a change of the NADH level and will be recorded as a fluorescence change. The NAD/NADH system is therefore an indicator for nonsynchronous changes of fluxes at the enzymes glyceraldehyde-3-phosphate dehydrogenase and alcohol dehydrogenase.

The left part of the diagram represents the phosphokinase reactions. Under the conditions of continuous oscillation, 2 moles of ATP are split per mole of hexose. This is indeed the case shown in the energy balance of Figure 1.

How are oscillations of metabolite concentrations and

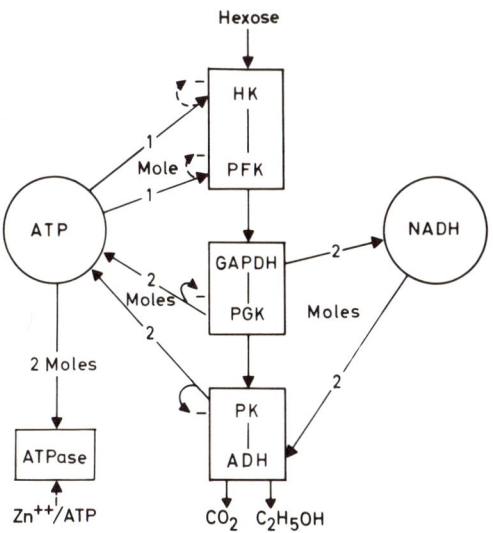

Figure 5. Metabolic control diagram.

fluxes generated in such a system? On the basis of its known allosteric properties, as well as the crossover behavior of phosphofructokinase, it has been suggested as the oscillophor by Ghosh and Chance (4) and Higgins (5). Furthermore, the oscillophor function of phosphofructokinase has been directly demonstrated by the substrate injection technique developed in our laboratory (2,6). Fructose-6-phosphate is indeed a minimum substrate for the induction of glycolytic oscillations. This is further supported by the fact that additions of fructose diphosphate, as substrate, to the complete system only generate an increase of the CO_2 production, but no oscillations of any metabolite concentration.

The periodic activity change of phosphofructokinase by allosteric self-excitation yields a pulsed output of fructose diphosphate and ADP, as illustrated in Figures 2 and 3. The fructose diphosphate output cannot be responsible for the propagation of the periodic flux change. Experiments in our laboratory (6), confirmed by Pye (7), show that fructose diphosphate additions within the oscillating range have no influence on the oscillation. However, it is known that additions of ADP strongly influence the phase state of

the oscillation, producing a shift from any state to the
NADH maximum. AMP additions have the same effect. In contrast, additions of ATP immediately shift the oscillation
from any state to the NADH minimum (2,8,9). Thus the control
function of the adenosine phosphates is evident. Since they
are in equilibrium by interconversion through the enzyme ATP:
AMP phosphotransferase, all the adenosine phosphates represent only one control variable from the functional point of
view.

Titration experiments with the oscillating system, as
well as with isolated enzyme preparations, prove that the
adenosine phosphate control variable is operating via the
kinases of the glycolytic pathway. Indeed, glycolysis is
controlled at any time by the adenosine phosphate variable.
The variable itself is a function of the velocities of all
ATP-generating and ATP-splitting reactions and of the adenosine phosphate pool. Any experimental change of the ATP-splitting activities or adenosine phosphate pool influences
the phase angle between the oscillophor and secondary oscillations of the other glycolytic intermediates. This has
been tested under a great variety of conditions in our laboratory (10).

In order to illustrate the control function of the
adenosine phosphate variable, the glycolytic flux rates at
their extreme values are schematically drawn in Figure 6.
The lower part of the picture shows the state of the glycolytic system at a high ADP:ATP ratio. The flux rate through
the kinases, which are then fully activated, is more rapid
than the ethanol production rate. So, pyruvate, acetaldehyde
and NADH pile up.

When the ATP synthesis overcomes the ATP breakdown,
the system is forced to the state shown in the upper part of
the figure, corresponding to a decreased ADP:ATP ratio. The
flux rate through the kinases is low because they are strongly
inhibited by ATP. Therefore, 3-phosphoglycerate, 2-phosphoglycerate and phosphoenolpyruvate pile up. The rate of
ethanol production is not inhibited and is only limited by
the substrate concentrations. This leads to a decrease in
the concentration of NADH, pyruvate and acetaldehyde. The
next change of the ADP:ATP ratio leads back to the lower
part of the figure, and so on.

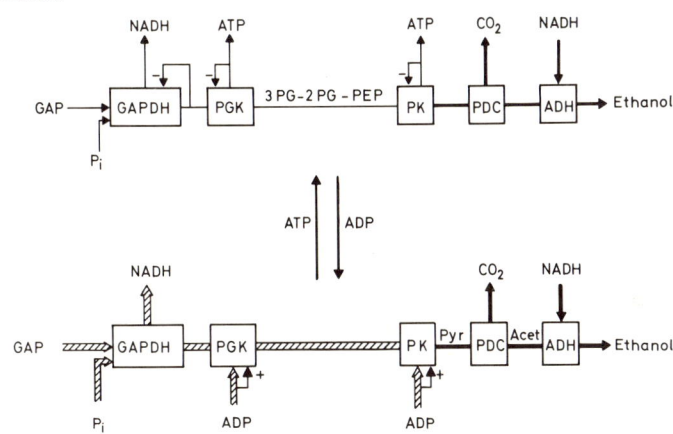

Figure 6. For explanation, see text.

In summary, glycolytic self-excitation is produced by the allosteric oscillophor phosphofructokinase and propagated with a dependent phase shift by means of the coupling variable adenosine phosphates via the kinases phosphoglycerate kinase and pyruvate kinase.

REFERENCES

1. Hess, B. and Boiteux, A. This volume, p. 229

2. Hess, B. and Boiteux, A. In <u>Regulatory Functions of Biological Membranes.</u> Ed. J. Jarnefelt. Biochem. Biophys. Acta Library, <u>11</u>, 148 (1968).

3. Hess, B. In <u>Systems Theory and Biology.</u> Ed. M.D. Mesarovic, Springer Verlag Inc., New York, 1968, p.88.

4. Ghosh, A. and Chance, B. Biochem. Biophys. Res. Commun. <u>16</u>, 174 (1964).

5. Higgins, J., Proc. Nat. Acad. Sci. 51, 989 (1964).

6. Boiteux, A. and Hess, B., Hoppe Seyler's Z. Physiol Chem. 348, 1228 (1967).

7. Pye, E.K., Fed. Proc. 26, 564 (1967).

8. Chance, B., Schoener, B. and Elsaesser, S., Proc. Nat. Acad. Sci. 52, 337 (1964).

9. Hess, B., Brand. K., and Gey, G., Internal. Cong. Biochem Abstr., New York, 1964, p. VI-41.

10. Boiteux, A. and Hess, B., in preparation.

Reference added in proof:

Hess, B. and Boiteux, A. Mechanism of Glycolytic Oscillation in Yeast (I), Hoppe-Seyler's Z. Physiol. Chem. 349, 1567 (1968).

COMPONENT STRUCTURE OF OSCILLATING GLYCOLYSIS

B. Hess, H. Kleinhans and D. Kuschmitz

Max-Planck-Institut für Ernährungsphysiologie,
Dortmund, Germany.

Considerable progress in our understanding of the mechanism of glycolytic self-oscillation has been made

1) by the elucidation of the input condition necessary for the induction of glycolytic oscillations,
2) by a detailed analysis of the concentration pattern of glycolytic intermediates,
3) by titration experiments with inhibitors, metabolites and enzymes,
4) by analysis of enzyme activities and enzyme kinetics mainly of PFK, PGK and PK under the conditions of oscillation,
5) by theoretical studies of oscillating chemical systems (for summary see references 1,2,3).

A more detailed understanding of the phenomena should be based on a complete knowledge of the catalytic parameters of the system, namely the kinetic constants and concentrations of the individual enzymes and metabolites. In this paper we will discuss the concentrations of the glycolytic enzymes and metabolites, and the concentration of the enzyme -

Abbreviations: ADH, alcohol dehydrogenase; ALD, aldolase; AK, adenylate kinase: ENO, enolase; GAPDH, glyceraldehyde-3-phosphate dehydrogenase; HK, hexokinase; PFK, phosphofructokinase; PGI, phosphoglucoisomerase; PGM, phosphoglycerate mutase; PGK, phosphoglycerate kinase; PK, pyruvate kinase, PDC, pyruvate decarboxylase; TIM, triosephosphate isomerase; G-6-P, glucose-6-phosphate; F-6-P, fructose-6-phosphate; DAP, dihydroxyacetone phosphate; GAP, glyceraldehyde-3-phosphate; 3-PG, 3-phosphoglycerate; 2-PG, 2-phosphoglycerate; PEP, phosphoenolpyruvate; Pyr, pyruvate.

coenzyme complexes of the dehydrogenases under oscillating conditions.

The enzyme concentration in an oscillating extract represents a total amount of roughly 20 to 100 mg protein per ml, varying from one experiment to the other. The concentrations of the individual enzymes are estimated as follows. The enzymes ALD, TIM, PGM, ENO, PDC and AK were estimated on the basis of the known values for Vmax, T.N. and the molecular weight of the purified enzymes and by assay of Vmax in the extract. The application of the Michaelis-Menten equation for enzyme-saturated conditions leads to values for molarities in the extract, or in reconstructed systems (4). Because of the large error involved in estimating the molarities of the enzymes HK, PFK, GAPDH, PGK, PK and ADH, we have obtained their molarities by titration with specific antibodies (4). GAPDH was, in addition, determined in the original extract by its GAPDH-NAD complex measured by difference spectroscopy in the absence and presence of H_2O_2 as shown in Figure 1 (5).

The results of these estimations are summarized in Figure 2. The data represent the molarities of the enzymes in a yeast extract of total protein concentration of 50 mg per ml. The order of magnitude of the molarities is high (3). With the exception of phosphofructokinase the molarities in the extracts are above 10^{-6}M. The protein in such an extract represents at least 65% of the total amount of soluble yeast cell protein. The molarity range agrees with the data calculated for enzymes in animal tissues by Bücher (6) as well as by Srere (7). We have recently pointed out that such a high accumulation of enzymes of one system should be considered as a functional macromolecular unit.

A more realistic figure has been obtained by correcting the molarities for the number of binding sites as far as they are known. Rather speculative is the number for PFK which was taken as being eight on the basis of its molecular weight (8). Relating the normalities to hexokinase as unity, the enzyme normalities can be ordered within a range of 1 : 8 (Figure 3). Indeed, if we round the normality figures up to full numbers, neglecting some deviation, we recognize that the enzyme normalities can be ordered in simple proportions.

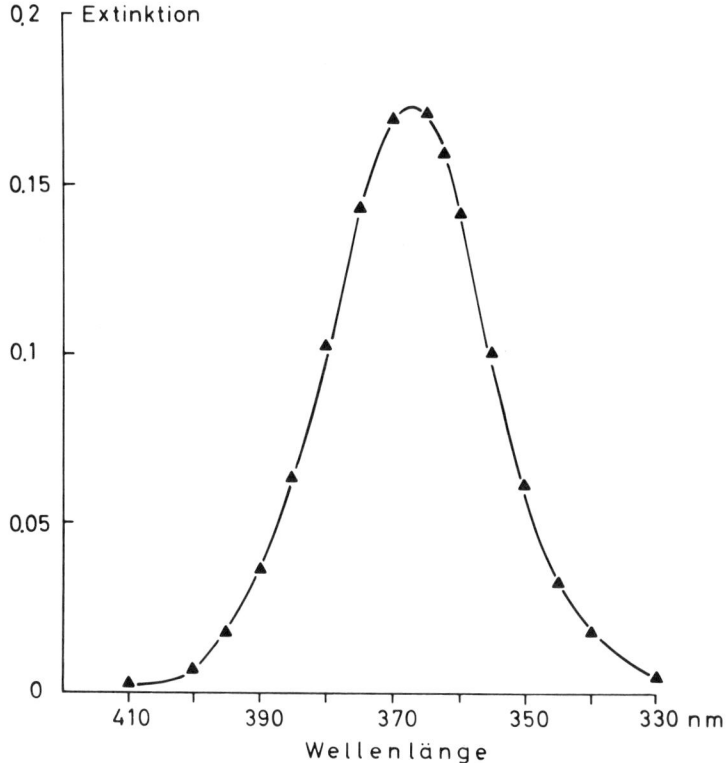

Figure 1. Difference spectrum of a NAD-saturated yeast extract (extract-H_2O_2 versus extract, 35 mg protein/ml, taken in a Johnson Foundation split-beam spectrophotometer, d = 1cm).

It should be mentioned that the normality values are multiplied by a dilution factor of 3 in order to represent the concentration activities per cytosol of the intact cell. This enzyme pattern raises a number of questions with respect to the character of interactions between individual components and the ratio of their respective subunit structure, problems which are out of the scope of this presentation.

Because of the uncertainties of the number of binding sites and of most of the dissociation constants involved, it

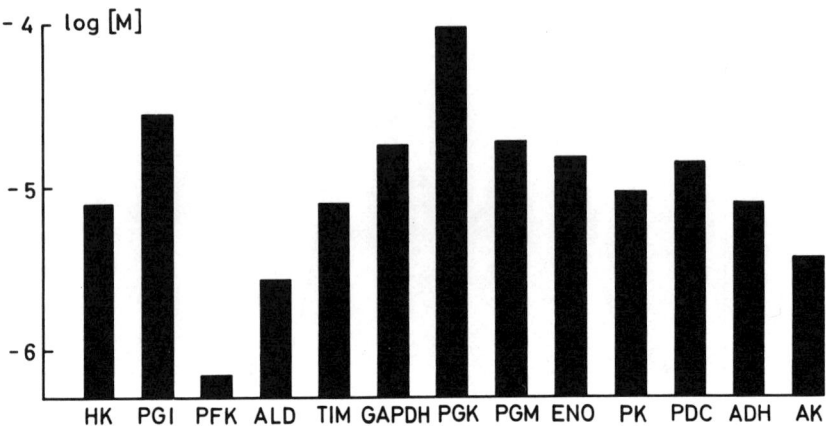

Figure 2. Molarities of glycolytic enzymes in a yeast extract (50 mg protein/ml).

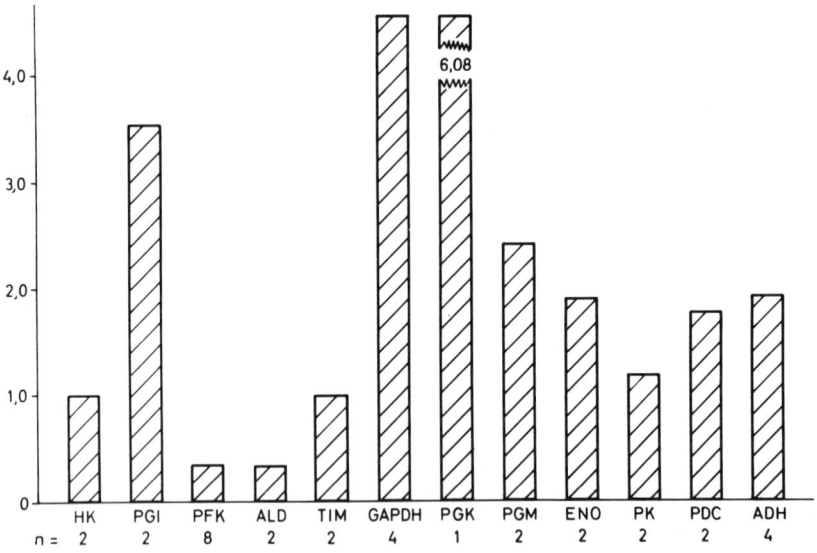

Figure 3. Relative normality of glycolytic enzymes in yeast (HK = 0.5×10^{-4}M = 1, n = number of active sites per molecule).

is difficult to relate the enzyme molarities to the substrate molarities under steady-state conditions on a quantitative basis and to evaluate the degree of saturation of most of the enzymes. A comparison of the molarities in the extract suggests that quite a number of the enzymes are not saturated with their respective substrates. At least their molarities are in an equal range (Figures 4 and 5)(3).

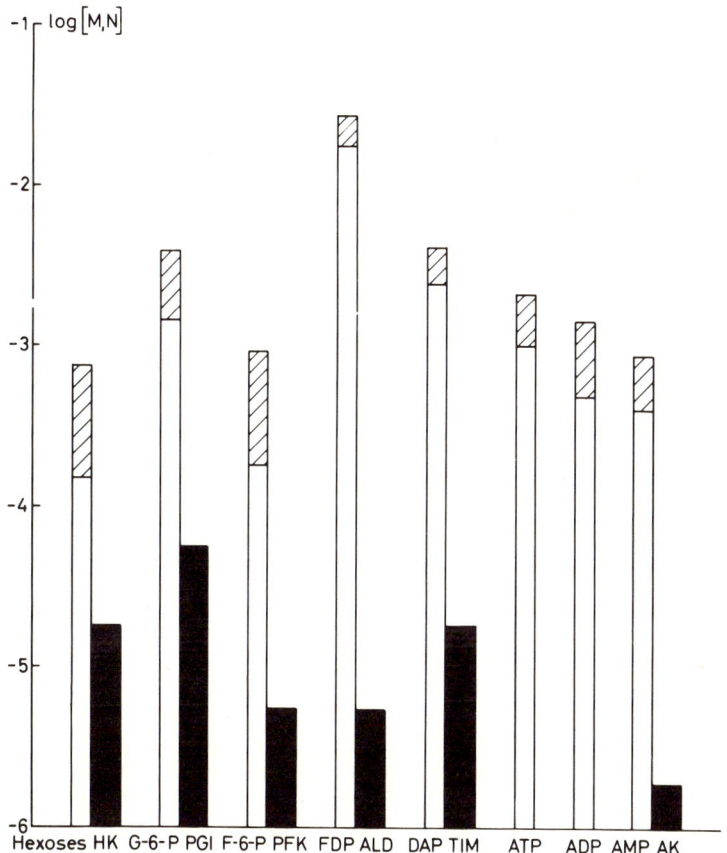

Figure 4. Concentration of glycolytic components in a yeast extract (50 mg/ml, enzymes are given in normalities, substrates and intermediates in molarities; hatched parts of columns signify range of oscillation).

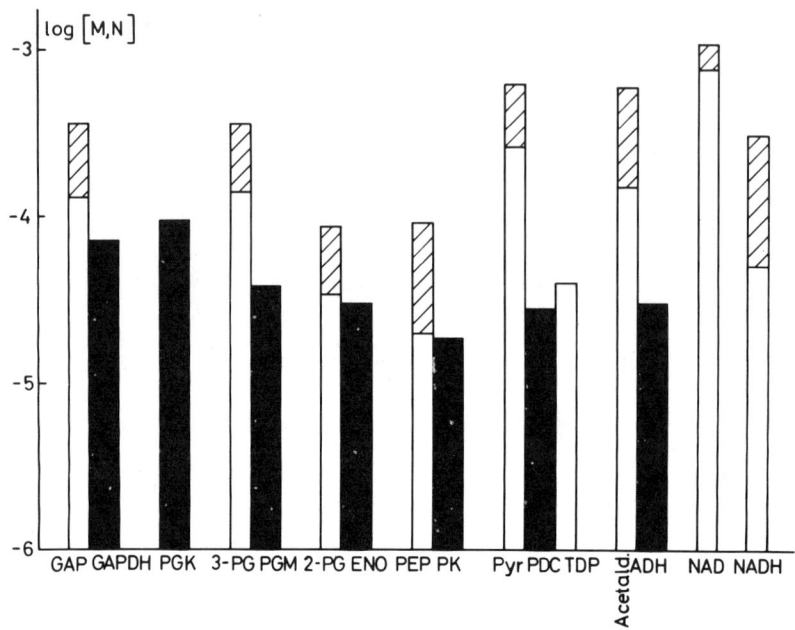

Figure 5. See legend of Figure 4.

In the case of the dehydrogenases, however, a reasonable estimate of the state of nucleotide saturation can be computed for the conditions of the oscillation. On the basis of the respective dissociation constants and the molarities of the components, the two equations summarized in Figure 6 can be solved, using a computer program, to yield the levels of the bound and free species. The percentage saturation of GAPDH with pyridine nucleotides is given in Figure 7 and accounts for 98.1% of bound NAD and 1.5% of bound NADH. The bound forms of NAD and NADH represent only 9% of the total NADH and NAD pool present in the system. Obviously the high percentage of NAD bound to GAPDH is functionally important to keep the enzyme always in the "R"-state and ready to go.

The reversed state of the ADH saturation with pyridine nucleotides is also shown in Figure 7. This enzyme is loaded to 85.4% with NADH and 13.8% with NAD and again only 18% of the total pool of pyridine nucleotides is used for ADH

$$\bar{K}_{NAD} = \frac{[E_{tot} - E \cdot NAD - E \cdot NADH] \cdot [NAD - E \cdot NAD]}{[E \cdot NAD]}$$

$$\bar{K}_{NADH} = \frac{[E_{tot} - E \cdot NAD - E \cdot NADH] \cdot [NADH - E \cdot NADH]}{[E \cdot NADH]}$$

Figure 6. Method for calculation of enzyme-bound pyridine nucleotides. $\bar{K}_{NAD}$ and $\bar{K}_{NADH}$ = equilibrium constants. E = GAPDH and ADH respectively.

Nucleotide saturation of GAPDH and ADH in Yeast

Enzym	total-E		free E		E-NAD		E-NADH	
	M	%*	M	%	M	%	M	%
GAPDH	2×10^{-4}	100	0.45×10^{-6}	0.2	2.1×10^{-4}	98	3.2×10^{-6}	1.5
ADH	1×10^{-4}	100	0.9×10^{-6}	0.89	0.14×10^{-4}	13.8	0.85×10^{-4}	85.4

* % of total

Figure 7. For explanation see text.

saturation. Here, NADH, to a large extent saturating ADH, keeps the enzyme in a state ready to go. From this calculation it can be seen that both enzymes bind up to 27% of the total free pyridine nucleotide pool available. Since we do not know how much of the rest of the free nucleotide pool is bound to other enzyme species, the availability of the free pyridine nucleotide pool is still a matter of speculation.

Let us consider the molarity values of the different components in more detail. The relevant molarities indicate, in Figure 7, 2.1×10^{-4}M GAPDH-NAD and 0.85×10^{-4}M ADH-NADH and, as shown in Figure 8, 2.4×10^{-3}M free NAD and 0.45×10^{-3} free NADH, as mean concentrations during the oscillation.

Concentration and ratio of free and bound NAD and NADH in Yeast

Nucleotide	bound		free	total
	GAPDH $\underline{M}$	ADH $\underline{M}$	$\underline{M}$	$\underline{M}$
NAD	0.2×10^{-3}	13.8×10^{-6}	2.4×10^{-3}	2.5×10^{-3}
NADH	3.2×10^{-6}	85.4×10^{-6}	0.45×10^{-3}	0.5×10^{-3}
NAD/NADH	Γ mV* 63 -268	Γ mV 0.162 -343	Γ mV 5.3 -299	Γ mV 5.0 -300

* $E'_{o(NADH/NAD)} = -320$ mV

Figure 8. For explanation see text.

The molarity range of these components raises the question as to whether the concentration of the complex-bound components also oscillate when glycolysis is in a state of self-excitation. The analytical data given above show that the total amount of NADH and NAD oscillates. Here we see that the bound forms of both nucleotides reach an appreciable level compared to the levels of their free forms. At least 20% of the free, available NADH is bound under steady-state conditions to ADH. Is it that only the levels of the free NADH and NAD oscillate, or is there indeed, as already suggested by Chance et al. (9,10), an oscillation of the enzyme-pyridine nucleotide complexes as well? We have tried to answer this question.

Indeed, the complexes GAPDH-NAD and ADH-NADH reach the concentration range which might well be detected spectroscopically during the oscillation. In 1954, B. Chance demon-

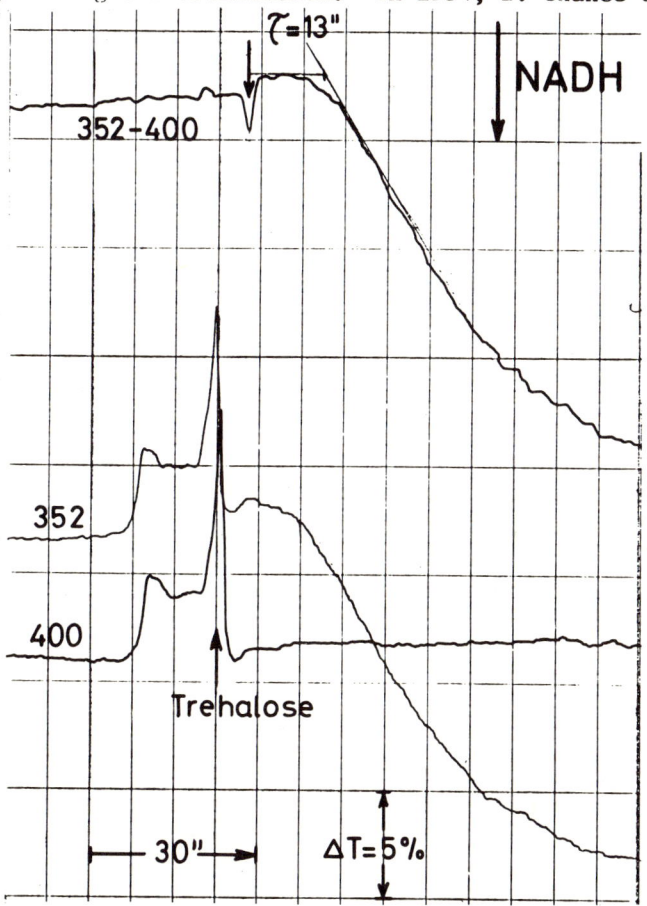

Figure 9. Double-beam record of the absorbancy changes induced by addition of trehalose to a yeast extract (50 mg protein/ml, d = 1cm). Trehalose was rapidly blown in to give a final concentration of 10 mM. The two bands are recorded by their respective single, as well as difference signals at 352 and 400 nm; the arrows indicate the addition of trehalose.

strated the presence of the GAPDH-NAD complex in a yeast suspension using the titration of the complex with iodoacetate in a double-beam spectrophotometer (10). As shown above, a complex with the same maximum can be identified by titration with H_2O_2 in the crude yeast extract which readily oscillates.

Given a molarity of 2.1×10^{-4}M for GAPDH-NAD and an extinction coefficient, at 366 nm, for the R-state of 0.95 $mM^{-1}cm^{-1}$ and for the T-state of 0.39 $mM^{-1}cm^{-1}$, at 400 nm, respectively, for half of each (11) we calculate an extinction for this compound of roughly 0.1 at 400 nm. This could readily be recorded if any change in the concentration of this compound occurred. However, as demonstrated in the next two figures, no oscillation can be detected. Figure 9 demonstrates the transient to oscillation following the addition of trehalose to a yeast extract, as recorded using a double-beam spectrophotometer set at the wave length pair of 352 and 400 nm. It can be seen that after addition of trehalose and a lag-period of 13 seconds, the oscillation of the NADH absorption is induced starting with an increase of NADH absorption. The record clearly shows that the 400 nm trace does not react to the addition of trehalose. Also, the full train of oscillations, as recorded in Figure 10, leaves the 410 nm band within the noise level. From this negative result we conclude that no oscillations of the GAPDH-NAD compound occur and furthermore that the enzyme is always NAD saturated and in excess under our conditions.

The identification of an ADH-NADH compound, which occurs in appreciable amounts in the extract, is difficult since the NADH-absorption is not influenced by ADH and the fluorescence enhancement of NADH by ADH cannot be detected so far for technical reasons. However, the action spectrum of the oscillation clearly points to an oscillating NADH-like compound which we would like to attribute to a NADH species bound to ADH. An action spectrum was taken using a double-beam spectrophotometer with the 340 nm absorption band serving as a reference, the measuring beam being varied between 330 and 380 nm, with the necessary correction for the non-linearity of the light source and the photomultiplier. A guard-filter eliminates from the measurement the strong fluorescence emission at 450 nm. Figure 11 summarizes the record of an extract displaying trehalose-induced oscillations. The extinctions were taken from the double amplitude of the

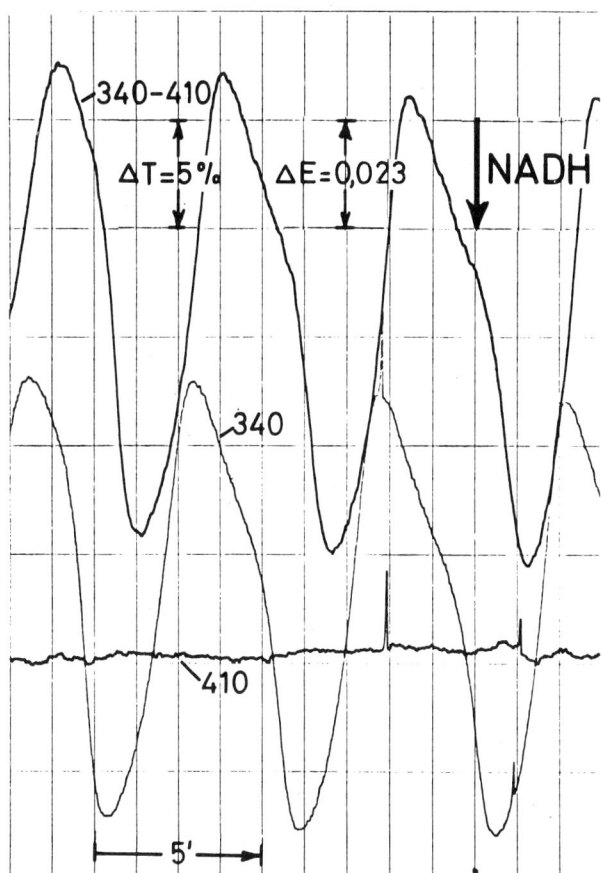

Figure 10. For explanation see legend Figure 9.

oscillations corresponding to an extinction of roughly 0.6 for the 340 nm band which was set equal to 100. The extinction ratios of the measured pairs of wave-lengths were plotted as shown in Figure 12 for two different experiments. The comparison of the pure NADH spectrum, drawn below, with the action spectrum of the extract shows that the oscillating compound, as measured by absorbancy changes, is not NADH, but a different compound with an absorption maximum at 355 nm. In parenthesis, it should again be recognized that no 400 or 410 nm compound is indicated by the spectrum.

Further identification of the "355" compound could be

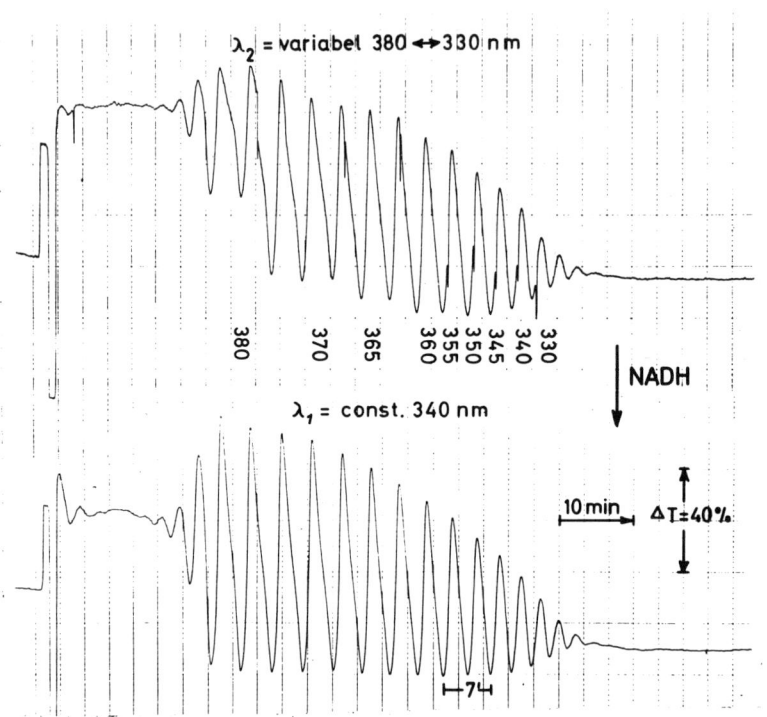

Figure 11. Double-beam record of the absorbancy changes induced by the addition of trehalose to a yeast extract (50 mg protein/ml). A guard filter selected the region between 325 and 380 nm. The band-width of the spectral resolution was 2.5 nm.

achieved by analysis of its kinetic behaviour relative to the NADH-fluorescence commonly used for the analysis of glycolytic oscillations. As reference for both indicators the pH-oscillation has been used, as demonstrated in Figure 13. As reported earlier the acidic maximum of the pH-oscillation preceeds or coincides with the maximum of the total NADH analysed analytically, as well as the NADH fluoresence maximum. The same is found with respect to the phase relationship between the pH and the absorbancy change of the "355" compound, as shown here. Thus, no phase shift between

BIOCHEMICAL OSCILLATORS

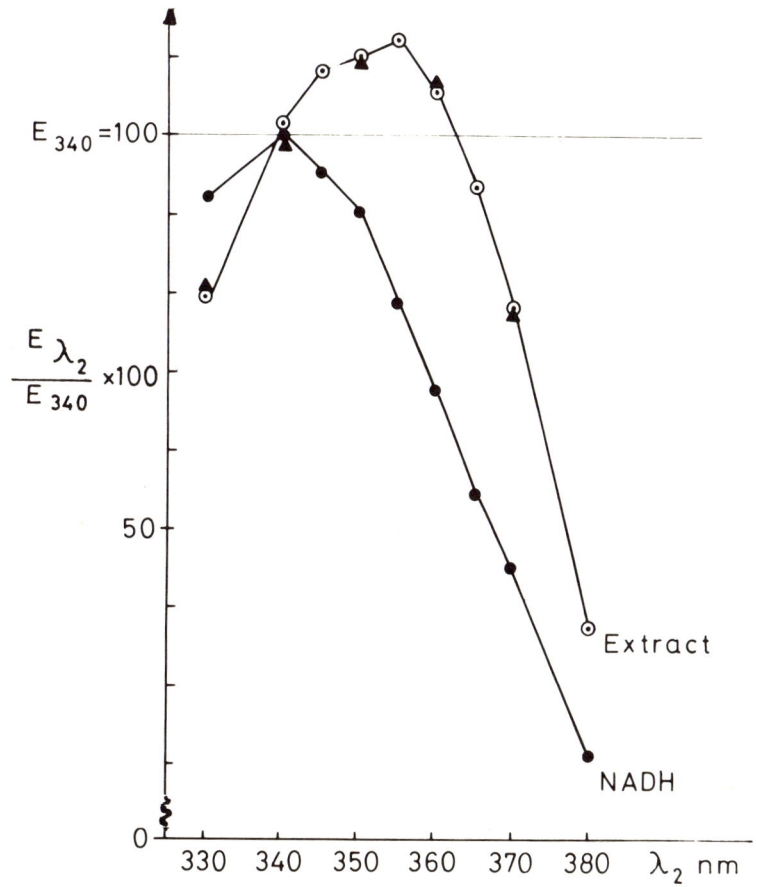

Figure 12. For explanation see text.

the fluorescent compound and the "355" compound is observed in these experiments. This confirms an earlier record reported by Chance et al. (9), in which oscillations were recorded at 350 - 380 nm. On the basis of the phase coincidence of the NADH-fluorescent compound and the "355" compound we cannot conclude that we are dealing with the same species. It might be the same compound, or there might be two compounds being rapidly equilibrated. Considering its phase relationship to the oscillation of the other glycolytic intermediates, the "355" compound most probably belongs to a new ADH-NADH

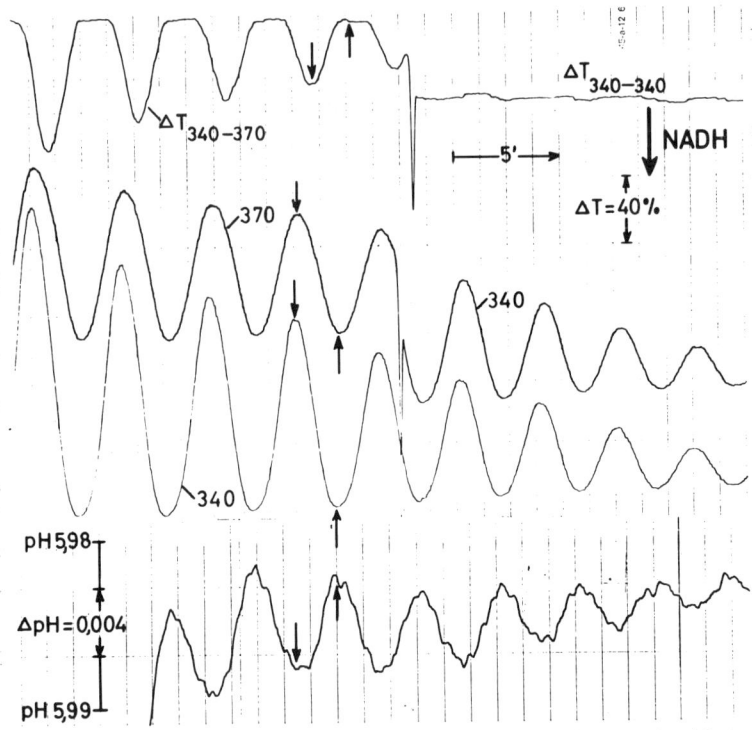

Figure 13. For explanation see legend of Figure 11. All traces are simultaneously recorded. Because of the small difference in pen positions of the three-pen recorder a constant deviation of the times given in the three spectrophotometric records is implied and indicated by the arrows which identify the points coinciding in time. After five cycles the reference wavelength was shifted from 370 to 340 nm, causing the disappearance of the difference signal.

species. Leaving the final identification of this compound open to further experimentation, our observations substantiate the suggestion that the oscillating NADH recorded here is not identical with free NADH and that we are dealing with a bound form of NADH. Thus we specify an enzyme-bound form which oscillates to a large extent. Research is currently being carried out to analyse in more detail the dynamics of these species under oscillating conditions using both an experimental approach as well as a computer model (in cooperation with E. Chance) based on the molarity data presented here.

ACKNOWLEDGEMENT

Part of this work was supported by a grant from the U.S. Department of Health, Education and Welfare. (CA 8255-01S).

REFERENCES

1. Hess, B. and Boiteux, A. In Regulatory Functions of Biological Membranes, (Ed. Johan Jarnefelt), Biochim. Biophys. Acta Library, 11, 148 (1968).
2. This volume.
3. Hess, B., Boiteux, A. and Krüger, J., Advances in Enzyme Regulation, 7, 149 (1969).
4. Hess, B., Wieker, H.J. and Krüger, J., Abstract No. 379, FEBS-Congress, Madrid, 1969.
5. Hess, B., Boiteux, A. and Krüger, J., Fed. Proc. Abstract No. 1556 (1969).
6. Bücher, T., Bolletino della Societa Italiana di Biologia Sperimentale, XXXVI, 1510 (1960).
7. Srere, P.A. In Metabolic Roles of Citrate, (Ed. T.W.Goodwin), p.11. Academic Press, London (1968).
8. Lindel, T.J. and Stellwagen, E., J.Biol. Chem. 243, 907 (1968).
9. Chance, B., Estabrook, R.W. and Ghosh, A., Proc. Nat. Acad. Sci. 150, 1244 (1964).
10. Chance, B., Harvey Lectures 40, 145 (1955).
11. Kirschner, K., Personal Communication.

GLYCOLYTIC OSCILLATIONS IN CELLS AND EXTRACTS
OF YEAST - SOME UNSOLVED PROBLEMS

E. Kendall Pye

Biochemistry Department, Medical School,
University of Pennsylvania,
Philadelphia, Pennsylvania 19104

INTRODUCTION

The phenomenon of autonomous oscillations in the concentration of glycolytic metabolites in yeast has been the subject of intense study for almost a decade (1-6). While the initial observation of this phenomenon can be traced back to the earlier studies of Duysens and Amesz (7), and probably even further to the work of Chance on the measurement of reduced pyridine nucleotides in living cells (8), it was not until the rediscovery of the phenomenon in 1964 (9) that a major impetus developed towards an understanding of the underlying basic mechanism. Since that time it has been shown that the oscillations originate through an alternating activation and deactivation of the key enzyme phosphofructokinase (PFK) (10), probably under the influence of positive allosteric effectors which are themselves either direct or indirect products of the PFK reaction. Furthermore, it has been shown that cell-free extracts of the yeast S. carlsbergensis can be produced which generate glycolytic oscillations of a type equivalent to those observed in intact cells (11). This discovery, in turn, led to the observation that these oscillations in cell-free extracts were, under certain conditions (12), essentially limit-cycle oscillations and could be sustained at close to constant frequency and amplitude for many hours (> 100 cycles). A section of one such experimental result is shown in Figure 1. The recognition that the glycolytic oscillations were not innately continuously damped came from the observation that the oscillations in cell-free extracts could be sustained by the addition of the reserve disaccharide, trehalose (12), or by the infusion of glucose, fructose or phosphorylated hexoses at the correct rates (4). These experiments strongly indicated that the rate of sugar entry into the cell was of crucial importance for the appearance of the oscillation phenomenon (5) and that this process limited the overall rate of glycolysis in these cells.

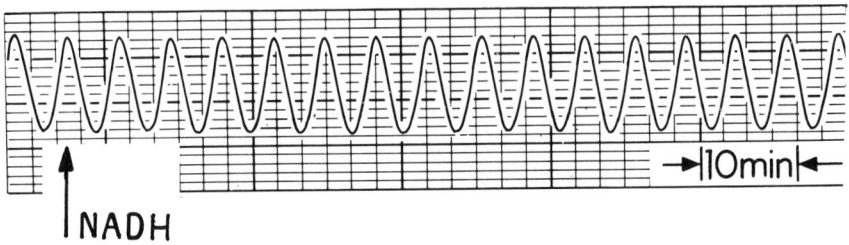

Figure 1. Portion of glycolytically-associated oscillations of NADH from a cell-free extract of the yeast S. carlsbergensis.

MECHANISM OF THE GLYCOLYTIC OSCILLATIONS IN YEAST.

With this knowledge and the results of previous experiments (5), and also drawing on the mathematical model developed by Sel'kov (13), it is possible to propose the mechanism shown in Figure 2. In this diagram, which depicts three stages in a single cycle, or pulse, the <u>relative</u> concentrations of intermediates are indicated by the intensity and size of their letter symbols. It is assumed that the rate of the sugar entry step and the reactions leading to fructose-6-phosphate (F6P) synthesis are constant throughout

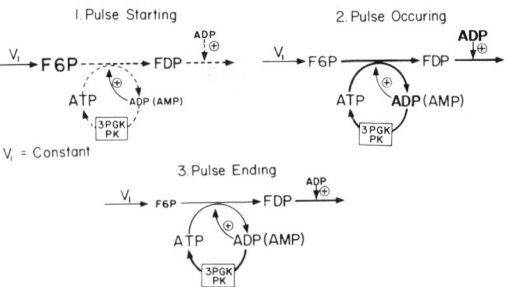

Figure 2. Mechanism of the Glycolytic Oscillations in Yeast. See Text for explanation.

and have a velocity V_1. However, this assumption might not be entirely correct since it is possible that the permease reaction may well be controlled by effectors which reflect the energy balance or the general metabolic state of the cell (see below).

As a matter of convenience the initial state depicted in Figure 2 is that where the velocity through the PFK reaction is low (broken arrow) even lower than V_1, because of the low concentration of the positive effectors ADP, and especially AMP (14), for PFK. This situation of F6P production rate being greater than its rate of utilization leads to an increasing level of F6P and, since F6P is not saturating PFK, this produces a simultaneous increase in the velocity of the PFK reaction. As the PFK reaction velocity increases the concentrations of its products, ADP and FDP will increase, together with AMP, which is in equilibrium with ADP and ATP through the myokinase reaction. AMP appears to be one of the strongest positive effectors of yeast PFK (14). Eventually the AMP concentration will approach the range where it will substantially activate PFK. At that point the activity of PFK will rapidly increase leading to an even greater concentration of its products and of AMP. This sharply increased velocity through the PFK reaction leads to a pulse of intermediates flowing through glycolysis which, in cell-free extracts at least, appears to represent a flux rate from five to seven times faster than that observed when PFK is in the deactivated state (low AMP).

During the time the pulse is occurring, the PFK activity is high, the ADP and AMP concentrations are relatively high, the velocity through the PFK reaction is high (solid arrow) and the velocity of the 3PGK and PK reactions (probably controlled by the ADP concentration) are high. The flux pulse is eventually limited by the decline in F6P concentration which occurs because the velocity of the PFK reaction is greater than V_1. The decline in F6P concentration will clearly cause the PFK velocity to decrease and thus to slow the rate of production of ADP and AMP. This, together with the still relatively high velocities of the 3PGK and PK reactions, will cause a continuing decline in ADP and AMP concentrations, which eventually will drop below their activation ranges for PFK. Thus the flux through the PFK reaction drops sharply, bringing an end to the pulse. At this point the velocity of the PFK reaction is below V_1 and the F6P concentration therefore starts to increase again. The system then returns to the condition initially described.

Several points are worth noting about this simply described mechanism, which can be characterised by the net flux diagram shown in Figure 3.

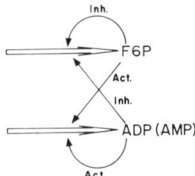

Figure 3. Net flux diagram of mechanism for the glycolytic oscillations of yeast.

1. The input velocity, V_1, is constant throughout and is poised between the maximum (activated) and minimum (unactivated) velocities of the PFK reaction.

2. The concentrations of ATP and FDP are <u>relatively</u> stable throughout. (Experiments indicate that both the FDP and ATP concentrations are relatively high and display only small <u>relative</u>-amplitude oscillations.)

3. The F6P, ADP and AMP concentrations display high <u>relative</u>-amplitude oscillations (confirmed by unpublished experiments).

4. The 3PGK reaction and possibly the PK reaction are under the control of the ADP concentration and the modifying action of Mg^{++} concentration.

5. The rate of removal of FDP is controlled by the ADP concentration via the 3PGK reaction. This reaction determines the flux through the latter part of glycolysis and is responsible for the reflection of the oscillations down the pathway and away from the PFK reaction.

6. Despite the fact that the oscillations monitored through the NADH concentration are smooth and sinusoidal in shape the flux through glycolysis represents a series of sharp transitions between high and low flux rates. This can clearly be demonstrated by measuring the output of end products of glycolysis during the oscillations.

While this mechanism satisfies and explains many of the observations made on the glycolytic oscillations in yeast, it by no means explains all of them. Furthermore, it leaves many questions unanswered. Before we can hope to modify the fine details of this simple mechanism to allow it to explain more of the behavior of the experimental system, it is first necessary to review those observations and aspects of the system in which we lack understanding. These include (a) the exact manner by which oscillations generated at the PFK reaction are reflected into the NADH concentration; (b) the possible modification of the oscillation generating system by interaction with the NAD/NADH system; (c) the factors which control frequency and amplitude; (d) the mode of generation of double-periodic oscillations; (e) the reasons for damping of the oscillations in intact cells; (f) the significance of metabolic coupling between individual cells on the appearance of oscillations in intact-cell populations, and (g) the mechanism of rapid phase shifting by ADP additions in cell-free extracts. In the remainder of this paper I would like to present experimental data on some of these areas in which we are clearly limited in our knowledge.

Reflection of the Oscillations.

The results of an experiment which sheds light on the characteristics of the mechanism by which oscillations generated at the PFK reaction are reflected into the NADH system are given in Figure 4. In this experiment equal amounts (0.35mM) of 3PGA, PEP, pyruvate and acetaldehyde were added separately to a cell-free extract of yeast which was generating stable oscillations. The point of addition in each case was at the NADH minimum in the cycle. This point is easily recognized and thus each addition was made while the system was in essentially an identical metabolic state.

The results show that 3PGA, 2PGA (not shown) and PEP additions have a significantly greater effect on the system than does the equivalent additions of pyruvate and acetaldehyde. The intermediates prior to the pyruvate kinase reaction cause an immediate and substantial decrease in the NADH level while equivalent pyruvate and acetaldehyde additions cause no decline in NADH but simply generate a short delay and a decreased amplitude for the next cycle. This clearly demonstrates that 3PGA, 2PGA and PEP represent an interacting intermediate pool which is separated from

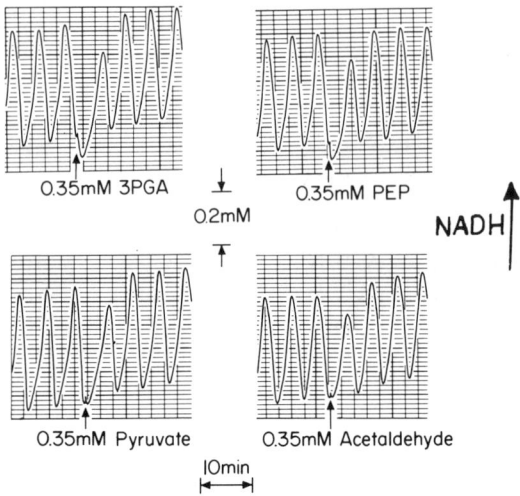

Figure 4. The additions, at the point of minimum NADH, of equal amounts of 3PGA, PEP, Pyruvate and Acetaldehyde to an oscillating cell-free extract of S. carlsbergensis.

pyruvate by the physiologically irreversible pyruvate kinase reaction. At first sight the oxidation of NADH caused by the 3PGA, 2PGA and PEP additions might be explained by the immediate conversion of these intermediates to acetaldehyde, via the later reactions of glycolysis, and the consequent stimulation of the alcohol dehydrogenase reaction. However, if this were the case, an equivalent addition of acetaldehyde should produce an equivalent or even greater effect.

The fact that acetaldehyde is less effective than 3PGA in causing an oxidation of NADH suggests two possibilities. One possibility is that the 3PGA, 2PGA or PEP is immediately converted to pyruvate, but the observed oxidation of NADH is not due to a stimulation of alcohol dehydrogenase by the increased availability of acetaldehyde but instead is brought about by a decrease in NADH production. This would probably be caused by the decrease in ADP concentration caused by the conversion of PEP to pyruvate, through the pyruvate kinase reaction. The decreased ADP would probably not affect the NADH level by any negative action on PFK since it is already known that FDP is very high in these extracts and almost certainly is not in control of aldolase (5). Thus the only other mechanism by which a decrease in ADP concentration would cause an oxidation of NADH would be by its action at the 3PGK step.

It is known (5) that ADP is at, or close to, its minimum concentration when NADH is at its minimum. Since the 3PGK reaction in general appears to be controlled by the ADP concentration it is only logical to assume that a further decrease of ADP concentration (caused by the addition of 3PGA, 2PGA or PEP) will produce an immediate decrease in the velocity of this reaction. If this occurs, the other substrate for the 3PGK reaction, 1:3 diphosphoglyceric acid (1:3DPGA), will build up. 1:3 DPGA is an effective inhibitor of glyceraldehyde phosphate dehydrogenase (GAPDH) and hence the result would be a decrease in the rate of production of NADH. In the experimental traces this would show up as a net oxidation of NADH, which is, in fact, observed.

The other possible explanation for the effective net oxidation of NADH by 3PGA addition is that 3PGA is not immediately converted to pyruvate but instead influences the 3PGK reaction directly. The 3PGK reaction is a readily reversible reaction and the enzyme is known to be in gross excess in yeast (15). Hence an increase in the concentration of the product, 3PGA, should cause an increase in the concentration of the substrate, 1:3 DPGA. Again this will inhibit GAPDH and result in a net oxidation of NADH.

Both of these possible explanations stress that the 3PGK reaction is of crucial importance in the reflection of the oscillations to the NADH concentration. The evidence points to the possibility that the oscillations in NADH arise because of the influence of an oscillating concentration of ADP (arising from the PFK reaction), which controls the 3PGK reaction, and the additional inhibitory action of 1:3 DPGA on GAPDH. It certainly eliminates the possibility that oscillations in the NADH concentration arise because of an oscillating substrate concentration for either the GAPDH or the alcohol dehydrogenase reactions. Supporting evidence for the theory that the action of 3PGA is mediated through a decrease in the ADP concentration comes from the fact that <u>addition</u> of ADP at the NADH minimum produces an exactly opposite effect to that caused by the addition of 3PGA.

Interaction of Oscillator with the NAD/NADH System.

One unexplained fact arising from this experiment is that it is possible to detect a phase-shift following the addition of 3PGA; (the interval between the peak prior to the addition and the next "normal" peak following the addition is not equal to an integral number of cycles). Such a phase shift is not observed with the addition of an equivalent amount of acetaldehyde. This might indicate that the phase-shift, which must represent an interaction with the oscillator mechanism, is generated by the ADP/ATP changes induced by 3PGA addition, influencing the PFK reaction.

However, this explanation disregards the possibility that the NAD/NADH changes themselves may interact in some indirect manner with the oscillator mechanism. Evidence that this may in fact occur comes from the results of an experiment shown in Figure 5. In this experiment increasing amounts of acetaldehyde (0.35 - 5.3 mM) were added to

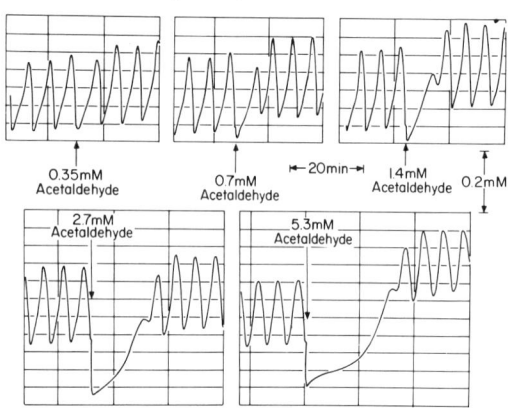

Figure 5. Effect of increasing additions of acetaldehyde on the oscillations of NADH in a cell-free extract of S. carlsbergensis. All additions were made at the NADH minimum.

an extract of S. carlsbergensis which was generating sustained oscillations. All additions were made at the NADH minimum for the reasons previously given. The results show that with the larger additions of acetaldehyde there is a larger net oxidation of NADH and with the two largest additions there is a substantial period during which the NADH oscillations are lost completely. This would indicate that acetaldehyde is normally very low in these extracts and that it controls, to a considerable extent, the alcohol dehydrogenase reaction. However, the close phase relationship of pyruvate and NADH (4) and the other arguments presented earlier, indicate that the NADH oscillations do not arise solely as a result of oscillating acetaldehyde levels but are much more dependent on the GAPDH reaction velocity. It must be assumed that the prolonged disappearance of the NADH oscillations following the addition of excess acetaldehyde (see Figure 5) represents a saturation of the alcohol dehydrogenase, as evidenced by the extreme lowering of the NADH level. Again, this indicates that under normal oscillating conditions there is a measure of control by acetaldehyde on alcohol dehydrogenase.

There is no reason to believe that the loss of NADH oscillations represents a total loss of oscillations within the system. It is most likely that the oscillations continue to be generated at the PFK reaction and that the loss of oscillations following acetaldehyde addition simply indicates interference with the NADH system by which the oscillations are monitored. However, there is evidence that either the acetaldehyde, or the changed NAD/NADH ratio, has influenced the oscillation generating reactions because significant phase shifts occur following the acetaldehyde additions. It is difficult to conceive of a direct influence of acetaldehyde on the PFK reaction and associated reactions. It is more likely that this phase-shifting occurs through the action of the changed NAD/NADH ratio on the GAPDH reaction and consequently the 3PGK reaction.

One unexplained result of the above experiment is the eventual return of the system to stable oscillations. The latter half of glycolysis is a closed loop system with NAD and NADH linking the two dehydrogenase reactions. Any intermediate added into this loop should cause a permanent change in the total concentration of intermediates within the loop. The fact that the system apparently returns to its previous condition indicates that there are reactions, other than the glycolytic reactions, occurring in the extract which can significantly alter the concentrations of intermediates in the latter half of glycolysis. Thus, cell-free extracts cannot be regarded as purely glycolytic systems.

Control of Frequency.

While it has long been known that the frequency of the glycolytic oscillations are temperature-dependent, with an approximate doubling of frequency for a 10°C rise in the physiologic range of temperature (16), it is not entirely clear which individual parameters in the oscillatory mechanism might control the frequency. Hess and Boiteux (4) have pointed to the importance of the substrate entry rate in the control of frequency in cell-free extracts. With intact cells it is more difficult to control the rate of entry of hexose into the cell except by the use of specific inhibitors for the permease, such as methylphenidate. Experiments with these inhibitors to determine the effects of decreased hexose entry on the oscillation frequency have proved to be inconclusive.

However, another possible approach to this problem is to use the specificity of the permease for the different sugars to achieve different hexose entry rates. It is known, for example, that fructose enters yeast cells slower than glucose (17) and that the rate of entry is highly concentration dependent even up to 10-20 mM.

The results of using this technique to vary hexose entry rates into intact yeast cells is shown in Figure 6. It can be seen that on adding 25 mM glucose to an intact cell suspension of S. carlsbergensis the resultant oscillations in NADH concentration had a frequency of close to 3 min^{-1}. When 25 mM fructose was added to an identical suspension of yeast under exactly the same conditions the frequency was less than 2 min^{-1}. Since it is known that fructose enters the yeast cell more slowly than glucose (17) this result would indicate that the rate of hexose entry into the cell is directly related to the oscillation frequency.

To confirm this finding for the intact cell different concentrations of glucose, mannose and fructose were added to relatively dilute yeast cell suspensions (to eliminate very rapid changes in sugar concentration) and the frequency of the initial oscillations was observed. Figure 7 shows the results of this experiment, which was carried out at 28°C, plotted as the frequency versus the initial sugar concentration. It can be seen that in all cases a saturation-type curve is produced. However, in the case of mannose

BIOCHEMICAL OSCILLATORS

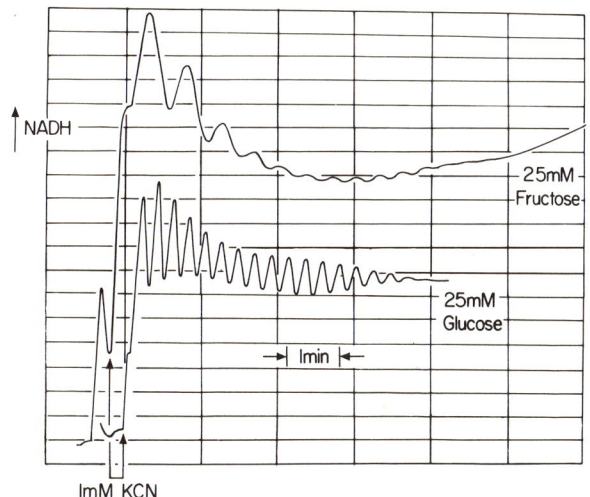

Figure 6. The effect of adding 25 mM glucose (bottom trace) or fructose (top trace) to intact cell suspensions of S. carlsbergensis under identical conditions. The traces were obtained by monitoring the fluorescence of intracellular NADH.

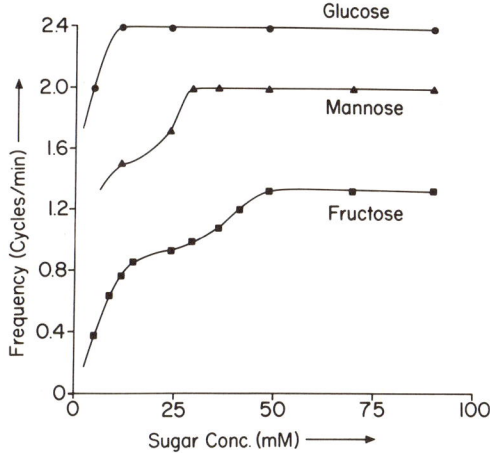

Figure 7. Plot of initial oscillation frequency against initial sugar concentration for glucose, mannose and fructose. The NADH oscillations were observed in intact cell suspensions of S. carlsbergensis at 28°C

and fructose at least, the curves produced are not the simple
rectangular hyperbolas usually associated with enzyme cata-
lysed reactions or permease mediated transport. The interpretation of these curves is not clear and they may well represent a complex output produced by multiple interactions
between cellular metabolism and the hexose transport system.

Multi-periodic waveforms.

Perhaps one of the most intriguing areas in which we
have little knowledge is the means by which double-periodic
and multi-periodic oscillations are produced. Generally
these are seen in cell-free extracts (12) but on occasions
double-periodic oscillations have been observed in intact
cells. The trace shown in Figure 8 is a typical example of
multi-periodic oscillations which were generated in a cell-
free extract of S. carlsbergensis supplemented with trehalose
(50 mM). The appearance of these oscillations has, in our
hands, been unpredictable but, as in this case, they often
degenerate to single frequency oscillations. Because it has

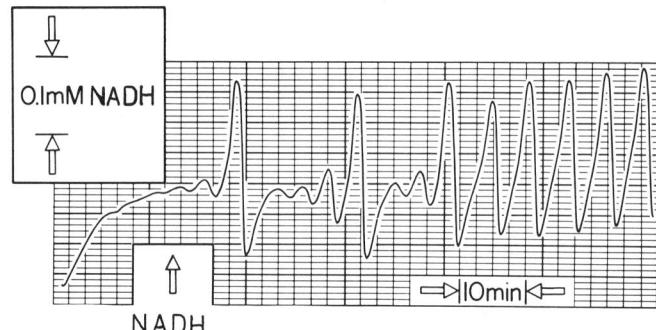

Figure 8. An example of a multi-periodic oscillation
which degenerated into single frequency oscillations.
This trace was obtained from a cell-free extract of
S. carlsbergensis supplemented with trehalose (50 mM).

been difficult to reproduce these phenomena at will they
have not been well studied and we can only speculate as to
the mechanism which generates them. Clearly it must be a
mechanism which generates the more common single frequency
oscillations but it can only be suggested that the time
taken to revert to the single frequency oscillations

represents the time required to build up or deplete a metabolic pool.

Damping of Oscillations in Intact-Cell Suspensions.

It is well known that the glycolytic oscillations in cell-free extracts can be sustained for many hours and for well over 100 cycles (12). While it is generally assumed that oscillations are generated by an equivalent mechanism in both the intact cell and the cell-free extract, it is of interest to note that the oscillations in intact cells have not been sustained for more than 30 minutes, nor for more than 48 cycles (5). This greater tendency for damping shown by the intact cell system may indicate a greater potential for secondary and more complex metabolic interactions which may, after a time, override and control the oscillation mechanism. Evidence for this possibility is given in Figure 9, which shows the rate of ethanol production by intact cells during glycolytic oscillations. The NADH oscillation pattern is superimposed on the graph for convenience. The striking observation, which has been repeated many times, is that when the oscillations damp out there is a sharp drop in the rate of ethanol production. This occurs despite the

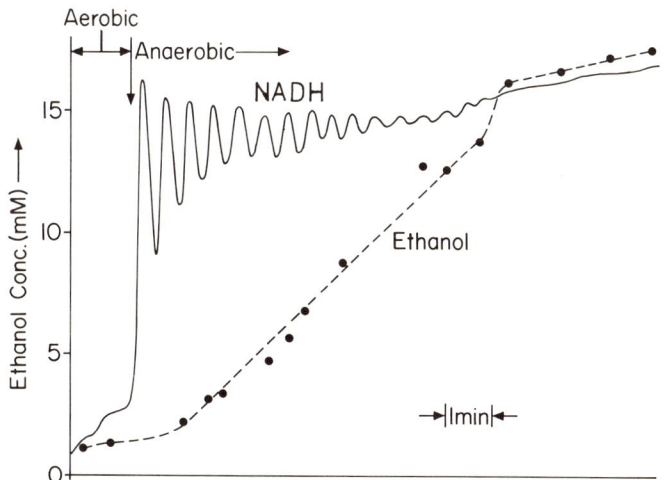

Figure 9. Ethanol production during glycolytic oscillations by an intact cell suspension of S. carlsbergensis. Initial glucose concentration was 100 mM. Both the ethanol and NADH data are on the same time scale.

fact that glucose is still present in gross excess (>50 mM). The only explanation for this is that the cell has switched to a new, more inhibited, metabolic state. What causes this switch is still unknown but it would appear to be due to a mechanism which overrides the oscillatory state. Possibly the inhibited state represents an "energy-rich" state which is not capable of being formed in cell-free extracts because of the less complex state of the system.

One other mechanism might produce damping in intact cell suspensions and that is if the individual cells become desynchronized. Should that happen the NADH fluorescence output from the entire population would be the average for all cells and would probably remain relatively stable even though individual cells in the population were still oscillating with full amplitude. To check this possibility a simple mixing experiment was performed (5) which indicated that the individual cells within the population remained metabolically coupled and synchronized. Later, more sophisticated experiments revealed many of the characteristics of this coupling phenomenon (18). In general, it appears that cells which are on the rising portion of their NADH oscillation can dominate and phase-shift cells which are on the descending portion of their NADH oscillations. Thus if two populations of cells are mixed, with one population having its NADH oscillation 180° out of phase with the other population, the new, mixed population will show full amplitude oscillations, the phase of which will be close to or identical with the phase of the dominant population.

CONCLUSION

The purpose of this paper has been to put into perspective our knowledge of the glycolytic oscillations shown by intact cells and cell-free extracts of the yeast S. carlsbergensis. An attempt has been made to identify those areas where our knowledge is limited in the hope that this will stimulate further research. While there is still much to be learned concerning the fine details of the underlying mechanism for the glycolytic oscillations, perhaps the most promising new line of study involves the search for the mechanism of intercellular metabolic communication which has been recognized through the use of oscillating cell suspensions.

ACKNOWLEDGEMENTS

The author is pleased to acknowledge the friendly interest, help and advice of Dr. Britton Chance throughout the period of these studies. This work was supported in part by USPHS GM 12202, 5K04 GM 46227 and NSF GB-8270.

REFERENCES

1. Ghosh, A. and Chance, B. Biochem. Biophys. Res. Commun. 16, 174 (1964).
2. Chance, B., Schoener, B. and Elsaesser, S. J. Biol. Chem. 240 3170 (1965).
3. Chance, B., Ghosh, A., Higgins, J.J. and Maitra, P.K. Ann. N.Y. Acad. Sci. 115, 1010 (1964).
4. Hess, B. and Boiteux, A. In Regulatory Functions of Biological Membranes. Ed. J. Jarnefelt. Biochim. Biophys. Acta Library, 11, 148 (1968).
5. Pye, E.K. Can. J. Botany. 47, 271 (1969).
6. This volume.
7. Duysens, L.N.M. and Amesz, J. Biochim. Biophys. Acta 24, 19 (1957).
8. Chance, B. The Harvey Lectures, Series XLIX Academic Press, New York, 1955, p.145.
9. Chance, B., Estabrook, R.W. and Ghosh, A. Proc. Natl. Acad. Sci. 51, 1244 (1964).
10. Higgins, J. Proc. Natl. Acad. Sci. 51, 989 (1964).
11. Chance, B., Hess, B. and Betz, A. Biochem. Biophys. Res. Commun. 16, 182 (1964).
12. Pye, E.K. and Chance, B. Proc. Natl. Acad. Sci. 55, 888 (1966).
13. Sel'kov, E.E. European J. Biochem. 4, 79 (1968).
14. Sel'kov, E.E. and Betz, A. This volume, p.197
15. Pye, E.K. Ph.D. thesis, Manchester University, 1964.
16. Betz, A. and Chance, B. Arch. Biochem. Biophys. 109, 579 (1965).

17. Lynen, F., Hartmann, G., Netter, K.F. and Schuegraf, A. In <u>Regulation of Cell Metabolism</u>. Ciba Foundation Symposium. Editors G.E.W. Wolstenholme and C.M. O'Connor. Churchill, London, 1959. p.256.

18. Ghosh, A.K., Chance, B., and Pye, E.K. Arch. Biochem. Biophys. <u>145</u>, 319 (1971).

SYNCHRONIZATION PHENOMENA IN OSCILLATIONS OF YEAST
CELLS AND ISOLATED MITOCHONDRIA.

B. Chance, Gary Williamson, I. Y. Lee, L. Mela,
D. DeVault, A. Ghosh and E. K. Pye

Johnson Foundation, University of Pennsylvania,
Philadelphia, Pennsylvania 19104.

The practically continuous nature of glycolytic oscillations observed in yeast cell suspensions (1-3) appears to require a remarkable similarity of the biochemical oscillators in individual cells of the population, or some internal synchronizing mechanism. Evidence for population heterogeneity, that would require a cell-synchronizing mechanism for continuous oscillations to be observed, should be afforded by direct fluorometric observations of the nature of the oscillations in a single cell, as compared with that in a population of cells.

This paper describes a microfluorometric method for observing glycolytic oscillations in single yeast cells in terms of periodic fluctuations of the NADH fluorescence.

Other methods for determining the synchrony of the oscillating populations have been evaluated. The first of these is provided by a small "chemically pure" (see below) temperature jump which can act as a synchronizer for the cell population. Alternatively, mixing of cell populations, in which the oscillations show varying degrees of phase shift with respect to one another may also be used to test not only the cell synchronizing properties of the preparation, but also to evaluate the important point as to whether the singular point of the phase plane profile is stable or unstable. Lastly, similar phenomena are investigated in the oscillations of K^+ and H^+ in suspensions of pigeon heart mitochondria.

<u>Microfluorometric observation of NADH oscillations in single yeast cells</u>. Microfluorometric techniques are

available which permit the measurement of the oxidation-reduction state of mitochondrial reduced pyridine nucleotide. This technique, which has been employed successfully for the measurement of the oxygen-nitrogen transition in single Baker's yeast cells (4), has not been previously tested for the detection of redox changes in the cytosol of the anaerobic cell. However, measurement of suspensions of cells by fluorometry on a mass scale clearly show changes of fluorescence due to cytosolic NADH. On this basis, it would seem reasonable that the microfluorometric method would prove suitable for recording the oscillations of cytosolic NADH in single cells.

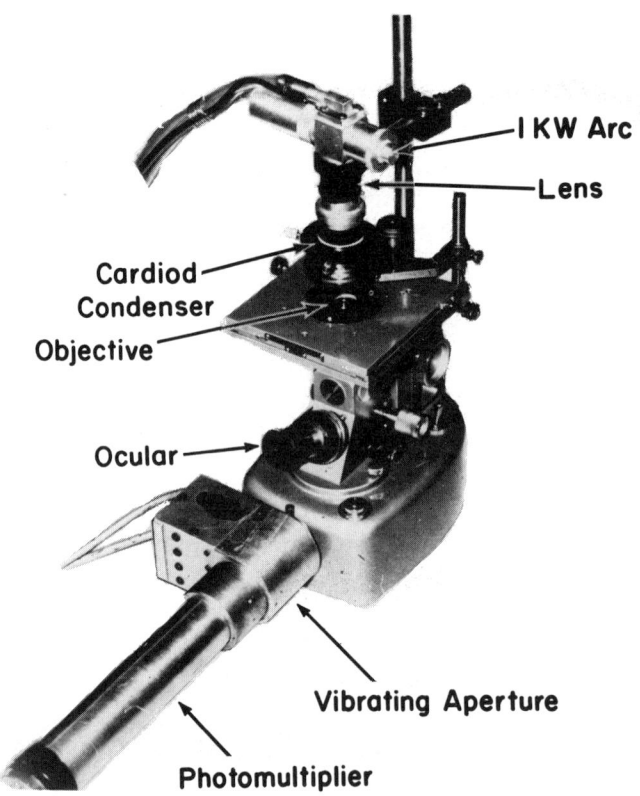

Figure 1. Fluorescence microscope set-up suitable for recording oscillations of cytosolic NADH in single yeast cells.

Figure 1 shows the microscope, which is similar to that employed in our earlier experiments (5); this geometry of the optical components has proved satisfactory for many types of experiments (6,7).

A technical problem in these studies has been one of manipulation: in order to observe the early phases of the oscillations it is necessary to minimize the time between initiation of the oscillations by consecutive additions of glucose and cyanide, and the first observation under the fluorescence microscope. Thus, the manipulations involved in preparing and mounting the yeast cells, aligning and focusing the microscope, and starting the recording have been simplified and, with practice, it is possible to make the first observations within two minutes of the moment of starting the oscillations.

A microfluorometric recording of oscillations in a single cell of S. carlsbergensis two minutes after adding glucose and cyanide is illustrated in Figure 2.

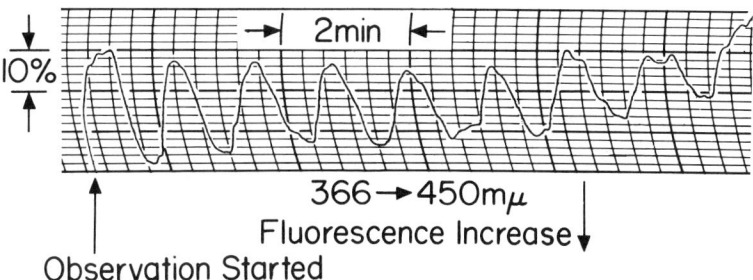

Figure 2. An example of a series of continuous oscillations of cytosolic NADH in a single yeast cell, recorded with the microfluorometer.

The important features of this record are the reasonably large number of oscillatory cycles and the relatively constant amplitude (small damping factor). In fact, in this case, as in a number of other experiments, the recording of the oscillations was terminated by the Brownian movement of the cell away from the 4 μ aperture of the microfluorometer, rather than by an actual cessation of oscillatory glycolytic reactions.

While Figure 2 provides good evidence for the oscillatory phenomenon in a single member of the yeast cell population, the relationship of the oscillations of the single cell to those of the population must also be considered. Visual observation of the field indicated the absence of non-oscillating or weakly oscillating cells, so that no gross heterogeneities of the population were observed. The relative phase angles of the oscillations in various cells were monitored by shifting from one cell to another in the field of the microscope. No clearcut phase shifts were observed.

A more effective way of determining the population heterogeneity is to compare the damping factor of the cell population from which the individual cell of Figure 2 was taken. If, indeed, the damping factor of the individual is less than that of the population, asynchrony of the cellular oscillators may be involved. The results of such an experiment are indicated in Figure 3 where it can readily be seen that the single cell (lower trace) has a somewhat lower damping factor than that of the population (upper trace).

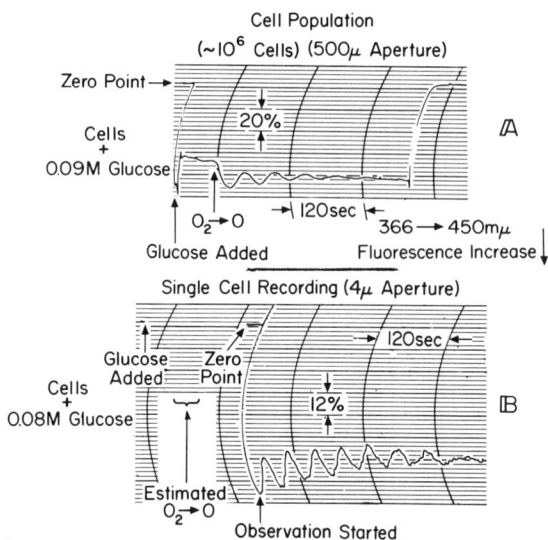

Figure 3. An illustration of the damping factor of a single cell (lower trace) from a population of cells whose damping factor is indicated on the upper trace.

In order to determine roughly the heterogeneity indicated by the larger damping factor of the cell population, one of us (DeVault) has derived a simple equation in which the difference of damping factors of the two cell populations is related to the frequency dispersion of the oscillator $\Delta\omega/\omega_o$.

$$\ln \frac{D_e}{D_o} = \frac{2\pi(\Delta\omega)^2}{\omega_o}\left(t_o + \frac{\pi}{\omega_o}\right) + \frac{4\pi^2(\Delta\omega)^2}{\omega_o^2} n \cong D_e - D_o$$

For the ensemble, the frequencies (ω) are assumed to be distributed according to:

$$P(\omega) = \frac{1}{\sqrt{2\pi(\Delta\omega)}} \exp\left\{-\frac{(\omega-\omega_o)^2}{2(\Delta\omega)^2}\right\}$$

The damping factor = $\frac{\text{amplitude of nth peak}}{\text{amplitude of (n + 1)th peak}}$

D_o = damping factor of individuals

D_e = damping factor of ensemble

t_o = time at which 0th peak occurs (at t = 0, all are synchronized).

When the damping factor is plotted against the cycle number (Figure 4A), the difference between the cell population and the individual is clearly indicated. The difference of damping factor is plotted in Figure 4B and extrapolates to the zeroth cycle with a $\Delta\omega/\omega_o$ value of 7%, a remarkably small value in view of the heterogeneous population of cell types to be expected in cells harvested at the end of the logarithmic growth phase. This value is, perhaps, best interpreted as a small lack of synchrony within the population, which, when viewed in regard to the large physical and physiological heterogeneity of the cells, would strongly suggest that an intercellular synchronization process is in operation in the oscillating yeast cell suspensions.

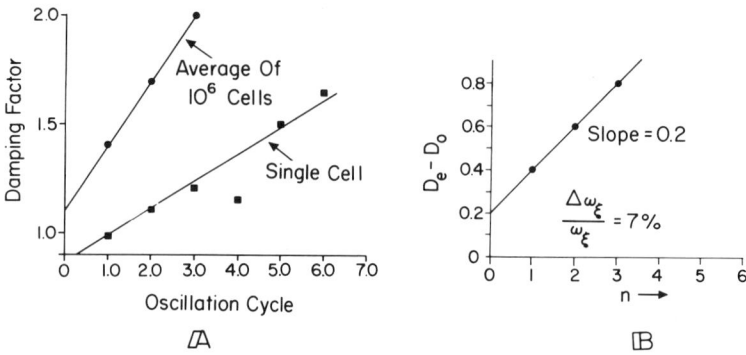

Figure 4. Comparison of the damping factor of NADH oscillations in a single cell and cell populations of S. carlsbergensis.

Effect of a Temperature Jump. In order to determine whether or not synchronization phenomena are sometimes inadequately operative in the yeast cells, and whether synchrony may be established by a sudden shock or jump in one of the parameters of the oscillatory phenomena, we have developed a simple apparatus for applying a temperature transient to the yeast cell suspension.

Here, the most important consideration is the achievement of a "chemically pure" temperature perturbation. Unlike the use of the temperature jump for rapid reactions, as perfected by Eigen and his co-workers (8), which involves the diffusion of electrolysis products from "Joule-heating", we desired a temperature perturbation that would have no long-term effects upon the cell suspension. This method, employed for the study of bioluminescence phenomena by Kreezer and Kreezer (9) is employed by us in the form indicated in Figure 5. Heat is generated by a 200 joule flash lamp which surrounds a blackened silvered cuvette containing 0.7 ml of the cell suspension which is rapidly stirred, magnetically. Excitation for the fluorometer illuminates the surface of the suspension and the emitted fluorescence is imaged upon a photomultiplier. A light-tight box avoids any "cross-talk" between the light flash and the fluorometer. Thus, observations can be made throughout the interval of the flash and continuously thereafter.

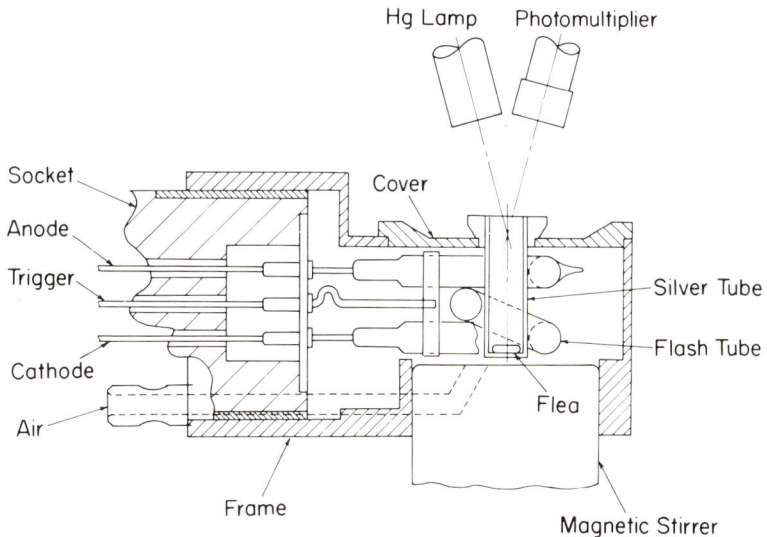

Figure 5. The flash heating apparatus employed for applying the temperature jump to an oscillating yeast cell suspension.

Figure 6 indicates the application of this technique to the initiation of oscillations in a suspension of S. cerevisiae, rendered anaerobic through additions of cyanide and glucose, but showing no development of persistent oscillations following the first cycle or the first temperature jump. However, the subsequent temperature jump gives a long train of damped oscillations while small oscillations appear following the third and none appear after the fourth temperature jump. These data suggest that the condition of the cells was such that their synchronizing mechanism was not operating effectively and that the temperature transient synchronized an otherwise non-oscillating preparation. The possibility that the temperature transient initiated an otherwise non-existent oscillation in the cells is not eliminated by these experiments, but is rendered less likely in view of previous observations which indicated that the individual cell oscillated for a longer period than did the population.

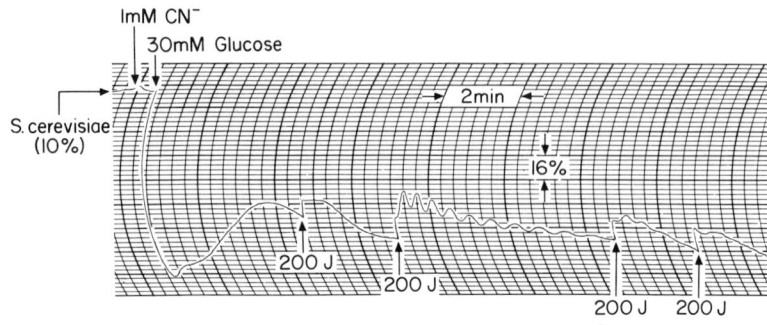

Figure 6. The application of temperature-jump and glucose pulses to the activation and de-activation of oscillations in suspensions of S. cerevisiae.

Mixing experiments. As reported by Pye (this volume) preliminary experiments on the response of the yeast cell oscillations to out-of-phase mixing suggests that the synchronization process between cells is rapid and effective. In order to pursue this phenomenon further, a mixing apparatus has been devised. The apparatus consists of two cuvettes which are mechanically moved rapidly in and out of the fluorescence excitation light. Thus, practically simultaneous readout of the oscillatory waveforms from yeast suspensions in the two cuvettes is possible. The oscillations in the two populations are initiated with a time difference of one-half cycle, or any other desired phase shift. In order to mix the contents of the two cuvettes the yeast suspensions are drawn up through a jet mixing chamber into a syringe and then discharged back again so that equal volumes of the thoroughly mixed contents fill the two cuvettes. The mixing time of a fraction of a second is short compared to the approximately three seconds required for mechanical emptying and refilling the cuvettes. Even so, this is only a small part of the oscillation cycle and, as the following records show, it does not interfere with the recording; see (10).

Figure 7 indicates, on phase plane diagrams, the phase relationships of the two suspensions - before and after mixing. It is apparent that the approximately 90° phase

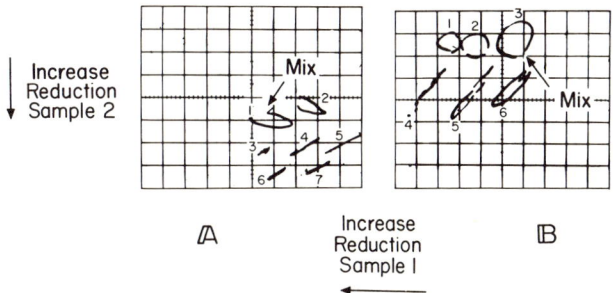

Figure 7. Phase plane plots of the fluorescence intensity of two yeast samples plotted one against the other. In A, the successive oscillations are displaced one from another manually by X-Y controls on the oscilloscope, so that each cycle is individually represented. The progression of cycles is 1 → 7. In B, the progression of cycles is similarly labelled 1 → 6.

shift between the two suspensions is reduced to zero and remains near this value, except for a small phase drift between the suspensions, with respect to each other, seen in Figure 7B, which is not completely understood at present.

A test of the response of the individual oscillating suspensions to mixing is shown in Figure 8. On the right-hand side is indicated the mixing of two suspensions in which the phase difference is 70°. It is apparent that there is no substantial change in the amplitude of the oscillations prior to and following mixing. (It should be noted that the amplification of the two channels is not identical, but nevertheless, the ratio of the amplitudes before and after mixing is not greatly altered.)

In the left-hand portion of the figure, however, the phase difference is 210°. At the point of mixing, one phase angle is 80°, and the other is 290°. After mixing, there is "practical silence" for at least two periods; then, at about the time that the third or fourth period would be expected, oscillations are observed. The phase angle of these oscillations is not obviously related to that of either of the parent oscillations; both have been considerably shifted in this case.

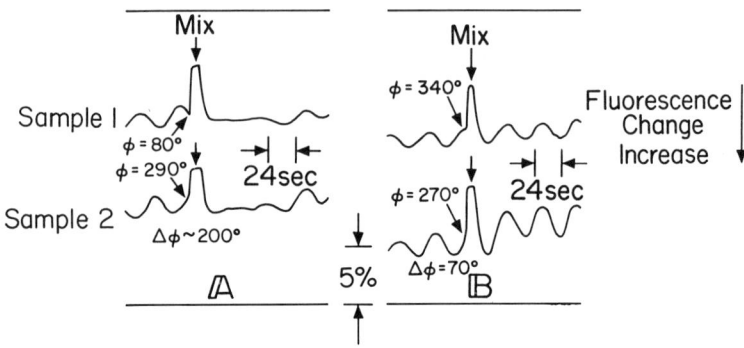

Figure 8. A plot of the individual oscillations following mixing of S. carlsbergensis cell suspensions at a phase difference of 70° (right-hand side) or mixing at a phase difference of 210° (left-hand side). In this case, two fluorometers record separately the oscillations of their contents. At the point labelled "mix" on the diagram the contents are intermixed and then equal portions are returned to the two cuvettes. The suspensions of S. carlsbergensis were at a concentration of 80 mg wet weight/ml in 50 mM TRA buffer, pH 7.5. 90 mM glucose and 0.9 mM KCN were added sequentially to initiate the oscillations.

Of the various agents that might readily permeate the cell, and therefore act as metabolic synchronizers, the most likely candidates would be pyruvate and acetaldehyde. Experiments reported by Betz at this meeting suggest that either would be suitable to advance the phase of the oscillation, even at rather low concentrations. These results, although carried out on a different strain of yeast, are suggestive of a synchronization mechanism.

Synchronization of mitochondrial oscillations. Figure 9 illustrates the continuous oscillations that can be readily observed in terms of a K^+/H^+ oscillation in pigeon heart mitochondria in the presence of valinomycin (11). The continuous nature of this oscillation is indicated by the increase in amplitude of the sinusoids.

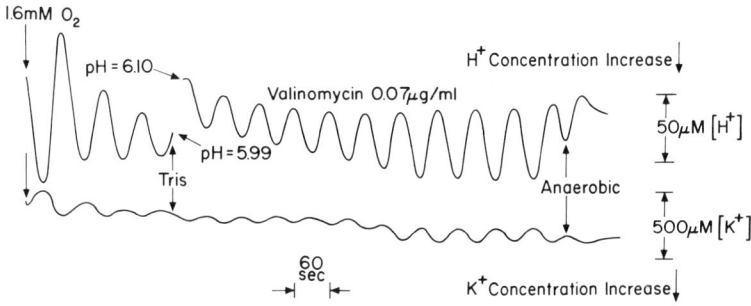

Figure 9. An example of continuous oscillations of K^+ and H^+ obtained in pigeon heart mitochondria. The mitochondria were at a concentration of 2.5 mg protein/ml in a medium, at pH 6.11, containing 0.225 M mannitol, 0.075 M sucrose, 2 mM malate-tris, 2 mM glutamate-tris, 2 mM phosphate-tris and 7 mM KCl. Valinomycin was added to give 0.07 µg/ml.

The phase angles of some of the constituents are shown in Figure 10. It is apparent that this material has at least two of its oscillatory parameters in the extra-mitochondrial space and, thus, should show large phase changes on mixing.

Figure 11 shows three conditions of mixing: 190° phase angle, 40° phase angle, and mixing with a non-oscillating suspension. It is apparent, from this record, that the out-of-phase mixing stops the oscillatory phenomenon, at least for an entire period (and for longer, as shown in other experiments). However, the mixing of suspensions with a 40° phase angle does not stop the oscillatory phenomenon, nor does mixing with a non-oscillatory suspension. This behavior seems to differ from that of the yeast cells.

DISCUSSION

Two general types of response are observed in the mixing experiments, and these presumably depend upon the nature of the singular point in the phase plane diagram for the oscillations. If, as in Figure 12A, the singular point

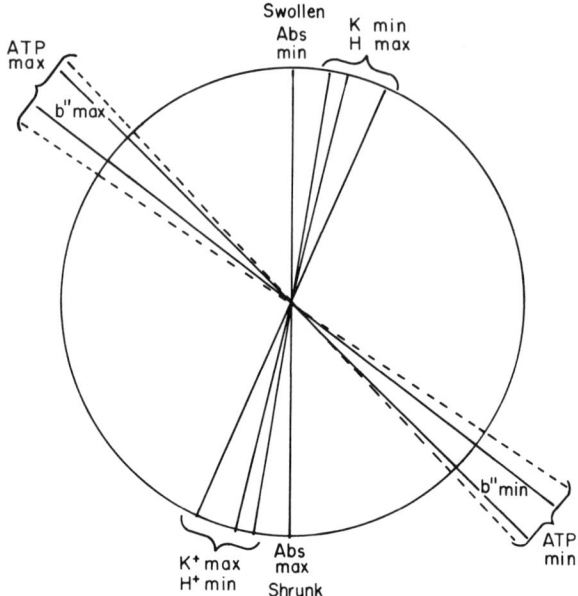

Figure 10. Experimentally determined phase angles for several components of the pigeon heart mitochondrial oscillations. The phase angles are plotted with respect to the absorption minimum at a phase angle of 0°; the K^+ minimum and the H^+ maximum are indicated, followed by the minima of the ATP and reduced cytochrome $\underline{b}$, and so forth around the circle, representing the relative phase angles of the oscillatory components. It should be noted that two of the parameters are measured in the extra-mitochondrial space, namely, the K^+ and H^+ changes.

is unstable, then mixing of oscillating suspensions, or even non-oscillating suspensions, will cause reactivation of the oscillations. Depending on the amplitude with which the phase angles are balanced, one against the other, the amplitude of the oscillations will drop to a low value, but maintain some vestiges of the phase relationships of the prior oscillation, particularly when one of the two populations has a phase angle from which it is not "shiftable". If, however, the mixing is appropriately out of phase, the oscillations may die away completely.

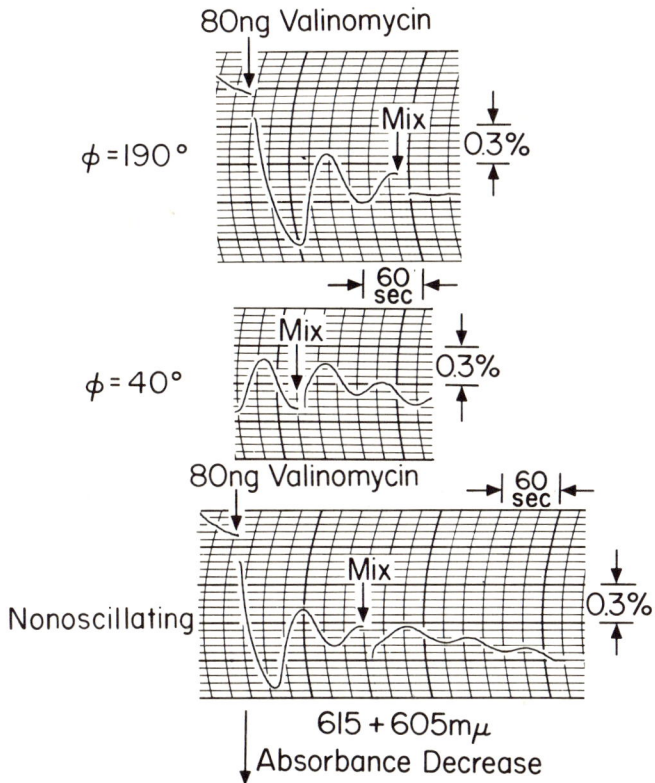

Figure 11. Effects upon the oscillations of mixing different mitochondrial suspension. Top trace, mixing at 190° phase angle; middle trace, mixing at 40° phase angle; bottom trace, mixing with non-oscillating suspensions. The oscillations are measured by changes of H^+ concentration, as observed by the change of bromthymol blue absorption measured at 605-615 nm. An absorbance decrease corresponds to decreasing alkalinity.

If the external synchronizer has a strong control over the oscillations, then an approach to the singular point may occur, and the restarting of the oscillations may depend upon "noise" phenomena which are usually involved in the initiation of oscillations. Under these conditions, it is probable that the phase angle of neither of the suspensions will be retained. This is apparently the case in Figure 8A. The observation of this phenomenon identifies an instability of the singular point of the yeast oscillations.

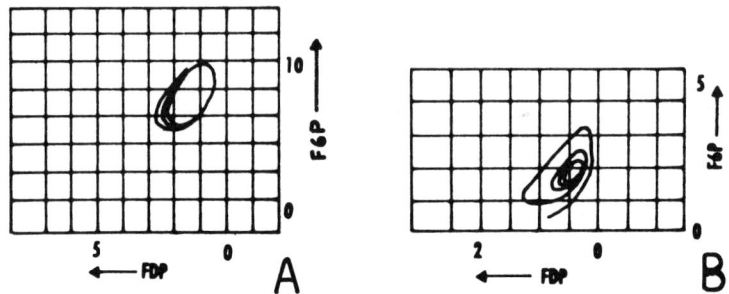

Figure 12. A. Limit cycle with an unstable singular point. B. Limit cycle with a stable singular point.

In Figure 12B there is a stable singular point, so that mixing of out-of-phase oscillations will yield stabilized, lower-amplitude oscillations; or, if the singular point is reached, no oscillations at all, with no possibility of restarting the oscillations spontaneously. This phenomenon, at least according to our preliminary results, is observed in the heart mitochondria.

Relation to general aspects of oscillatory control. Mixing of the out-of-phase cell suspensions immediately identifies whether the oscillations exist in two separate and miscible phases. For example, if the yeast cell oscillation existed only in a homogeneous phase within the cells, not communicating with the external medium, no alterations of oscillatory amplitude and phase would be detected on mixing. The relationship which the new phase angles would bear to the parent phases would, of course, be algebraic, instead of showing definite quadrant selectivity, as indicated above.

Mixing provides, automatically, a forcing function on as many variables as are in a common -- or external -- phase. This forcing function can identify the control properties of those variables which are in the external medium; and, as shown in the case of the yeast cell oscillation, a surprisingly great effectiveness is demonstrated for at least two variables which are present in adequate external concentrations for control, namely, pyruvate and acetaldehyde. Alternatively intercellular collisions may cause synchronization.

As clear as the experimental results for the yeast cells may be, two interpretations are possible in the case of the mitochondria, since insufficient data are presently available. The explanation above, that the singular point is reached by the mixing of out-of-phase mitochondria, must also be considered in relation to the possibility that the mitochondria do not communicate at all with the outside medium, and that the stopping of the oscillations is only apparent. While in the yeast cells this possibility was readily ruled out by an examination of the phase angles of the "daughter" oscillations, it cannot at present be ruled out for the mitochondria. However, the great sensitivity of the oscillations to the external H^+/K^+ ratios (11) suggests that a dependence of the oscillations on the external concentrations is involved.

The support of USPHS RG 12202 is gratefully acknowledged.

REFERENCES

1. Chance, B. Estabrook, R.W. and Ghosh, A., Proc. Nat. Acad. Sci. (U.S.) 51, 1244 (1964).
2. Chance, B. and Williamson, G., Fed. Eur. Biochem. Soc. 5th Meeting, Abst. 948 (1968).
3. Pye, E.K., Studia Biophysica, 1, 75 (1966).
4. Chance, B., Unpublished observations.
5. Chance, B. and Legallais, V., Rev. Sci. Instr., 30, 732 (1959).
6. Chance, B. and Thorell, B., J. Biol. Chem. 234, 3044 (1959).
7. Chance, B., Thurman, R.G. and Gosalvez, M., Förvarsmedicin, 5, 235 (1969).
8. Eigen, M. and DeMayer, L., In Techniques of Organic Chemistry, Vol VIII, Interscience, New York, 1963 p. 895.

9. Kreezer, G.L. and Kreezer, E.H., J. Cell. Comp. Physiol. 30, 173 (1967).

10. Ghosh, A., Chance, B. and Pye, E.K., Arch. Biochem. Biophys. 145, 319 (1971).

11. Chance, B. and Yoshioka, T., Arch. Biochem. Biophys. 117, 451 (1966).

IV
OSCILLATIONS IN TISSUES

OSCILLATING CONTRACTILE STRUCTURES FROM INSECT FIBRILLAR MUSCLE

J. C. Rüegg

Department of Cell-Physiology, Ruhr University, Bochum
and
Max-Planck-Institute for Medical Research, Heidelberg.

In recent years, much attention has been paid to oscillatory phenomena in insect fibrillar flight muscle (see the surveys of Pringle (1) and Rüegg (2)). The oscillatory contraction is an example of cell oscillation since it occurs according to a myogenic rhythm, which is largely independent of the rate of nervous stimulation and not in synchrony with the action potentials. The frequency is similar to the resonance frequency of the thorax wing system of the insect and may vary from about 1000 cycles sec^{-1} in small midges to 20 cycles sec^{-1} in large tropical waterbugs (Lethocerus maximum), which we used for the experiments reported in this paper.

<u>Myofibrillar oscillation</u>. The myofibrillar origin of the oscillatory performance of muscle cells was demonstrated by Jewell and Rüegg (3), who found that glycerol-extracted muscle fibers from the flight muscles of waterbugs would oscillate when suspended from a lever system with suitable resonance frequency and immersed in an ATP-salt solution. Glycerol-extracted fibers are essentially bundles of myofibrils since after the treatment with glycerol, the membrane becomes permeable and the sarcosomes and vesicles of the reticulum are largely disrupted (see (4), but cf (5)). It is noteworthy that the sequence of contraction and relaxation occurs at a constant, buffered, free calcium concentration in the ATP-salt solution. This shows that the switching on and off of contractile activity during oscillation does not require an increase and decrease in the level of ionized calcium. In these free oscillation experiments the myofibrils oscillated with a frequency near to the resonance frequency of the lever system used, but the oscillatory performance is

optimal at a particular frequency, the frequency optimum. This suggests that oscillation occurs because the oscillatory system responds to length changes after a delay with negative feedback. After a delay, a length increase evokes a contraction and the resulting length decrease leads to relaxation, so that the system will elongate once more because of the restoring pull of the spring lever.

The delayed rise in tension after extension can be demonstrated by suddenly stretching the extracted fibers in ATP-salt solution. This causes an increase in the ATPase activity also (6). It can be easily shown that the stretch activation phenomenon leads to the output of oscillatory power when the preparation is sinusoidally stretched or released, i.e. when it is driven to oscillate. In such a stretch/release cycle, extension leads also to an increase in tension after a delay, with the consequence that sinusoidal tension changes lag behind sinusoidal length changes. The size of the oscillatory work per cycle corresponds to the area enclosed by the tension length diagram in which tension is greater during release (shortening) than during stretch. This oscillatory performance requires calcium ions; at very low calcium concentrations, there is no stretch activation and no oscillatory power output.

<u>The frequency dependence of oscillatory performance</u>. This may be studied by observing the power output at various frequencies of driven oscillation, rather than by changing the resonance frequency of the recording lever system in free oscillation experiments. There is no phase angle between tension and length changes at very low frequencies of oscillation. As the frequency of sinusoidal stretch and release increases, the phase angle and the work per cycle (and the viscous modulus) increase likewise, pass through an optimum (at about 3 cycles sec^{-1}) and then decrease again. These experiments on fibers from <u>Lethocerus maximus</u> confirmed earlier experiments on <u>Lethocerus cordofanus</u> fibers (3), in which, however the frequency optimum was higher. Figure 1 summarizes experiments done in collaboration with Mr. G. Steiger. About 40 fiber bundles were sinusoidally stretched and released at an amplitude of about 2.5% fiber length and at the reference frequency of 2 cycles per second. They were then driven at one of the following frequencies: 0.1, 4, 10 or 60 cycles sec^{-1}, and the power output was expressed in percent of that at the reference frequency. (Mean and standard error of the mean.)

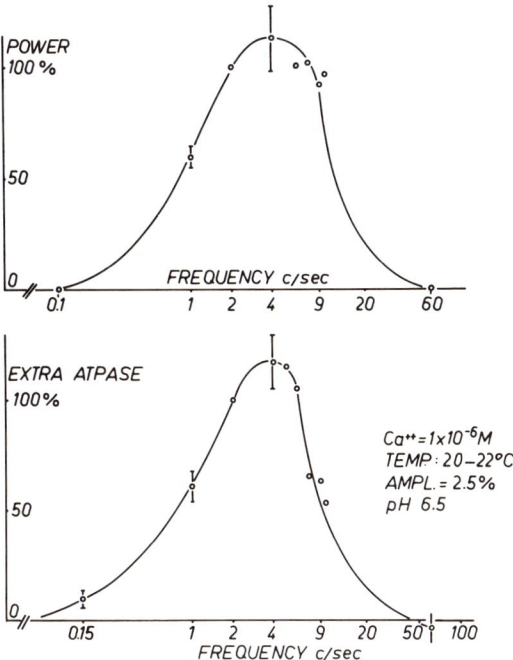

Figure 1. Coupling of oscillation induced ATPase activity of extracted muscle fiber bundles (below) and oscillatory performance or power output (above) at various frequencies of driven oscillation, at an amplitude of 2.5% fiber length. Ordinate: rate of work output (above) and rate of extra ATP-splitting (below) in percent of the rates at the reference frequency of 2 cycles sec^{-1}.

In each one of the 40 fiber bundles the ATPase activity was determined in 20 min periods before, during and after driven oscillation at the reference frequency (2 cycles sec^{-1}) in order to obtain a reference value (=100%) for the oscillation induced extra ATPase and power. This was compared with the oscillation effect at another frequency in the range of 0.1 to 60 cycles sec^{-1}.

The preparation (four to five fibers in a bundle of glycerol extracted dorsolongitudinal muscle of Lethocerus maximus, 0.5 cm long and containing about 10-20 µg protein) was glued to a tension transducer and a vibrator and suspended in 0.3 ml activating solution containing 30 mM KCl, 20 mM histidine buffer, 10 mM MgCl$_2$, 10 mM ATP, 4 mM calcium EGTA buffer, 10 mM Na azide, pH 6.5, pCa 6.0, 21-23°C and mostly o.1 mg/ml adenylate kinase (15).

305

In an attempt to determine whether the frequency dependence of oscillatory performance is associated with and possibly due to a frequency dependence in the rate of energy liberation, we determined the ATPase activity in fibers during periods of rest (each about 20 minutes) and periods of oscillation. The latter caused extra ATPase activity in addition to the basic ATPase observed during rest periods. The oscillation-induced extra ATPase activity is frequency-dependent and parallels the power output at the various frequencies. This suggests that the extra ATPase activity is indeed oscillation-coupled; it is an oscillatory ATPase. The close coupling of ATP-splitting and mechanical events implies also that the oscillation optimum occurs at that frequency at which the energy release is optimal; then the ATPase activity is maximal and the chemo-mechanical energy transformation occurs with the best efficiency (approaching about 3 kcal/mole of ATP split). The frequency dependence of the efficiency is not so obvious in the experiments shown in Figure 1, but it is rather striking at high calcium concentration (about 10^{-5}M). Under these conditions there is about 1 kcal of work produced per mole of ATP split at 2 cycles sec^{-1}, 2-3 kcal at 10 cycles and 0.5 kcal at 15 cycles.

The effect of temperature on the oscillatory performance.
In the experiments shown in Figure 1, the frequency optimum for power output is about 5 cycles sec^{-1} at 20°C while it must be about 20 cycles sec^{-1} in the living waterbug muscle *in situ*, since this is the wing beat frequency during flight of the large tropical waterbugs (7). It seems, however, that in living muscle during flight the temperature is about 30-35°C (8). In glycerinated fiber, an increase in temperature to 35°C increases the frequency optimum for oscillatory power output to about 20 cycles sec^{-1} at an amplitude of 2% muscle length, provided that ATP-regenerating systems are added; if not, ATP starvation occurs and the fibers enter the high tension state described by Jewell and Rüegg (3). At room temperature, the ATP supply is not rate-limiting (9). It may be mentioned in passing that an increase in free calcium concentration also increases the frequency optimum but only slightly so. Both the temperature and the calcium effects are due to a shortening of the delay between length changes and tension changes.

The dependence of oscillatory performance on the amplitude of oscillation. This phenomenon has been described in detail by Pringle (1). It is noteworthy that an increase in the amplitude of oscillation scarcely affects the frequency optimum of oscillation and energy release, suggesting that it is the increase in frequency per se and not the increase in the sliding rate of filaments which leads to a diminuition in ATPase activity and power above the optimum. At constant frequency, the work per cycle and the oscillatory power often increase in proportion to the square of the amplitude during driven oscillation (3) and so does the ATPase activity (11) which is apparently coupled to power output (6). The maximal possible amplitude is about 4% of the fiber length, which is more than one would expect on the hypothesis that oscillation occurs by means of synchronous beating of actinomyosin crossbridges in a one beat per cycle manner (cf discussion to reference (1)). This is also unlikely since more than 2 or 3 molecules of ATP may be split per crossbridge in a cycle of oscillation (10).

The mode of stretch activation. So far we have seen that oscillation occurs because changes in length are followed by delayed changes in tension. It is now pertinent to consider how the length signal input might be transformed in the myofibril into the tension signal output. The available evidence suggests that upon stretching, the length increase is sensed by the passive, tension-bearing strained myosin filament which is attached to the Z-lines in insect flight muscle (2). The result of strain is apparently an increase in the extent of actin-myosin interaction leading to an increase in both tension and ATPase activity of myofibrils (10). The activity increase is not due to an increase in the calcium affinity of the contractile system as suggested by Chaplain (12), since stretch activation is observed even at 'enzyme-saturating' concentrations of free calcium (10^{-5}M; see (11)). The implication of ADP in stretch activation has been discussed by Pringle (1) and discarded by Rüegg (2). It seems possible, though, that stretch activation involves the removal or displacement of an inhibition, analogous to that removed by Meinrenken (13) when he extracted from glycerinated insect fibers a 'contraction inhibitor' together with Ebashi's factor, with Tris buffer, pH 8.

Conclusion: Coupling of ATP-ase activity and mechanical oscillation in myofibrillar bundles. To summarize then the findings relevant to the problems of biochemical oscillations, we recall first the ATPase activation of insect myofibrils by stretch. At the muscle fiber length allowing optimal oscillatory performance of myofibrils (about 102% L_o), the ATPase activity changes by about 30% for 1% length change, implying that during mechanical oscillation the myofibrillar ATPase activity increases and decreases in an oscillatory fashion and in synchrony with the length changes. This point is, however, difficult to establish without using very fast recording methods. We should also recall the increase in the mean ATPase activity of myofibrils which occurs at the onset of mechanical oscillation and which apparently liberates the energy required for the output of mechanical power during oscillation. This 'oscillation induced' ATPase activity is obviously oscillation-coupled, since it depends on the oscillatory mechanical performance when the latter changes with frequency.

Finally, we might point out that the described oscillation phenomena occurring in a simple biochemical system (myofibrillar bundles, glycerol, extracted fiber) are not necessarily a unique oddity of insect flight muscle contractile systems. In view of some recent findings by Armstrong et al (14), oscillation may well be a more general phenomenon of muscle.

REFERENCES

1. Pringle, J.W.S., Progr. Biophys. Mol. Biol., <u>17</u>, 1 (1967).

2. Rüegg, J.C. Experientia, in press.

3. Jewell, B.R. and Rüegg, J.C., Proc. Roy. Soc. B., <u>164</u>, 428 (1966).

4. Zebe, E., Meinrenken, W and Rüegg, J.C., Z. Zellforschg., <u>87</u>, 603 (1968).

5. Abbot, R.H. and Chaplain, R.A., J. Cell. Sei., <u>1</u>, 311 (1966).

6. Rüegg, J.C. and Tregear, R.T., Proc. Roy. Soc. B., <u>165</u>, 497 (1966).

7. Barber, S.B. and Pringle, J.W.S., Proc. XII Int. Congr. Zool. London, p.185 (1964).

8. Leston, D., Pringle, J.W.S. and White, D.C.S., J. Exper. Biol., <u>31</u>, 525 (1964).

9. Mannherz, H.J., Pflügers Arch. Ges. Physiol., <u>291</u>, 94 (1966).

10. Rüegg, J.C. and Stumpf, H., Pflügers Arch. Ges. Physiol., in press.
11. Rüegg, J.C., Am. Zool., 7, 457 (1967).
12. Chaplain, R.A., Biochim. Biophys. Acta, 131, 385 (1967).
13. Meinrenken, W., Pflügers Arch. Ges. Physiol., 294, 45 (1967).
14. Armstrong, C.F., Huxley, A.F. and Julian, F.J., J. Physiol., 186, 26P (1966).
15. Rüegg, J.C., and Steiger, G., Unpublished experiments.

KINETIC MODEL OF MUSCLE CONTRACTION

V.I. Descherevsky

Institute of Biophysics of the
U.S.S.R. Academy of Sciences,
Puschino, Moscow Region, U.S.S.R.

The myosin-actin interaction is the necessary condition for striated muscle contraction (1). These two basic muscle proteins are located in the two systems of protofibrils, which are able to make contact with each other by means of myosin cross-bridges (2,3) at certain discrete points only. According to the sliding filament concept, it is the interaction of myosin bridges with the active sites of the actin protofibrils that provides the moving force of the contracting muscle (4-6). Each bridge must act cyclically (6) because the filament length does not change considerably during contraction (3).

The mechanical properties of the contracting muscle are probably determined at each moment by the distribution of its myosin bridges among the stages of an elementary working bridge cycle. If so, we may postulate on the series of stages of which the cycle consists, the probabilities of the transition between them and also try to describe muscle contraction by the methods of formal chemical kinetics.

Such an approach is a simplification of the idea given by Huxley (5).

Description of the Model

We shall simulate the behavior of a pair of protofibrils - thick and thin. Then the sarcomere will be considered as a parallel set of such identical pairs, the muscle fiber representing a series of identical sarcomeres. Protofibrils are regarded as rigid pivots with active sites placed periodically along them, i.e. with cross-bridges on myosin fibrils and with sites of their possible binding on actin fibrils. The following assumptions have been made with regard to the structure of an elementary working bridge cycle.

1. In an excited muscle a cross-bridge has a chance to be bound with a linking site on the actin filament. Binding makes the bridge, or, more probably, the remainder of a myosin molecule associated with the bridge (perhaps the L-meromyosin part), begin to shorten. It is essential that the myosin molecule be allowed to undergo the conformational rearrangement only if the filaments are sliding relative to each other. The speed of shortening of the molecule will be equal to the speed of this sliding. During this "molecular contraction" the cross-bridge will produce an active moving force.

2. Only when the conformational transition is over does the bridge have the chance of splitting. From this moment, up to the point of realizing this chance, the bridge is producing a hindering force due to a continuous sliding of the filaments.

3. Active and hindering forces are constant and do not depend on the velocity of sliding, their absolute values being identical.

4. The speed of association of the free bridges with the actin filaments is not limited by the frequency of their encountering the receptive sites. This process, as well as the splitting of the associated bridges, follows monomolecular kinetics.

We shall not concentrate on the possible nature of all these processes because it is not important for our model. But the qualitative picture given by Davies (7) may serve as a good illustration for this scheme.

Mathematical Formulation

According to our scheme, the bridge in an excited muscle can be in one of three possible states: a) a free state, b) an associated state, when the bridge produces an active force, i.e. the state of conformational rearrangement, and c) an associated state in which the bridge produces a hindering force, i.e. when the active conformational rearrangement is completed.

An elementary working cycle of the myosin bridge may be depicted as follows:

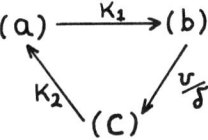

The symbols near the arrows signify the corresponding transition constants. The constant of transition from (b) to (c) is proportional to the speed of sliding of a thick filament relative to a thin filament, U, and inversely proportional to the mean value of the shortening of the myosin molecule during the active conformational rearrangement, δ.

Suppose that α is the total number of myosin bridges in the muscle layer of a half sarcomere length with the cross-section being 1 cm^2. If n and m represent the number of bridges in states (b) and (c) respectively, the following equations can be written:

$$\frac{dn}{dt} = K_1(\alpha - n - m) - \frac{U}{\delta} n \qquad [1]$$

$$\frac{dm}{dt} = \frac{U}{\delta} n - K_2 m \qquad [2]$$

The force developed by the muscle at any one time will be $f(n-m)$ where f is the absolute value of active and hindering forces of the associated myosin bridges. So, the equation governing the motion of the mechanical system connected to the muscle will be

$$M \frac{d^2 L}{dt^2} = f(n-m) - P(L) \qquad [3]$$

where M is the effective mass of the mechanical system, L is the shift of the end of the muscle associated with the mechanical system, and $P(L)$ is the external force dependent on the shift L. The muscle mass and the force of mechanical friction are regarded here as being negligible.

The shift L is expressed as follows:

$$L = \mathcal{N} \ell$$

where $\mathcal{N}$ is the number of sarcomeres along the muscle and ℓ is the shortening of a single sarcomere. Thus the whole system of equations will be:

$$\frac{dn}{dt} = K_1(\alpha(\ell)-n-m) - K_{-1}\upsilon n \qquad [4]$$

$$\frac{dm}{dt} = K_{-1}\upsilon n - K_2 m \qquad [5]$$

$$\frac{d\upsilon}{dt} = \frac{f}{2\sqrt{M}}(n-m) - \frac{P(\mathcal{N}\ell)}{2\sqrt{M}} \qquad [6]$$

$$\frac{d\ell}{dt} = 2\upsilon \qquad [7]$$

where $K_{-1} = \frac{1}{\mathcal{J}}$ and $\alpha(\ell)$ is the total number of myosin bridges in the filament overlap which depends upon the degree of shortening ℓ.

Stationary Contraction

The stationary contraction, at a constant speed, takes place under isotonic conditions if the muscle length is about the same length in situ. In this case $\alpha(\ell) = \alpha_0$ (8) and $P(\mathcal{N}\ell) = P$ are constant and do not depend on ℓ. Equations [4], [5] and [6] do not contain ℓ, so they form the exclusive system whose steady state ($\frac{dn}{dt}=0$, $\frac{dm}{dt}=0$, $\frac{d\upsilon}{dt}=0$) is single:

$$n = \frac{K_1(f\alpha_0 + P) + K_2 P}{f(2K_1 + K_2)}; \quad m = \frac{K_1(f\alpha_0 - P)}{f(2K_1 + K_2)}; \quad \upsilon = \frac{K_1 K_2}{(K_1+K_2)K_{-1}} \cdot \frac{f\alpha_0 - P}{\frac{K_1}{K_1+K_2}f\alpha_0 + P} \quad [8]$$

The expression for the stationary velocity, υ, may be changed in such a way that:

$$(P + a)\upsilon = b(P_0 - P) \qquad [9]$$

where $P_0 = f\alpha_0$

$$a = \frac{K_1}{K_1 + K_2} f\alpha_0 = \frac{K_1}{K_1 + K_2} P_0 \qquad [10]$$

and

$$b = \frac{K_2}{K_{-1}} \frac{K_1}{K_1 + K_2} = \frac{K_2}{K_{-1}} \frac{a}{P_0} = \upsilon_{max} \frac{a}{P_0} \qquad [11]$$

where υ_{max} is the contraction velocity at $P = 0$

Expressions [9], [10] and [11] coincide in detail with the experimental correlations discovered by Hill (9).

The rate of the total energy production will be

$$\frac{dE}{dt} = \mathcal{E} K_2 m = \frac{\mathcal{E} K_1 K_2 (f\alpha_0 - P)}{f(2K_1 + K_2)} = Const(P_0 - P) \quad [12]$$

where $\mathcal{E}$ is the energy of the chemical reactions occurring in each elementary cycle. It is probably equal to the energy of hydrolysis of one ATP molecule.

The rate of heat production can be obtained by subtracting mechanical power from equation [12] and by expressing $f\alpha_0 - P$ through the stationary v from equation [8]:

$$\frac{dQ}{dt} = \frac{dE}{dt} - PV = v\left[\frac{\eta K_1}{2K_1+K_2}P_0 + \frac{\eta(K_1+K_2)-(2K_1+K_2)}{2K_1+K_2}P\right] \quad [13]$$

where

$$\eta = \frac{\mathcal{E} K_{-1}}{f} = \frac{\mathcal{E}}{f\delta} \quad [14]$$

has the sense of an inverse efficiency of the elementary cycle because $f\delta$ is the mechanical work of the myosin bridge during the elementary cycle.

If $\eta = \frac{K_2 + 2K_1}{K_1 + K_2}$ then $\frac{dQ}{dt} = av$ this being in agreement with the experimental results of Hill (9).

Expression [13] is in conformity with the more recent results of this same author (10), if $\eta = 1.4 \pm 0.3$ and $\frac{K_2}{K_1} = 5.5 \pm 1$

It should be recognized that expression [13] does not contain the activation and maintenance heat, as they are probably due to the action of the muscle activation system rather than the contractile system (11).

Estimation of the Parameters

There are five parameters in our scheme connected with the intimate mechanism of muscle contraction: α, K_1, K_{-1}, K_2 and f. Comparison with Hill's equation gives three relationships between them:

$$f\alpha_0 = P_0, \quad \frac{K_1}{K_1+K_2} = \frac{a}{P_0}, \quad \frac{K_2}{K_{-1}} = V_{max}$$

α_0 may be estimated as 10^{13} from the structural data (6) and formula [14] gives the last relationship we need.

Assuming that $P_0 = 3\,kg$ (12), $\frac{\alpha}{P_0} = \frac{1}{4}$, $\eta \approx 1$
$V_{max} = \frac{4}{3}$ (a half of the sarcomere length)/sec = 1.5×10^{-4} cm/sec (9) and the energy of hydrolysis of the ATP molecule = 3×10^{-13} g cm^2/sec (13), we shall have
$f = 3 \times 10^{-7}$ g cm/sec^2, $K_{-1} = 10^6$ cm^{-1}, $K_2 = 150$ sec^{-1} and $K_1 = 50$ sec^{-1}.

Isotonic Contraction

The dynamics of movement towards the steady state [8] may be obtained by integrating the system of equations [4], [5] and [6] at $\alpha(\ell) = \alpha_0$ and $P(\mathcal{N}\ell) = P$. This system, however, has only a slight non-linearity and linear approximation gives quite satisfactory results in this case.

On introducing new variables such as $X = \frac{n}{\alpha_0}$, $\tau = K_1 t$, $y = \frac{m}{\alpha_0}$ and $u = \frac{K_{-1}}{K_1} v$, the system [4], [5] and [6] will be written as:

$$\frac{dx}{d\tau} = 1 - x - y - ux \qquad [15]$$

$$\frac{dy}{d\tau} = ux - 3y \qquad [16]$$

$$\frac{du}{d\tau} = B(x - y - A) \qquad [17]$$

where $A = \frac{P}{f\alpha_0} = \frac{P}{P_0}$ and $B = \frac{P_0}{M} \frac{K_{-1}}{2\mathcal{N} K_1^2}$

The steady state [8] in terms of these variables will be

$$X_0 = \frac{4A+1}{5} \ ; \quad y_0 = \frac{1-A}{5} \ ; \quad u_0 = \frac{3(1-A)}{4A+1}$$

The characteristic equation of system [15], [16] and [17] linearized near this steady state is

$$\lambda^3 + (3.25 + 3.75a)\lambda^2 + (15a + 0.4b)\lambda + b = 0 \quad [18]$$

Here $a = \frac{1}{4A+1}$ and $b = B(4A+1)$ have the following limits of change:

$0.2 \leq a \leq 1$ and $1 < b < 500$ if $\mathcal{N} \approx 1.5 \times 10^4$ and $\frac{P_0}{M}$ is allowed to change from 10^2 to 10^4 cm/sec^2 by using various isotonic levels.

The whole picture of the eigenvalues of equation [18] in the plane of the parameters a and b is shown in Figure 1. At almost all values of the parameters, one real negative root λ_1, and a pair of complex roots $\lambda_2 = P + i\omega$ and $\lambda_3 = -P - i\omega$ with the negative real part are present.

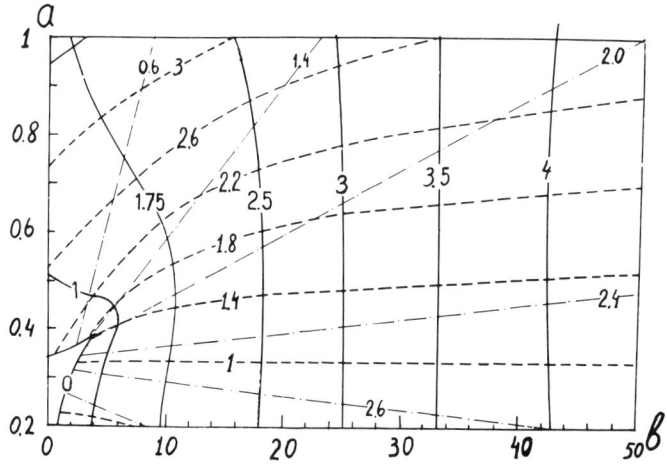

Figure 1. Parametrical plane of equation [8]. The levels of ω, P and λ_1 are given by solid, dashed and dashed-dotted lines, respectively.

This means that motion towards the steady state speed given by Hill's equation occurs as follows. The mean level of the contraction speed approaches the steady state value with a time constant of $\frac{1}{K_1 \lambda_1}$. Around this mean level there occur damped sinusoidal oscillations with a period of $\frac{2\pi}{\omega} K_1$ and a damping constant of $\frac{1}{K_1 P}$. The amplitude of the oscillations is determined by the initial perturbation.

317

It should be emphasized that in these calculations we do not take into account the elastic properties of the muscle. Therefore, the system [15]-[17] does not contain any resonance elements. The oscillatory mechanism may be explained as follows. The speed of transition of myosin bridges from state (b) to (c) varies directly with the velocity of contraction. If, at the first moment, the contraction velocity is zero, then the transition from (a) to (b) will prevail. The force developed by the muscle, the acceleration and the velocity of inertial loading will rise rapidly, this resulting in the predominance of the transition from (b) to (c). Because of the loading inertia this will cause the force to fall to a level insufficient for the steady state speed to be maintained, and the cycle will then repeat.

As an illustration, a numerical integration system for [15]-[17] is shown in Figure 2. The value of the parameter B corresponds to the case of a muscle lifting a weight of $P_0/M \approx 10^3$ cm/sec^2. The oscillation frequency coincides with that calculated using a linear approach.

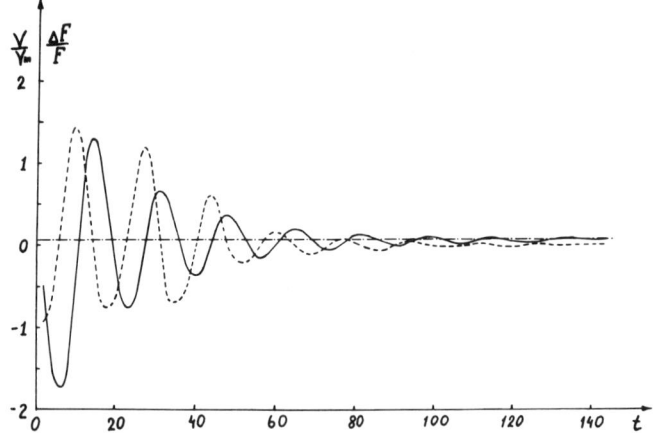

Figure 2. The initial portion of the isotonic contraction curves calculated from system [15]-[17] under the conditions: B=36; A=0.75; X(0)=0; Y(0)=0; U(0)=0. The solid line gives the speed of shortening as a fraction of the value in unloaded tetani. The dashed line gives the difference between muscle tension and load, divided by load. The dashed-dotted line gives the steady state value of contraction speed at A=0.75.

BIOCHEMICAL OSCILLATORS

Stretched Muscle Behavior

The behavior of stretched muscle under conditions of isotonic load is described by the system of equations [4]-[7] with $P(\mathcal{N}\ell) = P$ constant and with $\alpha(\ell) = \beta\ell$ being a linear function. From the structural and physiological data it follows that $\alpha(\ell) = \frac{\alpha_0}{1.4 \times 10^{-4}} \ell$ (14), if ℓ is a shortening of the sarcomere in cm from the initial length of 3.65×10^{-4} cm. Upon introducing these modifications, system [4]-[7] can, by substitution of the variables, be reduced to the form:

$$\frac{dx}{d\tau} = z - x - y - ux \qquad [19]$$

$$\frac{dy}{d\tau} = ux - 3y \qquad [20]$$

$$\frac{du}{d\tau} = B(x - y - A) \qquad [21]$$

$$\frac{dz}{d\tau} = \varepsilon u \qquad [22]$$

Here, $x = \frac{n}{\alpha_0}$, $y = \frac{m}{\alpha_0}$, $u = \frac{K_{-1}}{K_1} v$, $z = \frac{1}{7 \times 10^{-5}} \ell$,

$\tau = K_1 t$, $B = \frac{P_0}{M} \frac{K_{-1}}{2\sqrt{K_1^2}}$, $A = \frac{P}{P_0}$, $\varepsilon = \frac{2}{K_{-1} \times 1.4 \times 10^{-4}}$

This system has three natural time scales of order 1 for equations [19] and [20], for [21] and for [22]. Taking into account that $\varepsilon = 1/70$ and $5 < B < 100$ as usual, we may analyse this system in three steps.

As the first approach we may consider the group of equations [19]-[21] as "fast", while equation [22] is "slow". Then we may obtain the solution by means of Tikhonov's theorem (15). System [19]-[22] is stable at all values of the "slow" variable, z; therefore, the conditions of this theorem are satisfied. The solution can be obtained by substitution into the "slow" equation of the steady state value of the fast variable u, which can be derived from system [19]-[21] with z treated as a parameter. Coming back to the initial variables, we have:

$$\frac{1}{2}\frac{d\ell}{dt} = v = \frac{K_2}{K_{-1}} \frac{K_1}{K_1 + K_2} \frac{f \frac{\alpha_0}{1.4}\ell - P}{\frac{K_1}{K_1+K_2} f \frac{\alpha_0}{1.4}\ell + P} \qquad [23]$$

where ℓ is measured in microns.

319

Taking into consideration that $(f\alpha_0/1.4)\ell$ is the isometrical force $P_0'(\ell)$ developed by the sarcomere at a given shortening ℓ, and assuming, as usual, $K_1/(K_1+K_2) = 1/4$ and $K_2/K_{-1} = V_{max}$ we obtain:

$$\left[P + \frac{P_0'(\ell)}{4}\right]v = \frac{V_{max}}{4}\left[P_0'(\ell) - P\right] \quad [24]$$

This is analogous to the Hill equation. It describes the quasi-stationary isotonical contraction of stretched muscles (the sarcomere length ranges from 3.65 to 2μ).

The second approach to the solution of system [19]-[22] may be obtained by substitution of the algebraic expression x-y = A for equation [21].

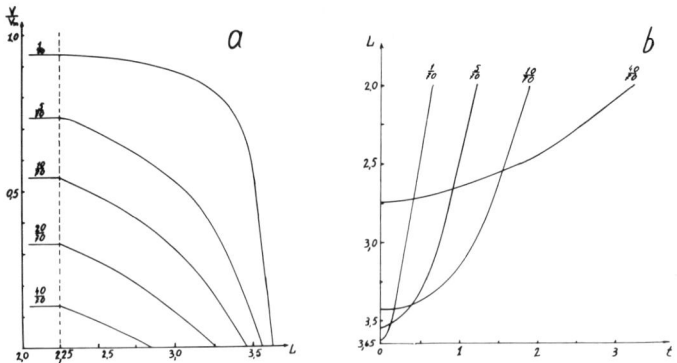

Figure 3. Isotonic contraction of stretched muscle. a) Mean of contraction speed as a function of sarcomere length, at various loads. Ordinate - speed of shortening as a function of the value in unloaded tetani. Abscissa - sarcomere length in microns. b) Shortening as a function of time at various loads. Ordinate - sarcomere length in microns. Abscissa - time in seconds. The load, as a fraction of isotonic tetani tension, is indicated above each curve.

Numerical integration of such a system of equations at several values of the parameter A is shown in Figure 3. It differs slightly from the calculations from formula [23] only at a very low load.

The whole system [19]-[22] has a single unstable steady state

$$u_s = 0, \quad y_s = 0, \quad x_s = \Gamma, \quad z_s = \Gamma \qquad [25]$$

corresponding to the load being balanced precisely by the muscle force at a given lengthening. Investigation of this system as a linear approximation in the neighborhood of this steady state shows that its approach to the quasi-stationary speed of contraction [24] is approximately the same as in the case of unstretched muscle. But there is a region in the plane of the parameters where oscillations of the contraction speed are undamped. As an illustration, the integration of system [19]-[22] for such a case is shown in Figure 4.

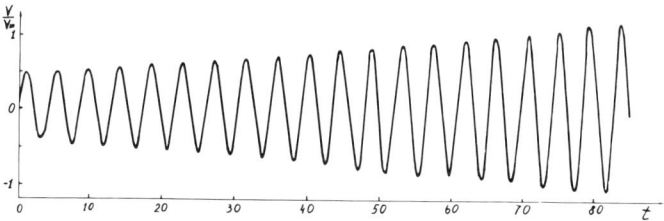

Figure 4. The initial region of stretched muscle isotonic contraction calculated from system [19]-[22] under the conditions: B=20, A=20/70, X(0)=22/70, Y(0)=0, U(0)=0, Z(0)=24/70. Ordinate - speed of shortening as a fraction of the value in unloaded tetani. Abscissa - time in milliseconds.

Isometric Contraction

To calculate the speed of the muscle-force developed under isometric conditions, it is necessary to know the load-extension curve $P(\ell)$ for elastic elements connected in series with contractile elements of the muscle. The data we need were taken from the work of Jewel and Wilkie (16).

The isometric contraction is described by system [4]-[7] at $\alpha(\ell) = \alpha_0$ and with equation [6] substituted by the algebraic correlation

$$f(n-m) = P(\ell) \qquad [26]$$

which expresses the equality of elastic and contractile forces. It allows v to be excluded from this system. After a linear substitution of the variables, it takes the form:

$$\frac{dx}{d\tau} = 1 - x - y - \beta \frac{1-x+2y}{\frac{dA}{dz} + 2\beta x} X \qquad [27]$$

$$\frac{dy}{d\tau} = -3y + \beta \frac{1-x-2y}{\frac{dA}{dz} + 2\beta x} X \qquad [28]$$

$$\frac{dz}{d\tau} = \frac{1-x+2y}{\frac{dA}{dz} + 2\beta x} \qquad [29]$$

Here $X = \frac{n}{\alpha_0}$, $y = \frac{m}{\alpha_0}$, $z = 100 \frac{\ell}{\delta}$, $\beta = \frac{K_{-1}\delta}{200}$; $\delta = 2.2 \times 10^{-4}$ cm is the length of a sarcomere, $A(z) = \frac{P(\ell)}{P_0}$ is the relationship between relative force and relative extension of an elastic component, which has the following analytical form:

$$A(z) \begin{cases} 0.0897 z^3 + 0.0348 z^2 + 0.2z & \text{for } z \leqslant 1.15 \\ 0.636 z - 0.319 & \text{for } z \geqslant 1.15 \end{cases} \qquad [30]$$

System [27]-[29], under initial conditions of X(0)=0, Y(0)=0, Z(0)=0 describes normal isometric tetani of the muscle. Tension redevelopment, after quick release is governed by the same system under initial conditions of X(0)=0.5, Y(0)=0.5, Z(0)=0. This may be explained as follows. Before the quick release the muscle develops the maximum isometrical force, all its bridges being in state (b). During the quick release the force falls and the bridges shift at the speed $K_{-1}v$ to state (c), but not to state (a), due to the limited value of the constant K_2. The fact that the force falls to zero means that half of the bridges are in state (b) and the other half are in state (c).

The speed of the force development can also be calculated easily from the load-extension dependence [30] and the stationary force-velocity relationship [9] (see ref.16). The results of our calculations are shown in Figure 5. For a comparison, the experimental and calculated curves from (16) are shown in Figure 6.

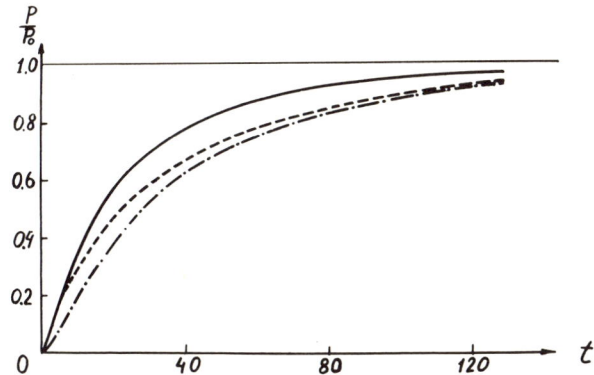

Figure 5. Isometric contraction. Ordinate - tension as a fraction of the maximum value. Abscissa - time in milliseconds. Solid line: calculation from the force-velocity [9] and load-extension [30] correlations. Dashed line: initial rise of tension calculated from system [27]-[29] under the conditions: $X(0)=0$, $Y(0)=0$, $Z(0)=0$. Dashed-dotted line: tension redevelopment after quick release, calculated from system [27]-[29], under the conditions: $X(0)=0.5$, $Y(0)=0.5$, $Z(0)=0$.

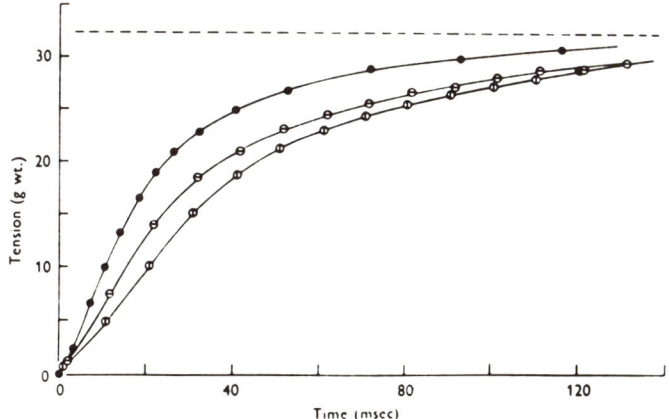

Figure 6. Isometric myograms from reference (16). Solid circles: curve of rise of tension calculated from the force-velocity and load-extension curves. Open circles: observed isometric myograms; ⊕ , initial rise of tension; ⊖ , redevelopment of tension after quick release.

Discussion

The sliding filament concept is at present the most valid scheme of muscle contraction. The first qualitative theory of muscle contraction based on it was advanced by Huxley (5).

The most essential feature of the work presented here is that it postulates a discrete spectrum of states in which the bridges are allowed to exist. This makes possible the simplification of mathematical formalism by using ordinary differential equations instead of partial ones.

The formalism of partial differential equations employed by Huxley(5) permits the analysis of only the simplest case of the stationary isotonic contraction. In this work, the partial equation has been derived for the fraction n(X,t) of myosin bridges combined with actin acceptors, where X is the deviation of the myosin bridge equilibrium position from the nearest actin binding site. The equation has the form:

$$\frac{\partial n}{\partial t} = (1-n)f(x) - n g(x)$$

where f(x) and g(x) are the rate constants of bridge binding and splitting, respectively. This equation should contain the term $\frac{\partial n}{\partial x}\frac{dx}{dt}$ governing the variation of n(x,t) conditioned by the bridge current due to filament sliding. It is probably the lack of this term that leads to some of the discrepancies mentioned by the author, himself (5).

The constant parameters of our model K_1, K_{-1}, K_2 and f can be treated as the average values of some arbitrary limited functions of the variables characterizing the myosin bridge state. We suppose, however, that good physical grounds exist for some of them to be truly constant.

The independence of these parameters from the speed of muscle contraction arises from the following fact. The rates of free macromolecular transitions exceed by several orders the maximum rates of conformational transitions permissible for myosin molecules when the bridges are combined with the actin filaments. So, in this case, any change of its state may be regarded as an equilibrium process. This leads to independence of K_{-1}, K_2 and f from v. Independence of K_1 from v results directly from assumption 4 made at the beginning of this paper. The validity of this assumption is increased by the fact that all g-actin monomers consist of a thin filament capable of combining with H-meromyosin, thus being the potential binding sites for myosin bridges (17).

Some of these actin monomers, however, due to a periodic organization of the thin filament, may occupy a more suitable position relative to the given thick filament. This is not important for the case of skeletal muscles of vertebrates in view of the lack of synchronization of periodicity of thin filaments across the sarcomere (3). But it does become essential in the case with insect flight muscles, due to their crystalloid organization (18). A slight modification of our model is sufficient to describe an oscillatory contraction of these muscles.

The independence of f from the degree of myosin molecule conformation rearrangement, ξ, may be explained as follows. Evidently, $f = d\epsilon/d\xi$, where $\epsilon(\xi)$ is an energy change of the myosin molecule during its conformational rearrangement. If a large number of weak hydrogen or van der Waals bonds are formed, or broken, in this process, then $d\epsilon/d\xi$ may be a constant. This would be so, for example, if the conformational rearrangement were similar to the wave propagation along the uniform molecular backbone.

In our model we suggest that the rate constant for splitting of the bridge in state (b) should be zero. A good reason for this assumption is given by the independence of the isometric muscle force, P_0, from temperature (19). In our model P_0 does not depend on temperature since α_0, the structural characteristics of the muscle, does not depend on it. If the rate constant, K_2', of splitting for the bridge in state (b) were comparable to K_1, then P_0 would be:

$$P_0 = \frac{f \alpha_0 K_1}{K_1 + K_2'}$$

In this case P_0 would be independent of temperature only if very special assumptions concerning the K_1 and K_2' temperature dependence were made.

The present form of our model fails to describe the process of muscle relaxation. For this purpose it is necessary to postulate a fourth possible state of the myosin bridge. The bridge is allowed to change to this state from states (a) and (b) when the muscle is being stretched $(v<0)$. The force produced by the bridge in this state may be equal to f and the rate constant of its splitting may be about $K_2/6$. Such values of the parameters are in accordance with the findings of Katz (20):

$$\left|\frac{dv}{dP}\right|_{P<P_0} \approx 6 \left|\frac{dv}{dP}\right|_{P>P_0}$$

According to this modification of our model, relaxation is possible only if the muscle is subjected to an external force. It should be noted that under some conditions the process of relaxation is a very slow one. The high frequency Young's modulus of frog sartorius muscle maintains an increased value for 1 sec after the muscle stimulation has ended (21).

When describing muscle relaxation it is necessary also to take into account the rate of Ca^{++} removal from the contractile system of the muscle. This should also be done if the twitch and the first moments of tetani are of interest. For this purpose it is sufficient to assume that the constant K_1 is dependent of time.

The parameters of our model were estimated by checking with the Hill equations. Thus, the model is able to predict the dynamics of muscle contraction under various conditions if the stationary characteristics of muscle contraction are known.

This model satisfies the Hill equations exactly, with all parameters having reasonable values. It is in good conformity with the data on the isometric tetani force development and explains the distinctions between the normal isometric contraction and the force redevelopment after quick release.

A wide range of behavior patterns of the model under various conditions gives the probability of good experimental varification.

Some experimental data indicate that the oscillatory modes of ordinary skeletal muscle contraction are possible (22,23). In reference (16) the authors explain oscillations of the speed of isotonic contraction, following controlled release, as an artifact arising from oscillations of the release relay, - though this is not always the case.

It should be noted that oscillatory modes of contraction are more efficient than monotonous ones.

Acknowledgement

I thank Dr. A.M. Zhabotinsky for useful discussions and I am indebted to Prof. A.M. Molchanov and Dr. E.E. Sel'kov for their helpful advice in the mathematical treatment.

References

1. Сент-Джиордъи, А., О мышечной деятелъности, Москва (1947).
2. Hanson, J. and Huxley, H.E. Nature, 172, 530 (1953).
3. Huxley, H.E. and Brown, W. J. Mol. Biol., 30, 383 (1967).
4. Huxley, A.F. and Niedergerke, R. Nature, 173, 971 (1954).
5. Huxley, A.F. Progr. Biophys. and Biophys. Chem., 7, 255, (1957).
6. Хаксли, Х., сб. Молекулярная биология, Москва (1963).
7. Davies, R.E. Nature, 199, 1068 (1963).
8. Page, S. and Huxley, H.E. J. Cell Biol., 19, 369 (1963).
9. Hill, A.V. Proc. Roy. Soc., B126, 136 (1938).
10. Hill, A.V. Proc. Roy. Soc., B159, 297 (1964).
11. Хилл, А., сб. Молекулярная биология, Москва (1963).
12. Ramsey, R.W. and Street, S.F. J. Cel. Comp. Physiol., 15, 11 (1940).
13. Benzinger, T.H. and Hens, R. Proc. Natl. Acad. Sci., 42, 896 (1956).
14. Gordon, A.M., Huxley, A.F. and Julian, F.I. J. Physiol., 184, 170 (1966).

15. Тихонов, А.Н., Математический сборник, <u>22</u>, 193 (1948).
16. Jewel, B.R. and Wilkie, D.R.   J. Physiol., <u>143</u>, 515 (1958).
17. Huxley, H.E.   Proc. Roy. Soc., <u>B160</u>, 442 (1964).
18. Reedy, M.K.   J. Mol. Biol., <u>31</u>, 155 (1968).
19. Ernst, E., <u>Biophysics of the Striated Muscle</u>, Budapest (1963).
20. Katz, B.   J. Physiol., <u>96</u>, 45 (1939).
21. Сарвазян, А.П., Пасечник, В.И., Труды Тбилисского симпозиума (1969).
22. Frank, G.M.   Proc. Roy. Soc., <u>B160</u>, 473 (1964).
23. Armstrong, C.F., Huxley, A.F. and Julian, F.I.   J. Physiol., <u>186</u>, 26P (1966).

EXCITATION WAVE PROPAGATION DURING HEART FIBRILLATION

V. I. Krinsky

Institute of Biological Physics
Academy of Sciences of the USSR
Puschino, Moscow Region, USSR

A theory of wave propagation in an excitable medium is proposed. The medium under consideration is a two-dimensional model of heart tissue. The results obtained can also be used to explain the space behaviour of some oscillatory chemical systems.

1. **Fibrillation of the heart.** Under normal conditions the heart contracts periodically due to a special cardiac "pacemaker". Fibrillation disturbs regular pulse propagation and the smallest pieces of the heart contract chaotically and almost independently. Fibrillation can occur as a result of some pathologic changes in the heart muscle. Ventricular fibrillation is rapidly fatal; auricular fibrillation is not so dangerous and it can be self-terminating.

2. **Experimental fibrillation (1,2).** Fibrillation can be observed in experiments performed on animals. If stimulation electrodes are connected to the heart of a dog, the heart contracts in response to each pulse. The higher the pulse frequency of the stimulator, the higher the heart contraction frequency.

a. When the pulse frequency becomes too high, fibrillation is induced. Once initiated fibrillation persists even if the stimulator is switched off.

b. Fibrillation can usually be terminated by a much stronger electrical pulse (the necessary voltage from a defibrillator being about 1 KV and the pulse duration about 10 msec). The pulse must be that strong in order to excite relatively refractory tissue. It is possible to initiate and terminate fibrillation many times with the help of a stimulator and defibrillator.

<u>c</u>. The pieces which fire independently during fibrillation are as small as 0.5 mm, as is shown with the help of micro-electrode recordings and a high-speed cinematograph.

<u>d</u>. Fibrillation is possible in an isolated heart segment if its size is big enough and if its mass is greater than a certain value called the "critical mass". For a rabbit atrium treated with ACh the critical mass is about 30 mg. If the segment mass is less than the critical mass, fibrillation is impossible.

3. <u>Physiological theories of fibrillation</u>. The mechanism of fibrillation has been a subject of study for more than 100 years. There are two principal theories of fibrillation; – one deals with spontaneous activity of cells and the other with the circular movement of an excitation pulse. Neither of them can provide an adequate explanation of the arhythmia.

The first principal theory suggests that fibrillatory agents cause the cells to fire spontaneously at different frequencies. It can explain the small size of independently firing pieces during fibrillation, but defibrillation and the critical mass phenomenon remain unexplained. This hypothesis has been criticized by I. M. Gelfand and coworkers (3).

The second principal theory connects fibrillation with a circular movement of the excitation pulse. It is well known that the excitation pulse can circulate in a ring cut out of heart tissue. A similar process is supposed to take place in the intact heart. Actually, another kind of arhythmia, namely flutter, which is characterized by rapid but periodic contractions, has been explained in this way (4). Chaotic contractions during fibrillation could be explained by the interaction of a variety of such circulating waves, as was confirmed by G. K. Moe and coworkers (5) in experiments with a digital computer.

4. <u>The main targets of the analysis</u>. The major difficulty in explaining fibrillation by means of circular movement is as follows. It is commonly accepted (4), that a

pulse can circulate along a closed pathway if its length is greater than the wavelength. The wavelength $\lambda=RV$, where R is the duration of the refractory period and V is the velocity of the pulse. The paradox is that the minimum value of λ, estimated for a human and a dog heart, is about 30 - 50 mm, while pieces firing independently during fibrillation are about 0.5 mm (see Section 2c), *i.e.* 100 times smaller than λ.

Another question concerns the "critical mass phenomenon". The fibrillation problem is of great clinical importance. The theories in question must show which parameters should be regulated to render induction fibrillation more difficult. The existing theories are not quantitative. The conclusion inferred from them, namely that the greater the wavelength, λ, the more difficult the initiation of fibrillation, does not necessarily correlate with the action of antifibrillatory drugs. The investigation of the mechanism of the "critical mass phenomenon" gives rise to a hope of finding drugs that would enable us to enhance the critical mass value above the heart mass, thus making fibrillation impossible.

5. *The principal equations.* The problem in general is reduced to solving the three-dimensional Hodgkin-Huxley equations which govern pulse propagation in the heart. The Hodgkin-Huxley equations belong to the same class of partial equations which govern chemical reactions proceeding in space:

$$\frac{\partial u_i}{\partial t} = f_i(u_1,\ldots,u_n) + d_i \frac{\partial^2 u_i}{\partial x^2} \quad (i=1,\ldots,n) \quad (1)$$

where u_i are concentrations, and d_i are diffusion coefficients of ith substances. If the concentrations, u_i, are equal in all points of space, then $\frac{\partial^2 u_i}{\partial x^2} = 0$ and Equation (1) becomes a system of ordinary differential equations (2) governing the behaviour of an isolated point. System (2) usually

$$\frac{du_i}{dt} = f_i \ (u_1,\ldots,u_n) \quad (2)$$

serves as a model for chemical systems. The behaviour of an actual system which is governed by Equation (1) may greatly differ from that of the model (2).

The Hodgkin-Huxley equations represent a system of four equations of the same type as (1) where the singular diffusion coefficient d_i is not equal to zero. The variables are:- the membrane voltage, the K^+ - permeability of the membrane and two variables describing its permeability to Na^+. A greatly simplified version of the Hodgkin-Huxley equations is a system of the Van der Pol type with two variables.

$$\frac{\partial u_1}{\partial t} = f(u_1, u_2) + \mathcal{Z}$$

$$\frac{du_2}{dt} = \Psi(u_1, u_2)$$
(3)

where $\mathcal{Z} = d_1 \frac{\partial^2 u_1}{\partial x^2} + U_s$ is an external influence from the neighbouring membrane areas ($d_1 \frac{\partial^2 u_1}{\partial x^2}$) and from an electrical stimulator (U_s).

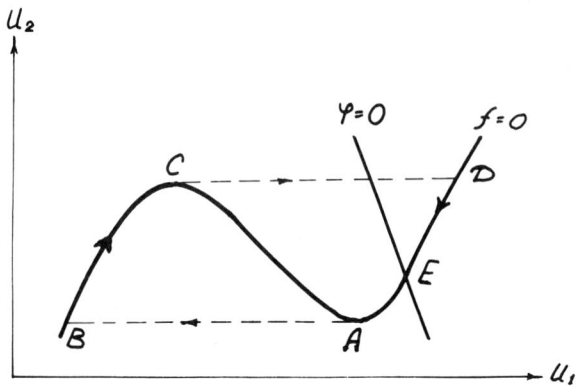

Figure 1. Phase plane of system (3)

On the phase plane shown in Figure 1, the regions corresponding to the quiescent (E), excited (BC) and refractory states can be identified (6).

6. Approach to the problem. Analysis of the Hodgkin-

Huxley equations is just too difficult. This paper deals with another approach to the problem. A simplified model system is analyzed below. We shall analyze the space behaviour of the system taking as axioms a) the response of an isolated point, and b) the manner of excitation propagation from one point to another.

Each point of the system is a chemical single-shot flip-flop. An isolated point is not oscillatory. The external force makes it generate a single pulse and return to the quiescent state.

7. <u>Axioms</u>. <u>a</u>. Each point of the medium may be in one of three states: quiescence, excitation and absolute refractoriness. Under an external influence the quiescent

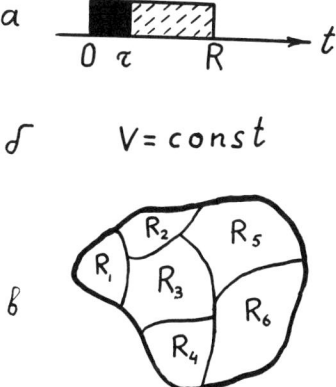

Figure 2. The properties of the excitable medium.

point switches first to the excited state, then to the refractory state and finally returns to the initial quiescent state.

The excited state has a non-zero duration, τ. The total duration of the excited and refractory states is R; R being called the refractoriness. During the whole period, τ, an excited point retains the ability to excite every adjacent quiescent point (see Figure 3).

<u>b</u>. In a quiescent area excitation propagates at a constant velocity, V.

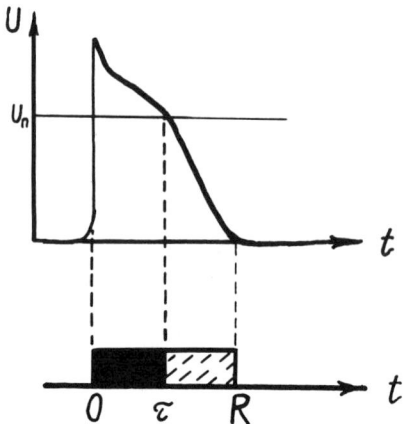

Figure 3. The relation between action potential and the states of the excitable medium. U_n = threshold voltage.

c. The medium is a two-dimensional, simply-connected region. The refractoriness, R, is a step-function of the coordinates. The other parameters, τ and V, are constant.

8. One-dimensional excitation propagation (8). This question will be considered to demonstrate the application of the adopted axioms. When a point of a quiescent fiber is excited, the pulse propagates from it in both directions at the velocity, V. The wavelength is $\lambda = RV$. The front part of the wave, of length τV, consists of excited points, while the points of the back part, of length $(R - \tau)V$ are in the refractory state (Figure 2a). If two waves move towards each other, they collide and die away due to the refractory tails (4).

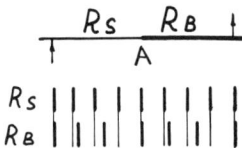

Figure 4. Rhythm transformation.

Now let us consider pulse propagation along a fiber which is nonhomogeneous with respect to refractoriness. If the pulse frequency is high enough some of the pulses die away in the heterogeneous parts. This phenomenon is called rhythm transformation (or Wenkebach's transformation). Suppose that the refractoriness in the left half of a fiber is R_s, while in the right half it is R_B, and $R_B > R_s$ (see Figure 4). Suppose the heterogeneity is small ($R_B - R_s < \tau$). Let us excite periodically the left end of the fiber with a period, T, such that $R_s < T < R_B$. Then some of the pulses will not enter the right half of the fiber. Designate the middle point of the fiber as A. The second pulse comes close to it before its refractory state is completed. If τ were zero, the pulse would die away at A. But according to the axioms the excitation of a point lasts for $\tau > 0$, during which time the refractoriness in A ends and the pulse proceeds on its way. On the right side of A the fiber can be excited at the times 0, R_B, $2R_B$, $3R_B$, ... and so on, but the pulse comes to it at the times 0, T, 2T, 3T, ... and so on. This means that the first pulse passes A without delay, the second pulse is delayed at A for $R_B - T$, the delay of the 3rd pulse is $2(R_B - T)$, ..., the delay of the k-th pulse being

$$\Delta_k = (k-1)(R_B - T) \quad (5)$$

A pulse can pass A if the delay is less than τ. The number N of the pulse which dies away at A, is determined from the inequalities: $\Delta_N > \tau$, $\Delta_{N-1} < \tau$ and

$$N = 2 + \left[\frac{\tau}{R_B - T}\right] \quad (6)$$

where square brackets mean an integer part.

When the (N+1)th pulse comes to A, the point A is already in the quiescent state and the pulse passes through the whole fiber. The picture is periodic. For every N pulses, (N-1) pass the fiber and the N-th pulse dies away at A. This situation holds for $\tau < 1/2\ R_B$. If $\tau > 1/2\ R_B$, the delay of the (N+1)th pulse is not zero, and the picture is a little more complicated.

9. Results. The following results concerning the two-dimensional medium are given without proof. A more detailed presentation can be found elsewhere (7-11).

a) If a two-dimensional medium is stimulated with a sufficiently high frequency ($R_{min} < T < R_{max}$) then some of the pulses do not enter high-refractory domains ($R>T$) (rhythm transformation) but rather bypass them. Then, in the non-homogeneous regions, whirls are formed and the wave pattern produced resembles a turbulent flow. These whirls represent temporarily closed pathways for excitation conduction which we call reverberators.

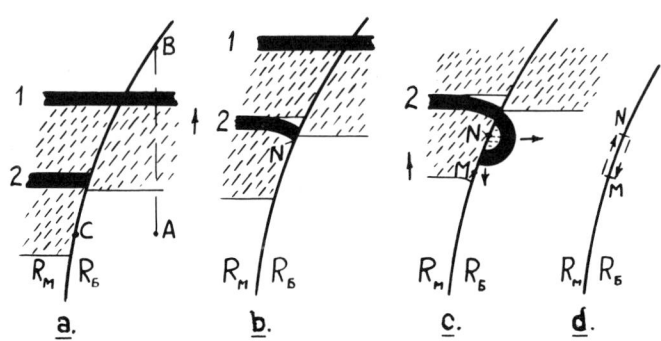

Figure 5. Formation of a reverberator. Refractoriness is smaller on the right of BC than it is on the left. Excited areas are black, refractory areas are hatched. a) Wave 2 moves through the left domain only. On the right the excited front of wave 2 borders on the longer refractory tail of the preceeding wave 1. b) Wave 2 falls behind wave 1 (because wave 1 comes to point B along the shortest path, i.e., AB, and wave 2 follows along BC). At point N the excited front of wave 2 comes into contact with the quiescent point of the right domain. c) Excitation from wave 2 propagates in the right domain. At point M it can come back to the left domain. d) Reverberator.

During interaction of the reverberators, some of them disappear while others are formed and this leads to the appearance of a chaotic wave pattern which looks like fibrillation.

We shall now describe the properties of reverberators.

a) The most curious fact is that the size of a reverberator can be much smaller than the wavelength, λ, (8). The main conflict between the large value of λ (3-5cm) and the small size of independently firing pieces during fibrillation (see section 2) is thus removed. The minimum length of reverberation is

$$L_k = \lambda(1 - k\ \tau/R) \qquad (7)$$

where k is the number of rhythm transforming points on that closed pathway (7). Figure 6 shows a closed pathway with k+4; its length can be infinitesimal for $\tau/R > 1/4$.

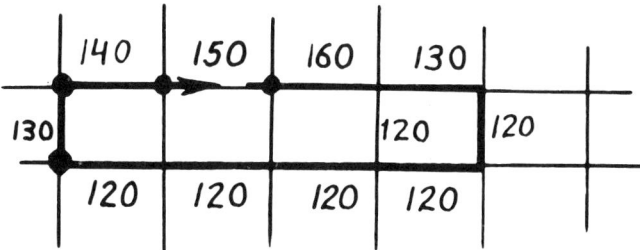

Figure 6. Closed path of conduction in a network. The numbers indicate refractoriness (in msec). Rhythm tranformation takes place at the 4 boldly marked points.

b) The points adjacent to a reverberator are excited at a maximum possible frequency (8). For this reason, reverberators in a heterogeneous medium send waves with different frequencies and they are not synchronized.

c) In a homogeneous medium, a reverberator exists for an infinitely long time. In a heterogeneous medium, the lifetime of a reverberator is finite (8) and after having sent several waves a reverberator disappears. Figure 5,d, shows a reverberator placed on the border of two domains with different values of refractoriness. The lifetime of such a reverberator (measured by the number of waves sent) is

$$T_L \approx \frac{\tau}{R_B \cdot \varepsilon} \qquad (8)$$

where $\varepsilon = 1 - R_s/R_B$, indicates the relative inhomogeneity of the region. The reverberator disappears because of the different frequency of impulsation of the left and right domains. (According to the previous property, the period of impulsation for the left domain is $T = R_s$, and for the right it is $T = R_B$). When a wave travelling in the left domain collides with the refractory phase in the right domain the reverberator dies away. It disappears for the same reason that it is created -- because of rhythm transformation. Equation (8) is an approximate expression for the number of pulses which die away during the rhythm transformation.

d) While a reverberator exists, it sends waves into the medium. Interaction of these waves can lead to the formation of new reverberators (8). Thus, a process similar to the reproduction of reverberators takes place.

The relationship of the rates of two opposing processes - the disappearance and reproduction of reverberators - controls the possibility of fibrillation in the medium. If the "death" rate is greater than the "birth" rate of reverberators, fibrillation is impossible. The opposite relationship creates conditions for the initiation of fibrillation. Such an interaction of reverberators determines the critical mass phenomenon in the medium.

The calculation for reverberator interaction is difficult and, in the general case, this task is not solved. For one special case ($\tau/R > 1/2$) a solution was found (9,11) and some analog of the critical mass was analyzed. It appeared to depend strongly on the ratio, τ/R. The reduction of τ/R increases the critical mass. This result is not surprising when equations (7) and (8) are considered. So, it appears that, in the medium considered, the effective way to render fibrillation impossible is to reduce τ/R. The process of mutual initiation of reverberators which is, essentially, fibrillation can last for an infinitely long time in the medium with a specific set of parameter values. An "arbitrary" set of values leads to the drift between the relative phases of reverberators. Generally, it results in the fact that a moment comes when all the reverberators die away almost simultaneously, the time difference being smaller than the time required for the formation of a new reverberator. After this no reverberator is formed and fibrillation is terminated. For the above-mentioned special case ($\tau/R > 1/2$) the mean lifetime of fibrillation is

$$T_f \approx \left(\frac{\tau}{R-\tau}\right)^m \qquad (9)$$

where m is the number of inhomogeneous domains in which the reverberators can be formed. The lifetime, T_f, is liable to increase rapidly when τ/R is increased.

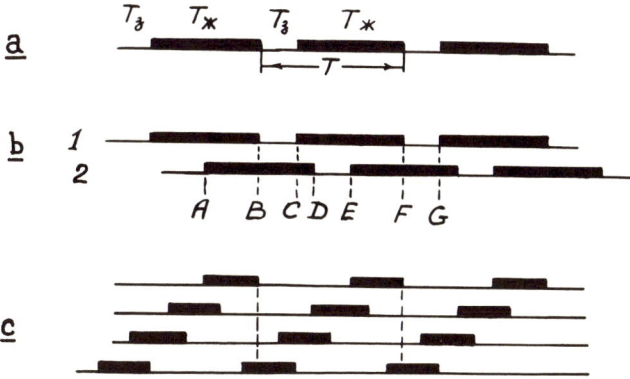

Figure 7. Calculation of critical mass and lifetime of fibrillation. a,b) Diagram of states of two heterogeneous domains where reverberators form and disappear. Line 1 represents the state of the first domain; line 2 the state of the second domain. The thick line represents the lifetime of a reverberator (e.g. AD and CF - lifetime of reverberators). At the moment B, the first reverberator disappears. The second reverberator sends waves until the moment D. During the time interval BC, its impulsation reinitiates the reverberator in the first domain. Now the first reverberator works until the moment F. During the time DE, the first reverberator creates a reverberator in the second domain, and so on.
c) Diminishing the ratio τ/R leads to a reduction of lifetime and an increase in the initiation time of a reverberator. In this figure the two reverberators cannot be mutually initiated. Fibrillation caused by four reverberators is shown, ($\tau/R > 1/2$).

10. Discussion. The results obtained on the simplest excitable medium suggest that the study of the parameter, τ/R, for heart, could be of great importance. It is highly probable that this parameter provides a more effective criterion for searching antifibrillatory drugs than does the wavelength, λ.

It is difficult to define precisely what is meant by

the "excited state duration", τ, for cardiac tissues. A more detailed analysis is necessary here.

SUMMARY

1. Wave propagation in a two-dimensional excitable medium can lead to the formation of turbulence. The whirls, or reverberators, are formed when the frequency of stimulation is sufficiently high.

2. The interaction of reverberators is similar to fibrillation of the heart. Reverberators send waves with different frequencies, and they are not synchronized. The length of a reverberator can be much shorter than the wavelength, λ.

3. In a heterogeneous medium, a reverberator has a finite lifetime. Reverberators can reproduce themselves. Two opposing processes, the reproduction and disappearance of reverberators, lead to the "critical mass phenomenon". Fibrillation is possible if the rate of reproduction is not less than the "death" rate of reverberators.

4. In an excitable medium, the most important parameter which influences fibrillation is the ratio, τ/R (τ - excited state duration, R - refractory period). The decrease of the ratio, τ/R, makes the initiation of fibrillation more difficult. It reduces the average lifetime of fibrillation, increases the critical mass of the medium and the length of a single reverberator, and cuts down the reverberator lifetime. Antifibrillatory drugs could have been found more effectively if examined for their ability to diminish the ratio, τ/R.

5. The results of this analysis allow prediction of the following effects not yet observed experimentally:

    a. A wave can circulate in a closed pathway which is much shorter than the wavelength, λ.

    b. In cardiac tissue the ratio, τ/R, is not too small in comparison with unity.

    c. The parameters of heart fibrillation must be strongly dependent on the ratio, τ/R.

d. The echo-phenomenon (9) is possible in fibers where $\tau/R > 1/2$.

REFERENCES

1. Hoffman, B.F. and Cranefield, P.F., <u>Electrophysiology of the Heart</u>, Blakiston, McGraw-Hill, New York, 1960.

2. Brooks, C.McC., Hoffman, B.F., Suckling, E.E. and Orias, O., <u>Excitability of the Heart</u>, Grune and Stratton, Inc., New York, 1955.

3. Gelfand, I.M., Kovalev, S.A. and Chailakhyan, L.M., ДАН СССР, <u>149</u>, 973 (1963).

4. Wiener, N. and Rosenblueth, A. Arch. Inst. Cardiologia de México, <u>16</u>, 205 (1946).

5. Moe, G.K., Rheinboldt, W.C. and Abildskov, J.A. Amer. Heart J. 67, 200 (1964).

6. Fitzhugh, R. Biophys. J. <u>I</u>, 446 (1961).

7. Krinsky, V.I. Проблемы кибернетики. No. 20 (1968).

8. Krinsky, V.I. Биофизика <u>II</u>, 676 (1966).

9. Krinkky, V.I. and Kholopov, A.V. Биофизика <u>12</u>, 524 (1967).

10. Krinsky, V.I. and Kholopov, A.V. Биофизика <u>12</u>, 669 (1967).

11. Krinsky, V.I., Fomin, S.V. and Kholopov, A.V. Биофизика <u>12</u>, 5 (1967).

CONFORMATIONAL OSCILLATIONS OF PROTEIN
MACROMOLECULES OF ACTOMYOSIN COMPLEX

S.E. Shnoll

Moscow State University, Faculty of Physics
Moscow Institute of Biophysics, USSR Academy of Sciences
Puschino, Moscow Region, USSR

In protein solutions of the actomyosin complex, spontaneous changes of the macromolecular states occur synchronously throughout the macrovolume. Such properties as ATPase activity, quantity of SH-groups titrated by $AgNO_3$, ability to react with I_2, ability to interact with diazosulfanilic acid and the bromphenol blue absorption also undergo changes. The protein macromolecules of the actomyosin complex exist in two (or four) states differing from each other by the features listed. The entire sample spontaneously converts from one state to another (and vice versa) with the synchronous character of such transitions in the macro-volume of the solution being proven (a) by experiments with simultaneous assays from different parts of the vessel, and (b) by experiments with simultaneous fixation of a solution previously taken from the total vessel.

Synchronization of these reversible changes of macromolecular states is preserved for dozens of minutes in isolated portions of the protein solution.

All the data available including, (a) the reversible changes of the functional properties of the macromolecule without any detectable changes in the main polypeptide chains and without changes in the effective macromolecular shape (light scattering measurements); (b) dependence of the oscillation amplitude on "nonchemical" factors such as ionic strength and on the type and concentration of nonpolar components in the medium, and (c) the specific dependence on temperature, taken together, these data allow the phenomenon described to be considered a result of conformational oscillations. These are synchronous for all macromolecules in the total sample, i.e. spontaneous reversible transitions of all macromolecules in the sample from one conformational state to another, and vice versa.

The process of conformational transition is similar to that of phase transition, for instance, where the process of crystallization of the whole sample is initiated by a particular macromolecular seeding crystal. The synchronization of this transition is achieved via macromolecular interactions through the medium, most probably by means of waves of structural transformation of both the medium and the macromolecules.

To carry out conformational oscillations in each separate macromolecule, the oxidation-reduction equilibrium of its functional groups is important. The amplitude of the observed oscillations decreases abruptly in the presence of oxidants and reductants such as glutathione, sulfite, hyposulfite and iodine in concentrations more than equimolar with that of the protein. The amplitude of the oscillations increases sharply at low concentrations of oxygen. The oscillations cease in formalin solutions, ATP solutions (oscillations of ATPase activity) and in 0.1 M ammonium hydroxide and ammonium nitrate solutions.

The amplitude of the conformational oscillations increases abruptly within narrow concentration ranges of some aliphatic alcohols (isopropanol, isopentanol). For most nonpolar substances, two more or less narrow concentration zones exist that either increase or decrease the amplitude of the oscillations (all the aliphatic alcohols from methanol to octanol, acetone, ethyleneglycol, glycerin and dichloroethane).

This summary paper is based on work published between 1958 and 1968 (1-7).

REFERENCES

1. С.Э. Шноль. Spontaneous synchronous transitions of actomyosin molecules in solution from one state to another. Вопросы медицинской химии. т. 4, вып. 6, стр. 443-454, 1958г.

2. С.Э. Шноль, Х.Ф. Шольц, О.А. Руднева. Change of protein adsorption ability with spontaneous changes of actomyosin state in solution. Вопросы медицинской химии. т. 5, вып. 4, стр. 259-264, в959г.

3. С.Э. Шноль, О.А. Руднева, Е.Л. Никольская, Т.А. Ревельская. Oscillation of spontaneous transition amplitude of actomyosin preparations from one state to another during storage. Биофизика, т. 6, вып. 2, стр. 166-171, 1961г.

4. С.Э. Шноль, Н.А. Смирнова. Oscillations of SH-group concentration in solutions of actomyosin, actin and myosin. Биофизика, т. 9, вып 4, стр. 532-534, 1964г.

5. С.Э. Шноль. Synchronous conformational oscillations of actin, myosin and actomyosin molecules in solution. В ст. Молекулярная биофизика. Изд. "Наука", Москва 1965г. ред. Г.М. Франк.

6. С.Э. Шноль. Conformational oscillations of macromolecules. В кн. Колебательные процессы в биологицеских и химических системах. стр. 22-41, Изд. "Наука", москва 1967г.

7. С.Э. Шноль. Influence of light and environmental properties on the conformational oscillation amplitude of actomyosin. Биофизика. т. 13, вып. 4, 1968г.

OSCILLATIONS IN MUSCLE CREATINE KINASE ACTIVITY

E. P. Chetverikova

Institute of Biological Physics,
Academy of Sciences of the USSR,
Puschino on Oka, Moscow Region, USSR

Some investigators, particularly in the last few years, have observed oscillatory kinetics in enzymic processes (1). Oscillations which occur in enzyme systems may be divided into two categories (2): kinetic or regulatory oscillations of the reaction rate which arise in operating systems as a result of substrate or product action (3-6), or the so-called conformational oscillations of isolated protein molecules which probably appear spontaneously (2,7-9).

Muscles have been found to manifest a number of processes of an oscillatory nature. Goodall (10) observed tension oscillations of a striated muscle fiber after the addition of phosphocreatine (PC). Oscillatory contractions have also been discovered in the glycerinated muscle fiber of insects (11). Chance and his colleagues (12) observed oscillations of the NADH content in a heart muscle. Vinogradov and Kondrashova (13) have described an oscillatory character in the changes in the NADH content of mitochondria isolated from the heart muscle. Frenkel (6) worked on a study of the mechanism of continuous oscillations in the NADH content of a heart muscle extract. This author considered the oscillations in the NADH content of the glycolytic system to occur as a result of activity changes of phosphofructokinase and glyceraldehyde-3-phosphate dehydrogenase due to the action of products and substrates of the reactions.

Shnoll and others (2,7-9) have succeeded in demonstrating oscillations in the purified contractile proteins of muscle: actin, myosin and their complex. The oscillations involved repeated synchronous changes of ATPase activity, the quantity of accessible thiol groups and the adsorption properties of the above mentioned proteins in solution.

Irreversible conformational changes of creatine kinase have been observed by optical rotatory dispersion and antibody action (14,15) as well as by the reaction with inhibitors (16,17). This paper deals with the results of a study of oscillations in the activity of creatine kinase, a globular muscle protein, during experiments on tissue extracts and with the purified enzyme.

Methods

Two methods were used for the purification of creatin kinase. The method of Kuby, Noda and Lardy (18), provided a specific activity of 60 units per mg protein, but the modified Kuby method (18) yielded about 200 units per mg. Our modification of the method will be published later.

The protein content of the enzyme preparations was established by the Lowry method (19), while the rate of the forward reaction was determined by the phosphocreatine (PC) formation and of the back reaction by creatine accumulation.

Tests for PC formation were carried out at $30^\circ C$. Incubation mixtures contained 1-3 µg of enzyme, ATP and magnesium acetate in amounts specified in the figure legends and 24 µmoles of creatine in 1 ml of 0.1 M glycine buffer at pH = 9.0. After a 2 min. incubation, the reaction was stopped by addition of 0.25 ml of 3.5% ammonium molybdate in 5N sulfuric acid (20). The amount of creatine formed was determined by the reaction with alkaline picrate.

For the measurement of the ATP-forming reaction using creatine assay the sample contained 1 µmole of ADP, 3 µmoles of phosphocreatine, 10 µmoles of magnesium acetate and 1 µg of enzyme in 1 ml of 0.05 M tris acetate buffer, pH 7.0. After a 2 min. incubation the reaction was stopped by addition of 1 ml of a 2 mM solution of p-chloromercuribenzoate (p-CMB) in 0.5 M NaOH. The creatine formed was determined using α-naphthol and diacetyl (21,22). Barium and sodium salts of phosphocreatine which did not contain either creatine or inorganic phosphate were synthesized according to the method of Ennor and Stocken (23). Before preparing the incubation mixture barium was removed with Dowex-50 resin in the sodium form.

In experiments with the muscle extract both the

creatine kinase and the ATPase activity were determined. 5.0 ml incubation mixtures were taken from the assay medium at 1 min intervals and placed in properly cooled tubes containing an equal volume of 5% trichloroacetic acid. The mixtures were then analyzed for PC with acid molybdate and alkaline picrate, and inorganic phosphate (Pi) by zero-time extrapolation according to the modified method of Lowry and Lopez (24,25).

Photo-oxidation of creatine kinase was performed using visible light, with methylene blue as a photosensitizer. The kinetics of photo-oxidation were studied by taking samples for activity determination at 5 min intervals. (For details see 26).

Oscillations of creatine kinase activity in the protein solution were followed by incubation in the buffer at $0.\pm 2^\circ C$. The enzyme, stored at $-10^\circ C$ in a frozen or lyophilized state, was diluted in Tris-acetate buffer, pH 7.0, in a tube held in a water and ice mixture. At 1 min or 3 min intervals, 0.1 ml samples of enzyme solution containing 5 mg protein/ml, were taken from the tube and diluted with Tris-acetate or glycine buffer. The diluted solution, containing 10 μg protein per ml, was transferred to the incubation sample for determination of activity.

The standard deviation (σ) of activity values from the arithmetic mean in certain samples served as a measure of the oscillation amplitude. The oscillation amplitude due to experimental error was determined in control tests in which the tubes contained either creatine or creatinine instead of enzyme. For these the same procedure used with the enzyme solution was repeated; the solution was diluted 250 times, then added to the incubation mixture and the content of the substances involved determined. The standard deviation values were 0.37 μg creatinine and 0.21 μg creatine (4.7% and 3.4% respectively). Each of these values was obtained as a result of 6 experiments involving 20 assays.

Results

In the sample containing ATP, creatine, magnesium ion and heart muscle extract the rate of PC formation changes, with the process going more or less rapidly and intervals of PC formation alternating with those of PC disintegration.

(Figure 1). The intensive ATPase reaction results in additional decomposition of creatine kinase substrate and produces the inhibitors ADP and Pi (27). At the same time myokinase converts ADP into ATP and AMP, the latter also being an inhibitor of the reaction in question. Therefore, oscillations of the rate and shifts in the direction of the creatine kinase reaction observed in the muscle extract could be connected with the described enzyme systems, involving alternate accumulation and consumption of the substrates and inhibitors of this reaction. At the same time they might depend upon oscillations in the creatine kinase system as well.

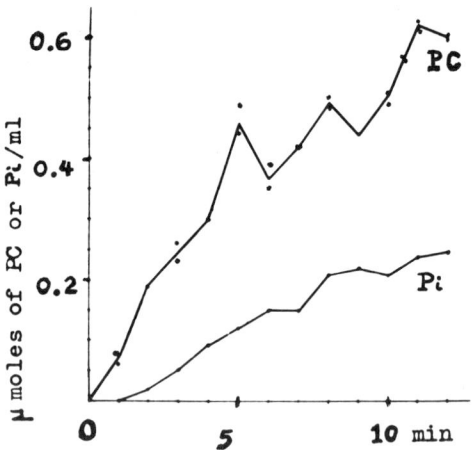

Figure 1. Kinetics of creatine kinase (PC) and ATPase (Pi) in the extract from rabbit heart muscle. The samples contained 5 µmoles ATP and 6 µmoles of magnesium acetate in 1 ml of incubation mixture.

Further experiments were carried out with enzyme purified by the method of Kuby (18) which contained no myokinase or ATPase. These experiments are illustrated in Figure 2, where curves 1 and 2 show the formation of PC with time. It can be seen from Figure 2 that when the enzyme is in excess and the rate is high, the PC formation is not monotonic even from the very beginning. The reaction rate undergoes repeated changes: periods of a positive rate are followed by those where the rate is less than or equal to zero. After 27 min on curve 1, or 15 min on curve 2, one can observe

changes not only in the rate but also in the direction of
reaction. PC formation frequently alternates with PC consumption, the changes in reaction direction occurring in
the state of "equilibrium" when the net reaction rate is
zero, which in Figure 2 corresponds to the straight line at
the 247 μg creatine level. If PC formation is studied over
a short time interval (30 sec or 15 sec) it can be seen that
in some cases the reaction rate in the "equilibrium" state
is 5 to 10 times higher than the initial reaction rate.

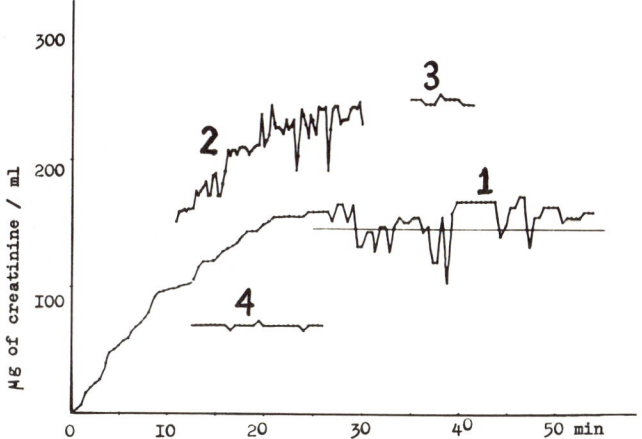

Figure 2. Kinetics of phosphocreatine (PC) formation by
purified creatine kinase. Samples contained 5 μmoles of
ATP and 6 μmoles of magnesium acetate. 1 - PC content
determined at 30 sec interval, 2 - same at 15 sec interval,
3 - an illustration of the experimental error, 250 μg of
PC sodium salt added to 1 ml of incubation mixture, 4 -
same, 1 mg of PC added.

Curve 1 (Figure 2) shows that the oscillations in the
reaction rate are different in shape. During the first 20
minutes of PC formation the oscillation shape resembles a
sinusoid with the amplitude being small and slightly exceeding the experimental error (curve 4). These oscillations will
be examined in the future and are therefore not considered
further in this paper. In the "equilibrium" state when the
rate of PC formation is zero (after 30 mins) the oscillations
take place more often. They have a higher amplitude and an

indeterminate shape. The amplitude of these oscillations considerably exceeds the experimental error (compare the upper part of curves 1 and 2 with curve 3). Assessing the oscillation amplitude in the "equilibrium" state by the standard deviation of certain samples from the arithmetic mean value of activity, we find it to be 9% while the error of the method in these experiments does not exceed 1%. We have failed to establish the oscillation period since a decrease in the sampling interval was followed by a decrease in the period. When taking samples at 1 min intervals, the period appeared to be 60 or 45 secs. Probably, the period is even shorter.

The study of the creatine kinase reaction kinetics using the purified enzyme revealed that oscillations in the reaction rate, and direction, occur during incubation of the enzyme with its substrates. Accumulation of reaction products, especially PC, is essential for the appearance of these creatine kinase oscillations (28). The investigation of the mechanism of creatine kinase activity oscillations in the presence of substrates is hindered by the fact that protein molecules, while left intact in solution will reveal oscillations of activity. Such oscillations have been observed during an investigation of the photo-oxidation of creatine kinase. It turned out that photo-inactivation of the enzyme in a slightly alkaline medium does not progress smoothly, the photo-inactivation rate undergoing repeated changes and sometimes dropping to zero (Figure 3A, curve 1).

Testing the freshly prepared enzyme, one can observe not only a decrease in oxidation rate but also a return of the creatine kinase activity either to its initial or even somewhat higher level (Figure 3B, curve 2). In this case the inactivation of the enzyme was occurring slowly. Conversely, in neutral and slightly acid media one can observe a smooth and complete photo-inactivation (Figure 3A, curve 2).

These observations induced us to study more thoroughly the dynamics of the changes arising in creatine kinase activity in solution. It turned out that over two hours the enzyme activity changed continually in all the preparations of creatine kinase obtained by the modified Kuby method (18). Figure 4 shows two experiments in which the samples were taken at 1 min intervals. It can be seen that the oscillations observed have a complicated shape. In general, two types of

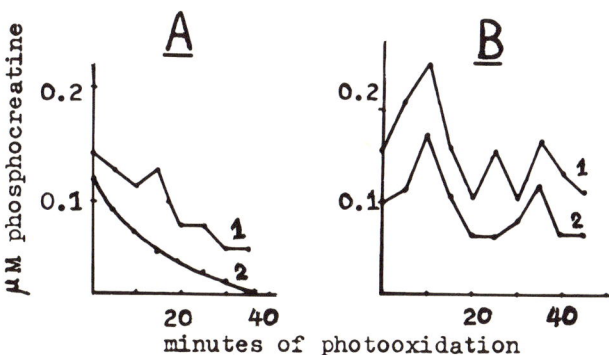

<u>Figure 3</u>. Kinetics of creatine kinase photo-inactivation. The samples contained 0.5 mg of protein in a cuvette. The activity assay contained 1 μmole of ATP and magnesium acetate in 1 ml. <u>A</u>. - stored enzyme preparation, 1 - photo-oxidation at pH 9.0, 2 - photo-oxidation at pH 7.0; <u>B</u>. - freshly prepared enzyme, 1 - activity in the control sample without methylene blue, 2 - same, with methylene blue.

oscillations have been detected: slow ones with a period of about one to two hours, and rapid oscillations, having a period considered to be less than one min, though a precise period for these oscillations has not yet been determined. The oscillation amplitude changes during the experiment, sometimes considerably. For instance, the standard deviation from the arithmetic mean in the experiment shown in Figure 4A is 35%. In another experiment, this value was 24% (Figure 4B).

The average oscillation amplitude which approximates to 30% for different experiments, actually varies from 10 to 100% which is several times, or even tens of times, higher than the experimental error (see Methods).

The distribution of the activity deviations from the arithmetic mean, analyzed by the statistical spectrum method, has been revealed to be substantially different from normal. This is illustrated in Figure 5 which shows the results of statistical handling of the experimental data given in Figure 4A. It can be seen that the most probable value of

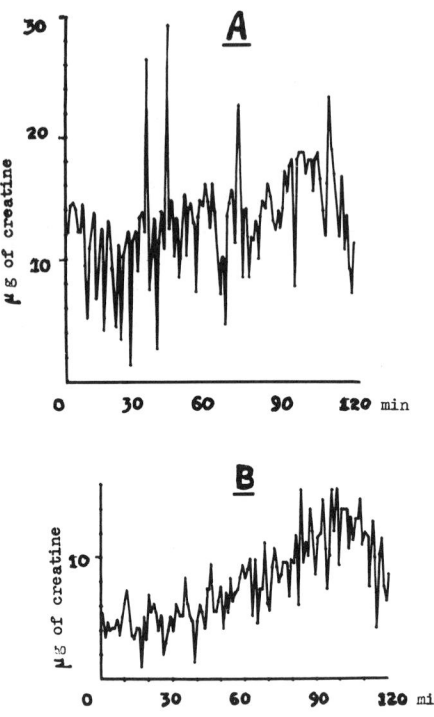

Figure 4. Oscillations of creatine kinase activity in a solution of Tris-acetate buffer at pH 7.0. The samples were taken at 1 min intervals and the activity estimated by creatine.

activity is expressed not by its arithmetic mean but by two specific values of "up" and "down" deviation from the mean which correspond to two more probable enzyme states.

The amplitude of the oscillation of enzyme activity in solution depends primarily on the quality of the enzyme preparation. Conditions affecting structure lessen the oscillation amplitude. Lyophilization causes a decrease in creatine kinase active and a two- or three-fold reduction of oscillation amplitude (Table I). Repeated freezing and thawing cause protein denaturation and disappearance of the oscillations. The oscillation amplitude for the same enzyme preparation depends on its concentration in solution. Relatively less concentrated solutions have an oscillation

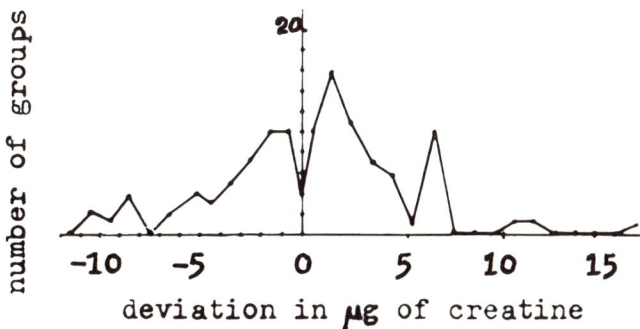

Figure 5. Statistical distribution spectrum of creatine kinase activity according to experimental results shown in Figure 4A. Deviations differing from each other by the experimental error x 4, and less, are united into groups. Abscissa: Absolute value of deviation of the groups having a given deviation from the arithmetic mean.

TABLE I

The Amplitude of Activity Oscillations of Creatine Kinase Stored in a Buffer Solution at $-10°C$ (Frozen Enzyme) and in a Lyophilized State.

Enzyme	Activity units/mg	Oscillation amplitude, σ	
		µg of creatine	% arithm. mean
Lyophilized	35	0.68	10
	40	0.86	12
	50	0.93	14
	40	0.63	12
Frozen	200	1.67	25
	160	1.91	35
	190	1.54	20
	212	2.06	30
	190	1.57	25
	190	2.37	30

amplitude of about 10%, while the amplitude of more concentrated solutions, (5 mg protein/ml), which correspond to the concentration of creatine kinase in muscles, is 20-30%. A further increase in protein concentration does not enhance the oscillations (Table II).

TABLE II

The Dependence of Oscillation Amplitude on the Creatine Kinase Concentration in Solution.

Protein concentration mg/ml	Oscillation amplitude, σ	
	µg of creatine	% of arith. mean
10.0	1.19	16
	1.36	16
	1.38	25
5.0	1.38	39
	1.34	21
	2.39	42
1.0	1.15	16
	1.36	20
	1.12	14
0.1	0.86	-
	0.65	9
	0.79	12

The oscillation amplitude does not depend on temperature. Though most of the experiments involving creatine kinase activity oscillations were made at thawing-ice temperatures in order to protect the protein from denaturation, the precaution was found to be unnecessary since the oscillations were the same at room temperature. After we found convincing evidence for repeated shifts of creatine kinase from one activity level to another in solution, we tried to find out those functional groups of protein that determine the oscillations. One of the factors causing oscillations of contractile protein molecules is the periodic oxidation of their thiol groups (9). Creatine kinase contains about 6 -SH groups per protein molecule (26,29). In each of two active centers of the dimer there is one thiol group that is strongly associated with the catalytic activity (30,31). Some experiments were carried out in which -SH groups were either bound by p-CMB or protected from oxidation with excessive cysteine. Table III shows that cysteine addition to the enzyme leads to a decrease in oscillation amplitude (molar ratio of enzyme to cysteine being 1 to 20). The absolute value of activity does not show considerable change.

TABLE III

Influence of p-CMB and Cysteine on the Oscillation Amplitude of Creatine Kinase Activity at pH 7.0 (Activity Estimated by Creatine Formation).

Experimental Variant	Standard Deviation, σ
1. Enzyme alone	2.4
	2.6
	2.3
Enzyme, 1×10^{-3}M cysteine	0.5
	0.7
	0.6
2. Enzyme alone	1.5
	1.3
	1.2
	1.3
	1.1
Enzyme, 5×10^{-5}M p-CMB	0.7
	0.8
	0.4
3. Enzyme alone	2.8
	5.0
	4.1
Enzyme, 1.6×10^{-4}M p-CMB	0.3
	0.2

p-CMB added to the enzyme in equimolar concentration (5×10^{-5}M) causes a decrease in the oscillation amplitude which was small in this series of experiments due to prolonged storage of the enzyme. This concentration of p-CMB could bind only a small fraction of the thiol groups since the ratio of the molar concentrations of the thiol groups and the inhibitor was 5 to 1. Since the p-CMB supresses the enzyme activity, 1×10^{-3}M cysteine was added to the incubation mixture to preserve the same arithmetic mean of activity. The cysteine addition, together with a proper dilution of the enzyme solution with inhibitor, removed the depressing action of p-CMB. During control tests it was found that cysteine addition to the incubation assay for determination of activity (unlike its addition to the oscillating enzyme) does not influence the oscillation amplitude. A higher concentration of p-CMB, 1.6×10^{-4}M (three times the enzyme concentration and approaching the concentration of the reactive -SH groups of the active center) completely removes the strong oscillations observed in the experiments without inhibitor (Table III and Figure 6).

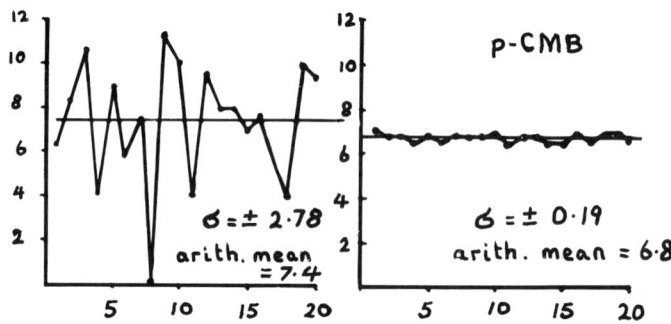

Figure 6. Influence of p-CMB on the oscillations of creatine kinase activity. A, oscillations in a control experiment without p-CMB; B, 1.6×10^{-4}M p-CMB added to the enzyme.

The study of oscillations in creatine kinase activity for different preparations and under various conditions helped to produce some data concerning the mechanism of the phenomenon in question. It was found that in order to have the capacity for oscillations the protein structure should necessarily be undamaged, with free thiol groups being present.

BIOCHEMICAL OSCILLATORS

Discussion

The experiments described here reveal that in enzyme solutions without substrates, oscillations in enzyme activity will occur, the only cause of their appearance residing in the conformational changes of the protein molecules. The oscillations of activity in enzyme solutions take place spontaneously and are not due to the influence of any substances that change their concentration. The enzyme solutions contained only buffer, (Tris-acetate or glycine) the nature of which had not been found to influence the oscillations. The oscillations of activity could not be caused by the changes in physical conditions. Thus heating and thawing of the preparation stored at $-10^{\circ}C$ had almost no significance, as the oscillation amplitude of freshly prepared enzyme was higher than that of the frozen one. The appearance of oscillations cannot be attributed to mechanical effects brought about by taking samples with a pipette, since the character of the oscillations does not change in experiments with intensive stirring of the enzyme solution. Since no effect on the oscillations of stirring was observed, it can be concluded that the oscillations of separate molecules of enzyme were rather well synchronized.

It is supposed that the spontaneous changes in the conformation of protein molecules are connected with interactions of their functional groups (9). One of these is the thiol group, the presence of which is necessary for initiation of creatine kinase oscillations. The group in question is also important for enzyme activity, its significance probably being associated with maintaining the catalytically active structure of the creatine kinase molecule (20).

Creatine kinase oscillations occur spontaneously in solution and are not caused by the alteration of either the chemical or the physical conditions of the experiment. The oscillations depend both on the enzyme concentration in the solution and on the state of its thiol groups, but they do not depend on temperature or on stirring of the solution. These oscillations may be classified as conformational, since according to Shnoll, _et al_ (2,7-9) it is just these features that are characteristic for myofibrillar proteins. Thus, the conformational oscillations are typical not only for the myofibrillar proteins but also for other muscle proteins of globular nature.

Oscillations in creatine kinase activity have been observed in operating creatine kinase systems during experiments with both purified enzyme and muscle extract. The mechanism of the oscillations in the reaction rate and direction, with the substrates and reaction products present, is more complicated. In this case, apparently, both kinetic and conformational oscillations are possible simultaneously, but when the system is in the equilibrium state, conformational oscillations probably prevail.

The amplitude of the oscillations in creatine kinase activity can be quite high. The activity may increase or drop two or three times over a short period of time (Figure 4), the highest level of oscillation being observed when the enzyme concentration corresponds to the physiological one. The enzyme in the oscillatory state reacts differently with other substances than it does in the absence of oscillations (Figure 3). Therefore, conformational oscillations of creatine kinase may be of physiological significance for the muscle contraction process and, probably, for the regulation of glycolysis.

Summary

1) The activity of purified creatine kinase in buffered solutions undergoes repeated changes, the oscillations observed having a complicated shape. The average standard deviation from the arithmetic mean of activity is 30%.

2) The creatine kinase oscillations may be classified as conformational ones since they occur spontaneously and depend on the enzyme concentration in the solution and the state of its thiol groups. They do not depend on the temperature or on stirring of the solution.

3) The oscillations in creatine kinase reaction rate and direction have been observed in operating enzyme systems in the presence of substrates and products of the reaction both with purified enzyme and muscle extract.

References

1. Sel'kov, E.E. In <u>Oscillating Processes in Biological and Chemical Systems</u>. Nauka, Moscow, 1967. p.7.

2. Shnoll, S.E. ibid, p.22

3. Betz, A. and Chance, B. Arch. Biochem. Biophys., 109, 585 (1965).

4. Chance, B., Hess, B. and Betz, A. Biochem. Biophys. Res. Commun., 16, 182 (1964).

5. Chance, B., Schoener, B. and Elsaesser, S. J. Biol. Chem., 240, 3170 (1965).

6. Frenkel, R. Arch. Biochem. Biophys., 115, 112 (1966).

7. Shnoll, S.E. Voprosy Med. Khim., 4, 443 (1958).

8. Shnoll, S.E., Rudneva, O.A., Nikoskaya, E.L. and Revelskaya, T.A. Biofizika, 6, 165 (1961).

9. Shnoll, S.E. In Molecular Biophysics, 1956, p.56.

10. Goodall, M.C. Nature, 177, 1238 (1956).

11. Jewell, B.R., Pringle, J.W.C. and Ruegg, J.C. J. Physiol. 173, 6 (1964)

12. Chance, B., Williamson, J.R., Jamieson, D. and Schoener, B. Biochem. Z., 341, 357 (1965).

13. Vinogradov, A.D. and Kondrashova, M.N. In Oscillating Processes in Biological and Chemical Systems, Nauka, Moscow, 1967, p.122.

14. Samuels, A.J. Ann. N.Y. Acad. Sci., 103, 858 (1963).

15. Samuels, A.J. Biophys. J., 1, 437 (1961).

16. Lui, N.S.T. and Cunningham, L. Biochemistry, 5, 144 (1966).

17. O'Sullivan, W.J. and Cohn, M. J. Biol. Chem. 241, 3104 (1966).

18. Kuby, S.A., Noda, L. and Lardy, H.A. J. Biol. Chem., 209, 191 (1954).

19. Lowry, O.H., Rosenbrough, N., Farr, A. and Randall, R. J. Biol. Chem., 193, 265 (1951).

20. Alekseeva, A.M. Biokhimiya, 16, 97 (1951).

21. Ennor, A.H. and Stocken, L.A. Biochem. J., 42, 557 (1948).

22. Ennor, A.H. and Rosenberg, H. Biochem. J., 51, 606 (1952).

23. Ennor, A.H. and Stocken, L.A. Biochem. J., 43, 190 (1948).

24. Lowry, O.H. and Lopez, G.A. J. Biol. Chem., 162, 421 (1946).

25. Peel, I.L. and Loughman, B.C. Biochem. J., 65, 709 (1957).

26. Chetverikova, E.P. and Alievskaya, L.L. Dokl. Akad. Nauk. SSSR, 181, 233 (1968).

27. Noda, L., Nihei, T. and Morales, M.F. J. Biochem., 235, 2830 (1960).

28. Chetverikova, E.P., Voronova, N.P. and Krinskaya, A.V. In Oscillating Processes in Biological and Chemical Systems, Nauka, Moscow, 1967, p.113.

29. Noda, L., Nihei, T. and Moor, E. Abstr. 5th Int. Congr. Biochem., 1961. p.277.

30. Rabin, B.R. and Watts, D.C. Nature, 188, 1163 (1960).

31. Watts, D.C. and Rabin, B.R. Biochem. J. 85, 507 (1962).

OSCILLATION OF SODIUM TRANSPORT ACROSS
A LIVING EPITHELIUM

James E. Allen and Howard Rasmussen

Department of Biochemistry
School of Medicine
University of Pennsylvania
Philadelphia, Pennsylvania 19104

Oscillations of ion flux have been observed across the mitochondrial membrane (1,2) but, somewhat surprisingly, there have been no reports of oscillations of ion transport at the level of the whole cell. Sodium transport across the bladder of the toad, Bufo marinus, can be easily and continuously monitored in vitro. We have observed large amplitude oscillations of sodium transport in this tissue with periods of from four to twenty minutes which are sustained for as long as ninety minutes.

The toad bladder serves in vivo as a reservoir of salt and water, drawn upon as needed under the influence of the hormones aldosterone and vasopressin. Its in vitro preparation is hardy and remains responsive to these hormones. Thus, it has served as one of the major tools in the investigation of the basic mechanism of action of these hormones.

The morphology of this tissue has been extensively described by Peachey and Rasmussen (3) and by Choi (4). Briefly, a layer on the urinary side of the bladder one cell thick is responsible for the salt and water transport. This cell layer is located on a connective tissue backing in which are located blood vessels, nerves, and an occasional smooth muscle bundle. On the serosal side of the membrane there is a thin discontinuous layer of peritoneal lining cells. The functional (in terms of sodium transport) mucosal layer is composed of three cell types: granular cells, mitochondrial-rich cells, and goblet cells, interconnected through tight junctions.

Data from this and other simple epithelia indicate that the transport of sodium occurs essentially as shown in the schematized cell of Figure 1. The sodium ion enters the cell at the mucosal, or urinary, border along an electrochemical gradient. The sodium diffuses across the cell to the serosal membrane and the Na^+-K^+ activated ATPase located in the lipid matrix of the membrane. This enzyme complex then translocates the sodium against an electrochemical gradient from the interior to the exterior of the cell, hydrolyzing ATP in the process. Some 30% of the ATP produced in this tissue seems to be used in transepithelial sodium transport (5).

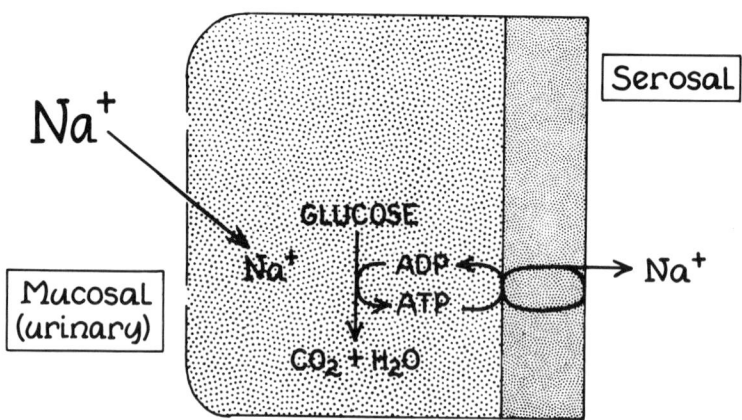

Figure 1. A schematized toad bladder cell showing the probable path of the actively transported sodium.

The transmembrane potential brought about in this sodium translocation can be used to electrically monitor the active ion transport, as described in the classic work of Ussing and Zerahn (6).

An experimental setup to carry this out is shown in Figure 2. It provides two separate circuits: one monitors the transmembrane potential, the other is adjusted to provide a current to drive the spontaneous potential to zero.

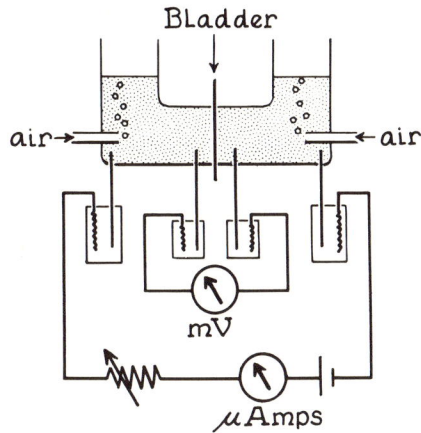

Figure 2. Apparatus for the determination of the short circuit current (SCC). Agar bridges lead from the buffer solution bathing the bladder to vials filled with saturated KCl. Ag - AgCl electrodes in the KCl lead to two separate circuits. One contains a millivoltmeter (mV). The other contains in series a variable resistor, a microammeter (µAmps), and a battery.

The current reading when the potential is zeroed is termed the short circuit current, and the bladder is said to be "clamped" under these circumstances. In this tissue, the short circuit current (SCC) is equivalent within about 5% to the rate of sodium transport. One feature to be noticed about the clamped condition is that less work need be performed in transporting sodium than in the unclamped state where sodium transposition is hindered by the transmembrane potential.

The oscillations of sodium transport were first seen when we began using a buffer solution described by Ling (7) to bathe these bladders instead of the standard amphibian Ringer's. The composition of the two is shown in Table I.

TABLE I

	Ringer's Solution	Ling Buffer
NaCl	113 mM	92.8mM
KCl	3.5	2.5
$CaCl_2$	0.9	1.0
$MgSO_4$	-	1.2
$NaHCO_3$	2.8	6.6
NaH_2PO_4	-	2.0
Na_2HPO_4	-	1.2

The initial observation is shown in Figure 3. The tracing demonstrates oscillatory behavior of short circuit current with a period of about 11 minutes. This was triggered by the addition of glucose to a bladder which had been under a high pressure of oxygen, a condition shown to drastically alter the pyridine nucleotide redox state, the levels of ATP, and to reversibly inhibit sodium transport (8).

We felt at that time that Mg^{++} was the critical factor in permitting or inducing oscillations because the Ling buffer in the absence of Mg^{++} would not support the oscillations (Figure 4). However, a subsequent observation using the Ringer's solution described above but with varying concentrations of calcium indicates that one of the critical factors is the total concentration of divalent cations. Figure 5 illustrates that the addition of vasopressin can trigger these oscillations, and that they depend on the presence of calcium in concentrations greater than 0.9mM. Notice that there is no other divalent cation in this solution. A train of oscillations is also triggered here by turning the voltage clamp apparatus off for several minutes,

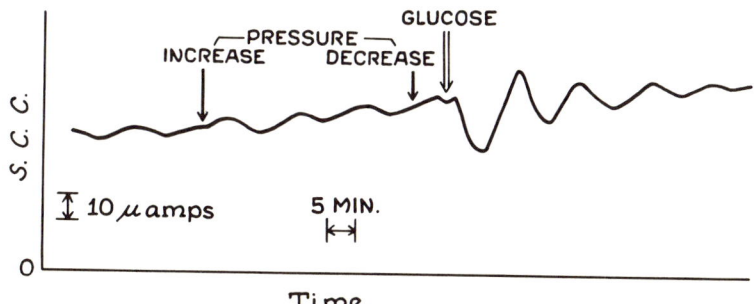

Figure 3. Oscillations in sodium transport induced by the addition of glucose (final concentration: 75 mg/100 ml) to a bladder deprived of substrate for five hours, and exposed to high oxygen tension. Pressure increased to 145 P.S.I.G. (pounds per square inch, gauge) from ambient. S.C.C.: short circuit current.

and then turning it back on. The difference in these two conditions, as we said earlier, is probably due to the decreased work of sodium transport occasioned by the clamped condition.

Investigations into the nature of these oscillations have been hampered by our inability to induce these oscillations at will. Recently, however, we have found that by deleting the HCO_3 from the Ling solution and adjusting the pH to 7.4 ± 0.1 (previous pH was 7.8), oscillations with periods of from 9 to 12 minutes can be regularly induced by reclamping the unclamped bladder.

Varying the amount of time that the bladder is unclamped gives rise to variations in the amplitude of oscillations as shown in Figure 6. Figure 7 is a graph of the time the clamp is off versus amplitude as a percent of baseline transport. This figure illustrates that one of the factors involved in the oscillations is a time-dependent alteration in intracellular composition as a result of the altered electromotive gradient across the tissue in the unclamped condition.

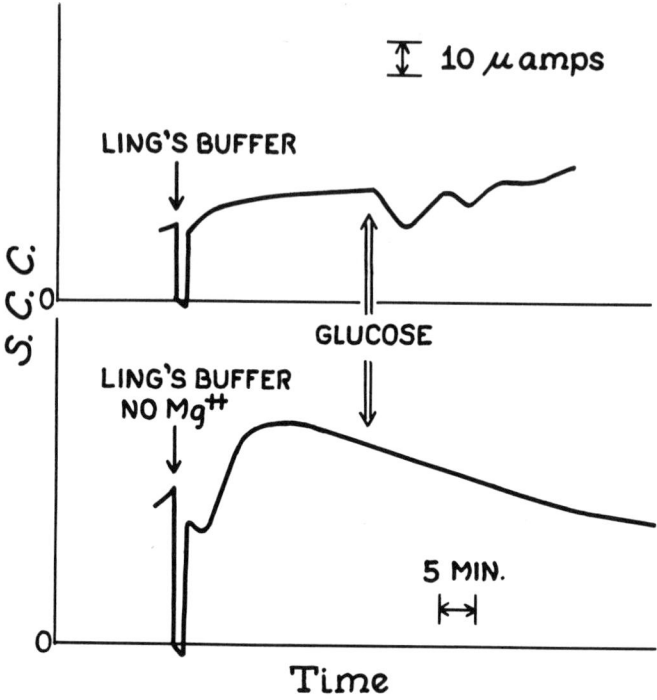

Figure 4. Tracing illustrating the absence of oscillations of sodium transport in the Ling buffer without magnesium. The control, with magnesium, oscillates upon the addition of glucose. This type of tracing is obtained by placing two of the chambers shown in Figure 2 side by side and clamping, with two separate circuits, two areas of the same half bladder.

Spontaneous oscillations have been observed several times. Figure 8 demonstrates a remarkable growth in amplitude, followed by a period of undamped oscillation, ending in a degeneration to a modified sawtooth waveform. This sort of spontaneous oscillation is always observed in a bladder whose S.C.C. is rising, presumably occurring when some threshold flux of either metabolite or ion is exceeded.

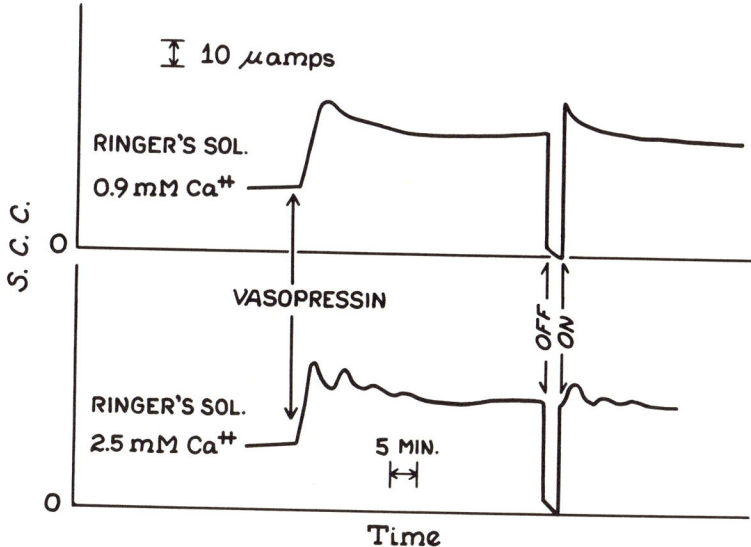

Figure 5. Oscillations triggered in ordinary Ringer's solution with a high concentration of calcium by the addition of vasopressin, and also by turning the voltage clamp off for several minutes, then back on.

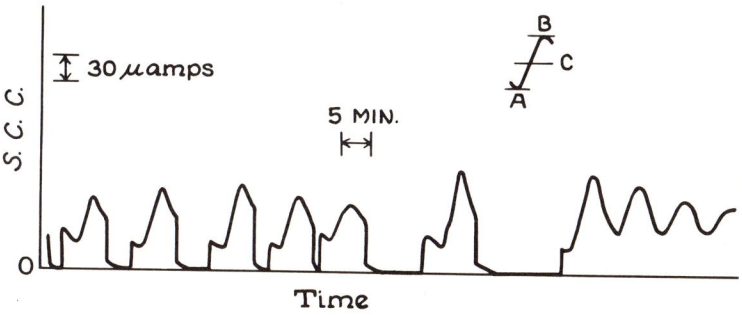

Figure 6. Variations in amplitude of the oscillations as a function of the time the clamp is off. The amplitudes of the first half-wave after turning the clamp on were quantified by relating the oscillations to the baseline value, thus: B-A/C x 100, see Figure 7.

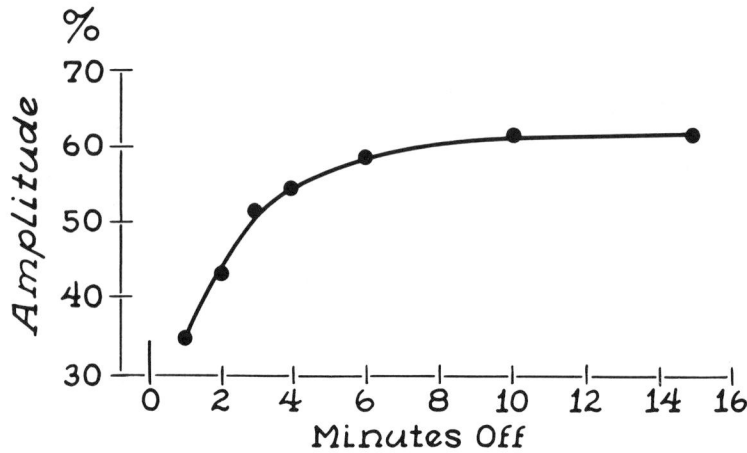

Figure 7. Amplitude of the first half-wave oscillation after turning the clamp on versus the time the clamp was off.

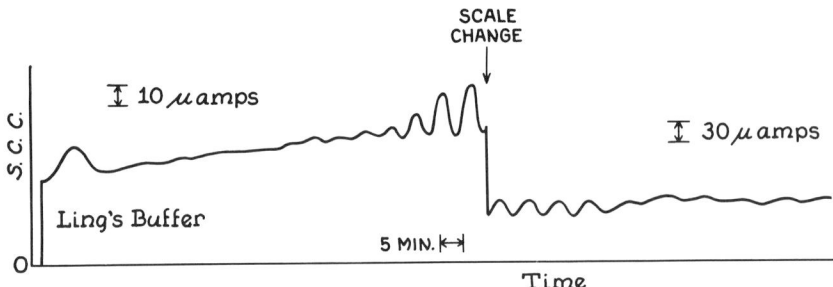

Figure 8. Spontaneous oscillations. Ling's Buffer.

A consideration of the model of sodium transport shown in Figure 1 indicates several possible mechanisms underlying the oscillations we have observed. They are obvious and need not be enumerated here. It should be pointed out, however, that Katchalsky and Spangler (9) have considered several of the theoretical aspects of such a system.

Thus, while it is not possible for us at this time to

clarify the mechanism underlying these oscillations, we feel that this observation presents a tremendous opportunity for furthering our understanding of the link between metabolism and ion flux, and also illustrates one way in which metabolic oscillations could drive rhythms in the remainder of the organism through an alteration of the ionic composition of extracellular fluids.

REFERENCES

1. Chance, B. and Yoshioka, T. Arch. Biochem. Biophys. 117, 451 (1966).

2. Graven, S.N., Lardy, H., and Rutter, A. Biochem. 5, 1735 (1966).

3. Peachey, L. and Rasmussen, H. J. Biophy. Biochem. Cytol. 10, 529 (1961).

4. Choi, J.K. J. Cell Biol. 16, 53 (1963).

5. Frazier, H.S. and Leaf, A. Medicine. 43, 281 (1964).

6. Ussing, H.H. and Zerahn, K. Acta Physiol. Scand. 23, 110 (1951).

7. Ling, G. A Physical Theory of the Living State. Blaisdell, New York, Appendix H. (1962).

8. Allen, J.E. and Rasmussen, H. Int. J. Clin. Pharm. In Press.

9. Katchalsky, A. and Spangler, R. Quart. Rev. Biophys. 1, 127 (1968).

ACKNOWLEDGEMENTS

Supported by a contract from the Office of Naval Research (NR 108-863) and a grant from the Public Health Service (AM 12013).

BIOCHEMICAL CYCLE OF EXCITATION
[Excitation is not so much a state as a process (1)]

M. N. Kondrashova

Institute of Biophysics of the
USSR Academy of Sciences,
Puschino, Moscow Region, USSR

Summary

Experiments are presented in favor of the suggestion that the metabolic state 3→4 transition represents an energy regulation of the physiological excitation-inhibition transition. An over-reduction of pyridine nucleotide, when compared to the initial level, was demonstrated in isolated mitochondria following ADP addition under conditions of limited substrate. This over-reduction of pyridine nucleotide, which may be due to endogenously generated succinate, provides regenerative compensatory processes after activity. The over-reduction of pyridine nucleotide is capable of summation which is quite typical for the hyperpolarization phase of physiological inhibition.

One of the reasons for the interest in oscillating reactions in biochemistry is the wish to discover their relationship to physiological reactions in cells and tissues of living organisms. There is one form of oscillating physiological reaction which is common to all living tissue; that is the oscillatory relationship between the intensity or duration of irritation and the tissue response, as shown in Figure 1 (1-9). Within the whole range of influences from the weak (subthreshold) to the strong (stress) ones, these relationships are represented by an M-shaped curve of a complicated waveform, given in Figure 1. The upward deflections of the curve correspond to an increase in external physiological reactions while the downward deflections correspond to a decrease of activity. The usual terms given for these phases are excitation and inhibition, respectively. I shall consider only the region of optimal threshold influences with well-pronounced physiological reactions which is shown in the right side of Figure 1 and which corresponds to the region a, b, c, on the left side of Figure 1.

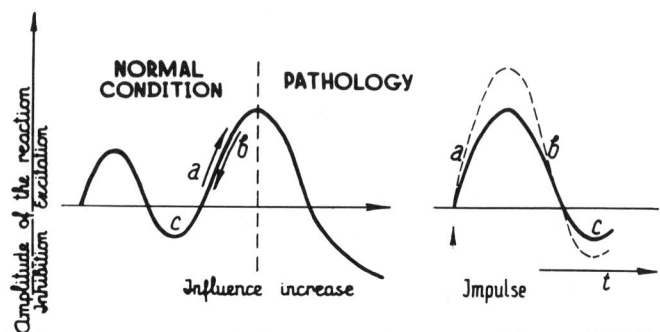

Figure 1. The cyclic relationship between physiological reactions and intensity (or duration) of irritation.

The following regularities of physiological excitation and inhibition development served as a guide for us in the analysis of the biochemical mechanism of these states. The essence of physiological inhibition is the compensation of changes caused by a previous period of activity and it provides a stabilization of the tissue steady state. It is quite typical of optimal physiological conditions that this compensation serves as a sure mechanism of autostabilization since it is switched on by the previous period of activity itself. This, inhibition is an obligatory result of excitation.

One piece of evidence for a close correlation between excitation and inhibition is the phenomenon of summation, given by the dotted line in the right side of Figure 1 (1-3, 12-14). More intensive excitation results in a more extensive compensation process. At last, from the physiological standpoint, inhibition becomes an active state of the tissue and one which is much more sensitive to injurious influences than is the first phase of excitation. This is a general phenomenon of many different forms of physiological activity which have only energy-dependence in common (10, 13, 15-17). Therefore it is logical to find the explanation in the regulation of energy metabolism. However, the kind of regulation of energy metabolism that is involved in the transformation from excitation to inhibition is completely unknown. The investigation of this problem demands a difficult combination of intact tissue observations coupled with detailed biochemical analysis.

The metabolic states of mitochondria may provide a new and rather useful approach to the understanding of biochemical mechanisms for the development of physiological excitation and inhibition phenomena. Of most importance in this respect is the ability of intact isolated mitochondria to maintain their steady-state and even actively restore it after the action of irritating influences has decreased the level of high energy compounds (18). This ability to develop the metabolic state 4 following state 3 is a great advantage of intact mitochondria, in short-term experiments, over other tissue preparations of oxidative phosphorylation systems. The existence of this regulation in mitochondria, which is in essence an autostabilization, led to the suggestion that one of the energy autostabilization mechanisms operating in living tissue is mainly preserved in intact, isolated mitochondria. If this is so, some metabolic states of mitochondria may serve as useful models of physiological excitation and inhibition.

The state 3→4 transition possesses many features common to the excitation-inhibition transition. Special attention must be paid to the remarkable parallelism between metabolic state 4 and physiological inhibition. Several features may be briefly mentioned, such as regenerative processes, the active character of respiratory inhibition, as expressed in the term "controlled state",* and the high sensitivity to damage when compared with state 3. According to Chance and Williams, milder conditions are necessary to obtain good respiratory control than those prescribed by Lardy and Wellman.

* The term "controlled state", introduced by B. Chance (19) is equivalent to the term "operative rest" used by A. Uchtomsky (3) for physiological inhibition. It is significant that in both of these cases, separated by a long time interval, by science and by territorial regions, the active term for the seemingly inactive state was preferred after the previous period of usage of more passive terms. One more significant point is that because of the greater analytical possibilities such a choice took considerably less time in biochemistry than in physiology.

But the 3→4 transition as viewed from the energy standpoint by the measurement of NADH fluorescence, does not reveal any signs of an overshoot which is so typical of physiological inhibition. The analysis led us to the conclusion that the absence of an overshoot in energy production in the State 3→4 transition is due to an excess of substrate under the conditions of the in vitro experiments. Some observations show that, in the cell, mitochondria exist under conditions of substrate deficiency, with pyridine nucleotides being incompletely reduced (20-22). The excess of substrate, especially the most active one, succinate, which is usually used in in vitro experiments for energy-linked NAD reduction measurements, results in a maximal level of NADH from the very start of the experiment. Such conditions do not permit the observation of any overshoot of energy production. Having used small concentrations of α-oxoglutarate, or higher concentrations of α-oxoglutarate, or even succinate with D-malate, Andrej Vinogradov obtained an NAD reduction overshoot in the State 4→3 transition, shown in Figure 2 (23). In some cases the course of the NADH change

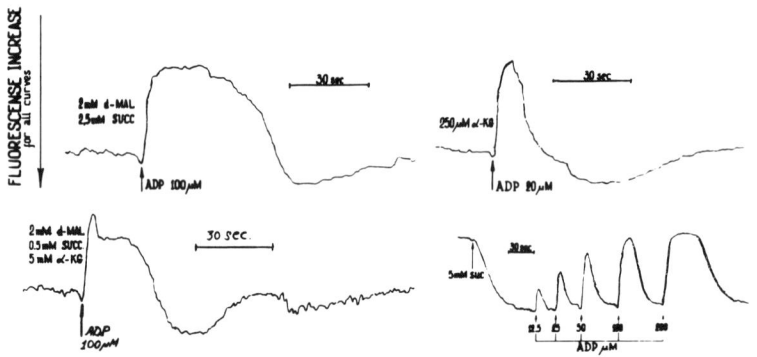

Figure 2. Various examples of NAD "over reduction" following ADP-initiated oxidation under conditions of low substrate supply in rabbit heart mitochondria. Experimental conditions are as previously described (23). In the lower right-hand section is given, for comparison, an example of the usual "linear" reaction for ADP.

has an oscillatory character. It is significant that even the details of the curves correlate closely with the NADH changes in intact muscle occasionally observed by Jobsis,

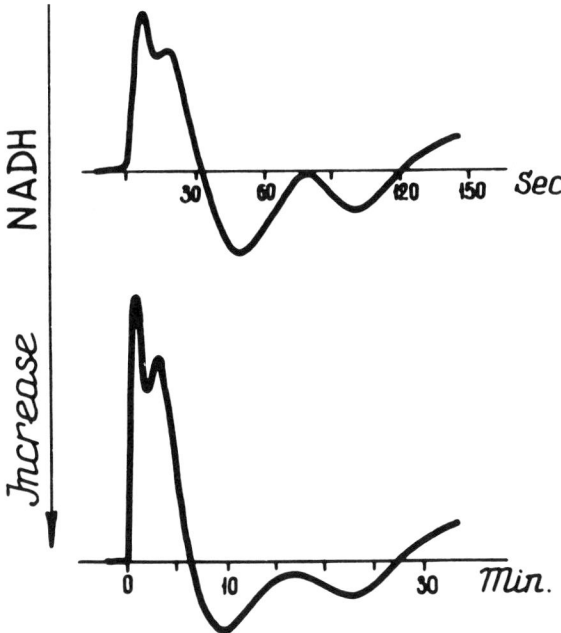

Figure 3. The comparison of NAD over-reduction in heart muscle mitochondria obtained in the experiments by Vinogradov and Kondrashova (23) (upper curve) with that obtained in intact muscle (lower curve). The original curve of Jobsis has been inverted.

as seen in Figure 3, (24), and are similar to the NADH curves from the electric organ (25). The phenomenon of summation, which is most typical of physiological overshoot, was also observed under these conditions in the two ways shown in Figure 4. The upper curve shows the effect of increasing ADP concentration while the lower curve shows repeated ADP additions at the peak level of NADH. The obtaining of summation at the mitochondrial level provides a biochemical basis for understanding such physiological phenomenon as facilitation, training, optimum physiological mobility and various others. The phenomenon of summation for reduced pyridine nucleotides is probably related to the remarkably stable oscillating muscle activity which continues for a long time with increased efficiency and without any signs of fatigue or muscle tetanus, both of which we obtained earlier (26,27).

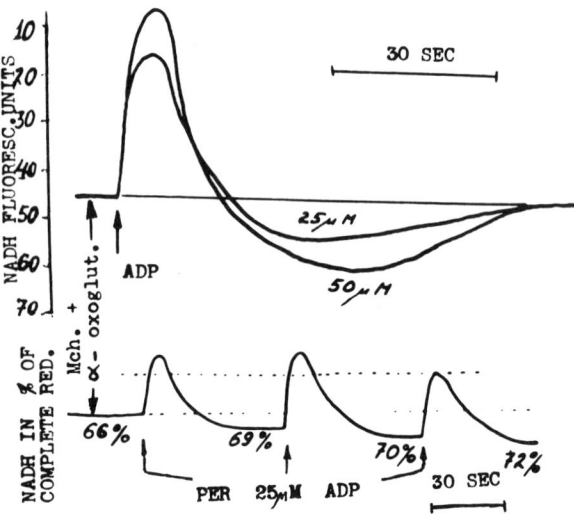

Figure 4. The phenomenon of summation in the mitochondrial NADH level, according to Vinogradov and Kondrashova (23).

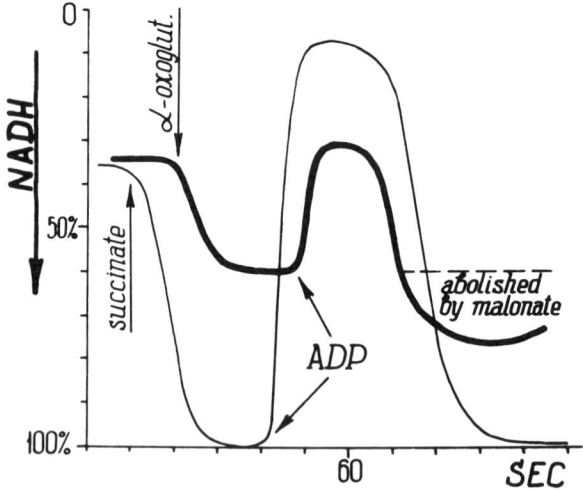

Figure 5. A comparison of the conditions of NAD reduction in the experiments of Vinogradov and Kondrashova (23) and in the usual experiments on succinate-linked NAD reduction.

The relationship between the conditions of our experiment and the usual conditions for measurement of energy-dependent NAD reduction is given in Figure 5. As is shown in this figure the initial level of NADH in our experiment before ADP addition was lower than usual, due to the substitution of α-oxoglutarate plus D-malate for succinate. Further analysis showed that over-reduction of NAD was completely abolished by malonate. It may be suggested that the over-reduction of NAD following ADP addition was due to endogeneously-generated succinate in the respiratory chain. Earlier we observed that following ADP addition the malonate-sensitivity of respiration and -SH group formation increased in mitochondria supplied with NAD-dependent substrates (28). All of those observations, together with some literature data showing the pronounced advantage of succinate in different energy-linked functions (29-32) (especially the data of Chance and Hollunger (33) on the even greater advantage of endogenously-generated succinate) led us to the following conclusion. An increase of compensatory regenerative processes under physiological inhibition may be due to a jump in the succinate-linked portion of the total oxidation reaction. Some considerations (see supplement to this paper) show that such a jump is quite possible under conditions of irritation when the ADP/ATP ratio increases. The well-known ability of succinate to monopolize the respiratory chain favors this possibility (34). Under optimal physiological conditions this switch to succinate oxidation must develop after each pulse of high-energy compound depletion. Consequently this biochemical mechanism is involved in each elementary act of excitation and; therefore, we propose to call it the "elementary biochemical cycle of excitation". Thus the switching from NAD-dependent substrates to succinate oxidation may represent the sought-after mechanism that is able to provide the over-restoration necessary following activity. This mechanism is only as certain as the biological system can be. In fact, such an elastic force for returning the system to the initial level, being created by the irritation itself, is generally safe within the physiological ranges of loading.

Some features of essential community between the controlled state in mitochondria and physiological inhibition have been listed above. The possibility of obtaining similar kinetics for their development and the phenomenon of summation prove the similarity between the two states. The detailed comparison of biochemical and physiological cycles of excitation have been given elsewhere (35) but some main points will be characterized briefly below.

Time correlation between physiological and biochemical cycles of excitation. It is often said that biochemical processes are considerably slower than the changes of physiological parameters following excitation. But, on the one hand, biochemical processes are not slow according to modern rapid-recording kinetic measurements. On the other hand, the velocity of a physiological cycle of excitation varies greatly in different tissues and depends on the parameter measured. Thus, the regenerative period of trace electropositivity for different nerves lies within the range of 40-100-1000 msec and is much longer in nonconductive tissues. The measurement of the velocity of the tricarboxylic acid cycle under conditions approaching the physiological ones with low substrate concentrations, gives a time of about one second for complete turnover (36). The switching to succinate oxidation takes less time than does the complete tricarboxylic acid cycle. The main part of the mechanism described above, reversed electron transfer, like other electron transport reactions is realized in milliseconds, especially on being activated by a high-energy compound acceptor (37). As most of the precise time measurements of physiological parameters are made on intact nervous tissue and most for biochemical parameters are made on isolated mitochondria of nonconductive tissues, it seems rather probable that for the same tissue a relatively close correlation exists between physiological and biochemical cycles with respect to time. The comparison of NAD reduction following previous oxidation in heart muscle mitochondria and in intact muscle, given in Figure 3, shows that the physiological cycle of excitation can take even longer than the biochemical one.

Correlation of the biochemical cycle with the course of excitability changes following irritation: decrease of excitability as the result of increased production of high-energy compound. The course of NADH reduction in intact heart mitochondria and in the intact heart, presented in Figure 3, when compared with the cycle of physiological parameter changes shown in Figure 6, suggests the following relationship between physiological function and energy supply. It is generally accepted that the functional ability of tissue depends on the energy supply. The complete loss of excitability in the absolute refractory phase is connected with a fall of the membrane ion gradient and potential, both of which are energy-dependent (10). The restoration of excitability during the relative refractory and exultation phases is usually related to the restoration of both of these energy-dependent parameters. It is unclear, however, why excitability falls again during progressive restoration of the energy supply and the energy-dependent parameters in the subnormal period, - the latter being considered as a model of physiological inhibition (10-16,38,39). The following explanation may be proposed. The diminution of excitability at hyperpolarization, in contrast to that at depolarization, is caused not by a decrease

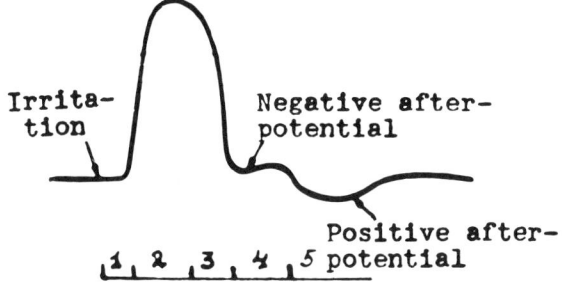

Figure 6. The course of various physiological parameter changes following irritation (10-14). Excitation involves depolarization and a fall in the level of high-energy compound. Inhibition involves hyperpolarization and a rise in the level of high-energy compound. 1 - Latent period, 2 - Absolute refractory period, 3 - Relative refractory period, 4 - Exultation phase, 5 - Subnormal period.

but rather by an increase of energy supply, the latter providing better resistance to irritation. Such an inhibitory mechanism must include a dependence upon the intensity of the irritation, which is in contrast to the impossibility of obtaining an external response to irritations of any force during the absolute refractory phase at hypopolarization. Indeed, with physiological inhibition, the external responses are abolished only at weak and moderate irritations, while the reactions to strong influences are preserved.

The reversed dependence of excitability on energy supply, which we propose for the initial and terminal phases of the excitation cycle, corresponds well to the fluorescence measurements of reduced pyridine nucleotide given in Figures 2, 3 and 4. The level of reduced pyridine nucleotide, which serves as an indicator of energy production reactions, falls at the beginning of the cycle and exceeds the initial level at the end of the cycle. This proposed mechanism permits the understanding of the transformation of hypopolarization into hyperpolarization (excitation into inhibition, respectively) which is an old mystery of physiology (15-17).

If succinate oxidation is responsible for the "overproduction" of high-energy compound and also for the phase of hyperpolarization (physiological inhibition), the addition of succinate might cause an increase in membrane potential and a decrease of the frequent spontaneous spikes because of the decrease of excitability (inhibition development). Such an effect of succinate was observed by Olga Jerelova in the giant neuron of mollusca, in situ. The data are presented in Figure 7. The addition of succinate

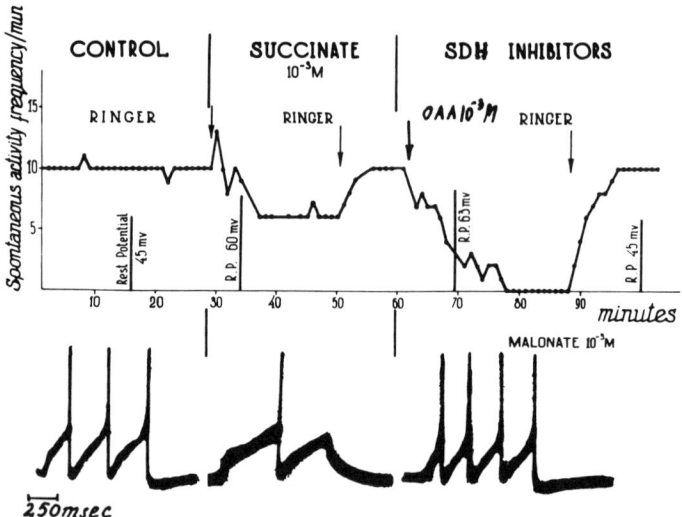

Figure 7. The influence of succinate, oxaloacetate and malonate on the spontaneous activity, membrane potential, and on the form and number of action potentials following a single pulse of an irritating current. Experiments by Olga Jerelova in the intact giant neuron of the mollusc Limneae stagnalis.

to the perfusion media, to give a final concentration of 10^{-3} M, resulted in a transition to the lower and stable ("controlled") level of frequency at the same time that hyperpolarization was developed. The application of "testing" pulses revealed a shift of the metabolic state of the neuron towards inhibition. The ascending part of the spike develops much more slowly and the number of spikes after a single pulse also decreases. The addition of oxaloacetate or malonate to prevent the utilization of endogenous succinate led to opposite changes in the neuron state. This was revealed more clearly following electrical stimulation of the neuron. (In spontaneous activity and in the resting state of neurons the difference between succinate and SDH inhibitors is not obvious in all cases, but exists nevertheless. This question will be analyzed in a special paper). Under the conditions of irritation both SDH inhibitors cause a facilitation of spike appearance, - the ascending part of the spike developing more rapidly and the number of spikes following a single pulse increasing. This burst of activity is unstable, however, and cannot be maintained for a long time. Non-stability of the cell state is rather typical of the action of SDH inhibitors, according to some other tests (see Figure 7a). In their presence the measured physiological parameters oscillate. The addition of succinate, on the other hand, leads to the stabilization of the metabolic state of the neuron.

The data given in Figure 7 show that succinate utilizing processes are responsible for the tissue resistance to the action of irritation, providing a control over the size and number of external responses. Succinate also gives a stability to the physiological state of the neuron. In other words succinate oxidation provides the usual function of physiological inhibition.

Our opinion is that Van-Rossum's observation with tissue slices (40) that succinate-initiated NAD reduction inhibits the sodium pump can also be interpreted as evidence for the dependence of physiological inhibition on succinate oxidation and reversed electron transport.

The considerations and data given above show that the whole sequence of changes of physiological parameters in an excitation cycle, including the most difficult point of inversion of membrane potential and of excitability changes

on the transition to inhibition, can be interpreted in terms
of mitochondrial respiration switching to succinate oxidation following irritation.

<u>Heat production in the cycle of excitation.</u> Heat production measurements are indicative of some biochemical
interconversions in tissue. The following phases of heat
production changes are now differentiated: the rapid, small
fall of heat production at the very beginning of irritation;
the well-known significant increase of heat production during activity; the pronounced and prolonged decrease of heat
production following work in a regenerative period, followed
by a slow return to the resting level (or one somewhat higher)
(41-45). The regenerative period has been linked to glycolysis in spite of the fact that this suggestion meets
some contradictions (see supplement). We believe that the
changes in heat production during the course of excitation
can be explained easily in terms of the "succinate concept"
considered here. Succinate-supported endergonic reactions
serve as good candidates for oxidative processes characterized by lower heat production. A wave-shaped curve of
heat production on excitation can be compared with succinate
oxidation and external tissue responses, as shown in Figure
8 (also see supplement). The irritation results in a
switch to succinate oxidation and a jump in the production
of high-energy compounds. Under the influence of weak
irritations, or at the very beginning of the action of
threshold ones, succinate oxidation provides tissue resistance and causes a latent period, or optimal work. Under
the action of stronger irritations succinate is formed more
extensively, the bulk of it being accumulated because of
the appearance of endogenous malonate and high concentrations of oxaloacetate. Under these conditions the energy
supply and increase in heat production may be due to fatty
acid oxidation (see supplement). Following the activity,
succinate-linked endergonic reactions predominate, causing
a pronounced temperature decrease. From the standpoint of
the biochemical cycle of excitation described here the
initial drop in heat production at the very beginning of
irritation is of most importance. This may be interpreted
as evidence that the total heat production with prolonged
activity is the result of summation of a number of elementary biochemical cycles of excitation (see supplement). Such
an approach allows the function to be broken down for easier
analysis. This approach also aids in the understanding that

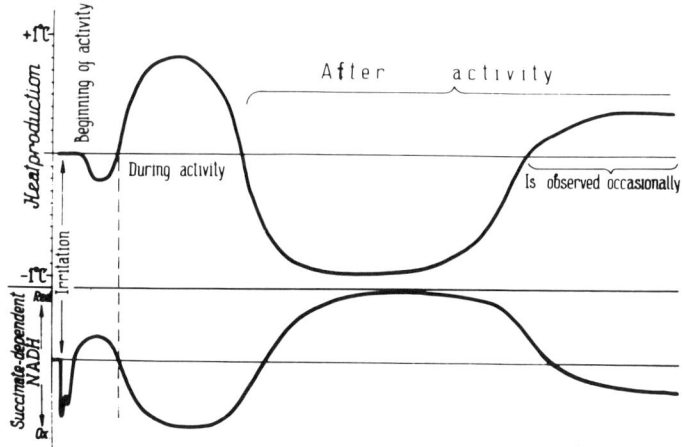

Figure 8. Comparison of heat production and succinate oxidation during and after the action of irritation.

even a physiological activity which seems constant and monotono s is a result of rapid oscillations of numerous elementary mechanisms which support a tissue steady-state.

REFERENCES

1. Насонов, Д.Н., Местная реакция протоплазмы и распространяющееся возбуждение. АН СССР, М-Л. 1962.

2. Введенский, Н.Е., О состояниях между раздражением и возбуждением при тетанусе. АН СССР, М-Л, 1960.

3. Ухтомский, А.А., Собрание сочинений, т. 2, АН СССР, М-Л, 1960.

4. Насонов, Д.Н., Александров, В.Л., Реакция живого вещества на внешние воздействия, М-Л, 1940.

5. Гельбрунн, Л., Динамика живой протоплазмы. ИЛ Москва, 1957.

6. Аршавский, И.Л., Введенский Н.Е. "Медгиз", М-Л, 1950.

7. Здродовский, П.Ф., Современное состояние экспериментальной иммунологии и ее задачи. М-Л, 1956.

8. Симонов, П.С., Три фазы в реакциях организма на возрастающий стимул.

9. Кулаев, Б.С., Лагутина, М.С., физиология и патология кровообращения и дыхания. Матер. Конф. Инт. Норм. и Патол. физиол. АМН СССР, М, стр. 46, 1960.

10. Ходжкин, А., Нервный импульс, "Мир", М, 1965.

11. Аккерман, Ю., Биофизика, "Мир", 1965.

12. Erlanger, J. and Gasser, H., Electrical Signs of Nervous Activity. Philadelphia, 1937.

13. Morgan, L.O., Physiological Psychology. McGraw Hill, N.Y., 1943.

14. Гуляев, П.И., В сб. Совр. Пробл. Электрофизиол. иссл. Нерв. Системы. Стр. 5, Парин, А.В., ред. "Медицина",1964.

15. Павлов, И.П., Лекции о работе больших полушарий головного мозга.

16. Eccles, J., The Neurophysiological Basis of Mind, Oxford, 1953.

17. Анохин, П.К., Внутреннее торножение как проблена физиологии. "Медгиз", 1958.

18. Chance, B. and Williams, G.R., Advances in Enzymol., $\underline{17}$, 65 (1965).

19. Chance, B. and Hollunger, J., J. Biol. Chem., $\underline{236}$, 1534 (1961).

20. Bellamy, D., Biochem J., $\underline{82}$, 218 (1961).

21. Chance, B., Schoener, B., Williamson, J. and Jamieson, D., Biochem. Z., $\underline{341}$, 357 (1965).

22. Terzuolo, C.A.T., Chance, B., Hadelman, E., Rossini, Z. and Schmelzer, P., Biochim. Biophys. Acta, $\underline{126}$, 361 (1966).

23. Виноградов, А.Д., Кондрашова, М.Н., В сб. Колебательные процессы в химии и в биологии. "Наука", 122, 1967.

24. Jobsis, F., J. Gen. Physiol., $\underline{50}$, 1009 (1967).

25. Aubert, X., Chance, B. and Keynes, R.D., Proc. Roy. Soc. B., $\underline{160}$, 211 (1964).

26. Кондрашова, М.Н., Корниенко, И.А., Биофизика, $\underline{10}$, 56 (1965).

27. Кондрашова, М.Н., Бюл. Эксп. биол. и мед., <u>7</u>, 43 (1956).
28. Кондрашова, М.Н., Озрина, Р.Д., Николаева, Л.В., Митохондрии, структура и функция. "Наука", М., 121,1966.
29. Chappell, J.B., Biochem. J., <u>64</u>, 212 (1962).
30. Chappell, J.B., Biochem. J., <u>90</u>, 237 (1964).
31. Höfer, M. and Pressman, B., Biochemistry, <u>5</u>, 3919 (1966).
32. Wenner, C., Laufenburger, J. and Hackey, J., Intnatl. Exch. Ir., No 1, 701 (1966).
33. Chance, B., <u>Energy-Linked Functions of Mitochondria</u>. Academic Press, New York, 1963.
34. Lehninger, A., <u>The Mitochondrion</u>, W. A. Benjamin, New York, 1964.
35. Кондрашова, М.Н., Сб. Митохондрии, "Наука", М., 1968.
36. Williams, G.R., Canad. Biochem., <u>43</u>, 603 (1965).
37. Chance, B. and Baltsheffsky, M., Biochem. J., <u>68</u>, 283 (1958).
38. Костюк, П.Г., Шаповалов, А.И., В сб. Совр.пробл.электрофизиол. иссл.нерв. системы, стр. 31, Парин, В.В., "Медицина", М., 1964.
39. Ройтбак, А.И., В сб. Совр.проблем.электрофизиол.иссл. нерв.системы. Стр.164, Парин, В.В., ред. "Медицина", М, 1964.
40. Van Rossum, G.D., Biochim. Biophys. Acta, <u>122</u>, 312 (1966).
41. Abbot, B., Hill, A. and Howarth, J., Proc. Roy. Soc. B., <u>148</u>, 149 (1958).
42. Путилин, Н.И., В сб. физиология нервных процессов. Киев, 337, 1955.
43. Путилин, Н.И., Вопросы физиологии, <u>7</u>, 44 (1954).
44. Березовский, В.А., Физиол. ж., <u>2</u>, 192 (1963).
45. Кондрашов, С.И., Динамика температуры мышцы и ее работоспособностъ. Автореф. дисс. канд., Киев, 1963.

POSSIBLE PATHWAYS FOR THE SUCCINATE CONCENTRATION BURST IN THE ACTIVE METABOLIC STATE.*

Y.V. Evtodienko and M.N. Kondrashova

Institute of Biophysics of the USSR Academy of Science
Puschino, Moscow Region, USSR

Abstract

Analysis of literature data shows two types of metabolic sequences leading to succinate accumulation in mitochondria in an active metabolic state (i.e., decreased ATP/ADP ratio). One of them occurs under aerobic conditions and is highly dependent upon fatty acid oxidation, while the other one occurs under anaerobic conditions and is dependent upon pyruvate accumulation. Here the suggestion is considered that the energy-providing function of mitochondria may also be realized at hypoxia and that succinate accumulated during activity is responsible for the bulk of the oxygen debt usually related to lactic acid.

The previous paper considered observations of an increase of malonate-sensitivity, NAD reduction, respiration and SH-group formation in mitochondria following the active metabolic state (1). These data may be interpreted as evidence for an increase in the succinate-linked portion of respiration after activity. Von Korff stated that in the active metabolic state a "cross-over point" appears at the succinate-dehydrogenase level (2). His measurements with labelled substrates showed that succinate accumulates in mitochondria in an active metabolic state and that this can be preferentially oxidized in a controlled metabolic state following the activity. These data confirm our conclusion on the succinate-dependency of NADH over-reduction during

---

* Supplement to the previous paper "Biochemical Cycle of Excitation" by M. N. Kondrashova.

the regenerative period after activity, which was considered earlier (1,3). An analysis of literature data shows that not just one but several reaction sequences may result in a succinate concentration "burst" in the respiratory chain under the conditions of depletion of high-energy compounds during active metabolism. Some of these possibilities are now considered.

1. Succinate concentration "burst" due to fatty acid oxidation under conditions of sufficient oxygen supply and CO_2 elimination (Scheme I). The utilization of ATP for physiological functions such as muscle contraction and ion transport is characterized by a low K_m of approximately 10^{-5}M. ATP utilization in endergonic synthesis seems to be characterized by a higher K_m. Therefore, when the ATP/ADP ratio is decreased during activity, ATP utilization in endergonic synthesis may be limited. In particular, lipid synthesis is inhibited under such conditions (Scheme 1-1), resulting in an accumulation of fatty acids and acetyl-CoA (1-2, 2-3). Acetyl-CoA is an allosteric inhibitor of pyruvate dehydrogenase (4). So, under conditions of acetyl-CoA accumulation the actual pathway for pyruvate transformation may be its carboxylation to form oxaloacetate (1-4). Oxaloacetate

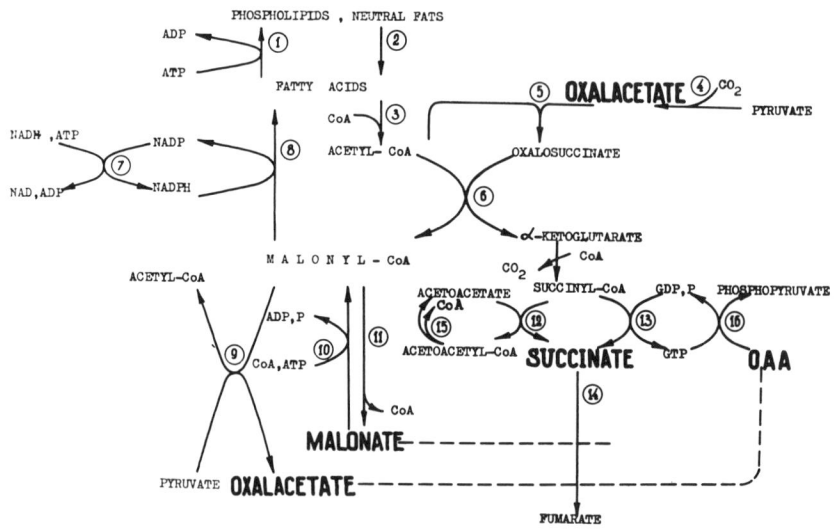

Metabolic Scheme I

could be involved in a condensation reaction with acetyl-CoA to form oxalosuccinate (1-5). Oxalosuccinate could participate in a transcarboxylation reaction with acetyl-CoA, with the formation of α-ketoglutarate and malonyl-CoA (1-6) - an intermediate of fatty acid synthesis. Under conditions of excitation, the ATP/ADP ratio and, as a result, the concentration of NADH and NADPH are decreased due to the inhibition of reversed electron transfer and energy-dependent transhydrogenase activity. Therefore, malonyl-CoA will not be utilized in fatty acid synthesis but will accumulate and an excess of pyruvate (due to the acetyl-CoA inhibition of pyruvate dehydrogenase) will favor oxaloacetate formation from malonyl-CoA and pyruvate (1-9).

Under conditions of low ATP concentration, different acylases are activated (1-6). On the one hand this results in malonate formation and, linked with it, an inhibition of succinic dehydrogenase (SDH) (broken lines indicate inhibition of SDH). On the other hand, due to deacylase activity, increased acetoacetate and CoA will be accumulated, resulting in a preferable conversion of α-ketoglutarate to succinate via reaction 1-12, the reactions 1-13 and 1-16 being inhibited. The inhibition of reaction 1-16 decreases the utilization of oxaloacetate, thus favoring its accumulation and giving rise to one more point of SDH inhibition (broken line - scheme 1). Therefore, as the ATP/ADP ratio decreases, the oxidation of fatty acids will increase with the accompanying formation of succinate and SDH inhibitors such as oxaloacetate and malonate (1-4, 1-9, 1-11, 1-16). Such a complex of conditions results in succinate accumulation.

After the physiological activity has ended, the ATP level rises, resulting in an elimination of malonate and oxaloacetate (1-10, 1-16). The preferential oxidation of accumulated succinate initiates reversed electron transfer and the restoration of both NADH and NADPH (1-7) and favors further utilization of malonyl-CoA in fatty acid synthesis (1-18).

According to these considerations, fatty acids rather than pyruvate are the main substrate of mitochondrial oxidation upon excitation. In connection with this possibility the following experiments are of interest. ADP causes a pronounced decrease in potassium ion transport through the mitochondrial membrane, the penetration of accompanying anions being correspondingly inhibited. Under conditions of ADP increase, the carnitine-dependent system of fatty acid transport that favors preferential utilization of lipids may not be inhibited. The switch from carbohydrate oxidation to lipid oxidation in the rest-activity transition has been shown in tissue slices (8). Under conditions of lipid utilization, succinate generation via the glyoxylate cycle seems rather probable (9), although the reality of its occurrence in mammalian tissues is still obscure.

2. **Pathways of succinate accumulation during prolonged activity under conditions of limited oxygen supply and CO_2 increase (Scheme 2)**. Under conditions of hypoxia due to activity, reduced intermediates, in particular NADH, accumulate and the ATP level is lowered. This rise in the NADH concentration results in an inhibition of oxidative reactions of glycolysis and the tricarboxylic acid cycle. However, in the presence of appropriate acceptors of electrons from NADH,

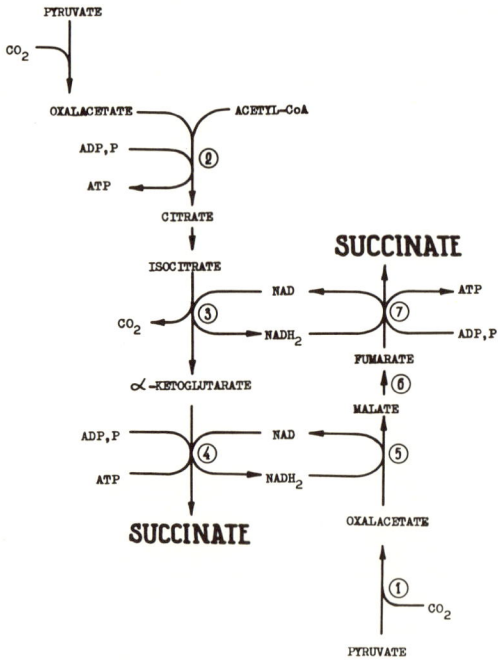

Metabolic Scheme 2.

glycolytic and tricarboxylic acid cycle reactions (9) may continue in such a way that ATP formation will occur in the system. Some of these reactions are given in Scheme 2. As was considered above, under conditions of metabolic activity the utilization of lipids increases with acetyl-CoA and pyruvate being accumulated. In the presence of CO_2 pyruvate will be carboxylated resulting in oxalacetate formation (2-1). Oxaloacetate and acetyl-CoA may be condensed by citrate synthetase into citrate, with the simultaneous formation of one molecule of ATP (2-2). Furthermore, citrate might be oxidized, via succinate, with the formation of ATP (2-3 and 2-4). Although these reactions are inhibited under conditions of hypoxia, in the presence of an appropriate amount of pyruvate (and oxaloacetate) the supplementary sequence of NADH-oxidizing transformations can occur allowing reactions 2-3 and 2-4 to be carried out. Thus oxaloacetate serves as an effective acceptor of electrons from NADH and oxidized NADH is transformed into malate (2-5). Malate is in equilibrium with fumarate. The latter, also a good acceptor of electrons from NADH, oxidizes NADH with the formation of succinate and ATP (2-7). This reaction is a reversal of reversed electron transfer and its equilibrium is shifted towards succinate formation in spite of ATP synthesis. The only additional succinate transformation possible, - into α-ketoglutarate, - is inhibited at low oxygen levels because of thermodynamic reasons.

In such a way, the described pathway of pyruvate transformation under hypoxia results in succinate accumulation and the formation of NAD and ATP. The high value of the total free energy favors this formation of succinate in spite of ATP synthesis. Using free energy values given by Burton (10), one can show the total free energy change of the following reaction to be about 12 kcal/mole.

$$\begin{matrix} CH_3 \\ | \\ C=O \\ | \\ COOH \end{matrix} + \begin{matrix} CH_3 \\ | \\ C=O \\ | \\ S-CoA \end{matrix} + 3\ ADP\ +\ 3\ P_i \longrightarrow \begin{matrix} COOH \\ | \\ CH_2 \\ | \\ CH_2 \\ | \\ COOH \end{matrix} + 3\ ATP\ +\ CoA$$

The described pathway of succinate formation seems to be possible during physiological excitation associated with

ATP depletion and oxygen exhaustion. After ATP depletion is completed, the oxygen concentration increases and the accumulated succinate will be preferentially utilized for some time until the attainment of new steady-state concentrations of tricarboxylic acid cycle metabolites corresponding to conditions of oxygen supply.

The considered pathways of succinate accumulation and ATP formation by mitochondria during hypoxia show the possibility for mitochondria to maintain their energy-providing system under such conditions. Our opinion is that glycolysis, with its considerably lower energy production power, does not provide an energy supply sufficient to support a really intact state of the cell. We believe that mitochondria are the only real candidates for this role. Glycolysis is included only during the action of excitations stronger than optimal. As was considered here, the respiratory chain can provide an energy supply even under anaerobiosis. Succinate accumulated in mitochondria during hypoxia can provide much more effective regenerative processes than can lactic acid, which is usually considered in this respect. If this is so, succinate is responsible for a great part of the so-called "oxygen debt" following work. This possibility is supported by observations that lactic acid disappearance can be responsible only for a small part of the oxygen debt; an oxygen debt develops in the muscle treated with monoiodacetate (11). Succinate accumulation has been demonstrated in mitochondria under hypoxia (12,13).

References

1. Kondrashova, M.N. This volume, p. 373

2. Von Korff, R.W. J. Biol. Chem. 240, 1351 (1965).

3. Kondrashova, M.N. and Severin, S.E. In Mitochondria, Nauka, Moscow, 1968.

4. Garland, P.B. and Randle, P.J. Biochem. J., 91, 218 (1964).

5. Hepp, D., Prüsse, E., Weiss, H. and Wieland, O. Biochem. Z., 344, 87 (1966).

6. Nakeda, H.J., Wolfe, J.B. and Wick, A.N. J. Biol. Chem., 226, 145 (1957).

7. Harris, E.J., Höfer, M.P. and Pressman, B.C. Biochemistry, 6, 1348 (1967).

8. Freinkel, N. In Metabolism and Physiological Significance of Lipids (Ed. R. Dawson and D. Rhodes), John Wiley & Sons, New York, 1964, p.455.

9. Krebs, H.A. and Lowenstein, J.M. In Metabolic Pathways (Ed. D.M. Greenberg), Academic Press, New York, 1960, p.129.

10. Burton, K. In Energy Transformations in Living Matter (Ed. H.A. Krebs and H.L. Kornberg), Springer Verlag, Berlin, 1957.

11. Belitser, V.A. In Chemical Changes in the Muscle, Moscow, 1940.

12. Randall, H.M. and Cohen, J.J. Am. J. Physiol., 211, 493 (1966).

13. Ozawa, K., Seta, K., Araki, H. and Handa, H. J. Biochem., 61, 512 (1967).

V

OSCILLATIONS IN GROWING CELL POPULATIONS

UNDAMPED OSCILLATIONS OCCURRING IN
CONTINUOUS CULTURES OF BACTERIA

D. E. F. Harrison[1]

Johnson Research Foundation,
University of Pennsylvania, Philadelphia, Pa. 19104.

Introduction

The chemostat is a technique for producing a steady-state in a growing culture of microorganisms under controlled conditions (1). The growth-rate of the organism is fixed by the feed rate of the medium, containing a simple limiting nutrient and, usually, the pH, temperature and oxygen supply are kept constant. Under these conditions the organism and substrate concentration should arrive at a stable level unique for the particular growth conditions involved (2). A simple mathematical model was proposed by Monod (3) to define this steady-state. Under the ideal conditions of the Monod model, with a well-controlled environment, continuing oscillations in population, substrate and respiration rate would not be expected although damped oscillations may occur (4). There have been several reports of damped oscillations in continuous cultures (5,6). Continuing undamped oscillations in well-regulated chemostat cultures are indicative of a more complex system and may occur through feedback interaction between cell metabolism and the environment or by the synchronization of oscillations based on wholly intracellular feed-back mechanisms.

The first report of stable oscillations in a well-regulated chemostat was of oscillations in oxygen tension in cultures of Klebsiella aerogenes (7). These oscillations of relatively long period (1-2 hours) could not be attributed to fluctuations in temperature, pH, medium feed-rate or

[1]. Present Address: Shell Res. Ltd., Sittingbourne,
                        Kent, England.

oxygen supply, but were shown to be a result of feed-back interaction between respiration rate and dissolved oxygen tension. More recently, oscillations of much higher frequencies (period 2-5 minutes) have been demonstrated in the NADH fluorescence of a chemostat culture of K. aerogenes (8) which also cannot be attributed to artifacts of the apparatus. These, possibly, represent oscillations arising from intracellular feedback mechanisms in individual cells which are synchronized in the chemostat culture. Studies of such oscillations should reveal much interesting information on the regulatory properties of growing organisms.

Low Frequency Oscillations in Respiration Rate

The oxygen tension in a continuous culture can be regulated by changing the partial pressure of oxygen in the gas supply, keeping all other conditions constant. Under these conditions the dissolved oxygen tension is given by:-

$$T_L = T_G - \frac{N}{K} \qquad (1)$$

where T_L is the oxygen tension in the liquid, T_G the partial pressure of oxygen in the gas, N is the oxygen uptake rate of the culture and K is a rate constant for oxygen transfer from gas to liquid phase (7). Therefore, if the oxygen uptake rate of the culture remains constant and K is unchanged, a straight-line relationship will exist between T_G and T_L. A plot of dissolved oxygen tension against partial pressure of oxygen in the gas phase for a chemostat culture of K. aerogenes is shown in Figure 1. At oxygen tensions above 10 mm Hg a linear relationship was obtained indicating that the oxygen uptake rate was constant and independent of dissolved oxygen tension. However, when the oxygen tension was reduced below about 5 mm Hg a steady oxygen tension trace could not be obtained but oscillations of oxygen tension and respiration rate occurred (7). During this oscillating phase some of the points in Figure 1 fall below the extrapolated straight line. From equation (1), this indicates that the expression $\frac{N}{K}$ has increased. In fact it was shown (7) that the oxygen uptake rate of the culture had increased by about 19% over the fully aerobic rate, when the oxygen tension was at a minimum during the oscillations. The amplitude and frequency of these oscillations varied with growth conditions but the waveform was similar, consisting of a sudden fall in

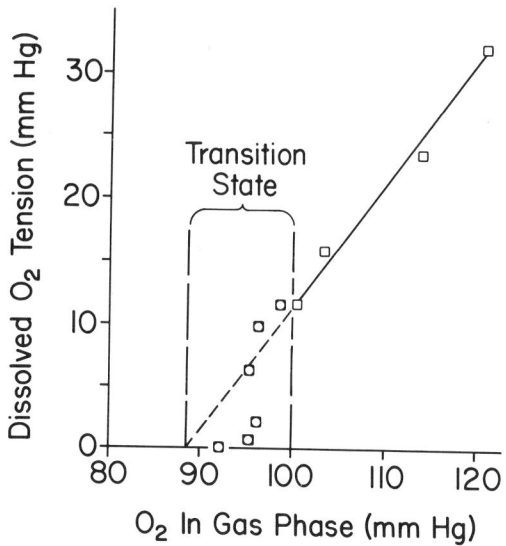

Figure 1. Relationship between dissolved oxygen tension and oxygen partial pressure in the gas phase of a culture of K. aerogenes. ▫: values obtained during fully aerobic steady-states. ◨: values obtained during the oscillating state. The culture was nitrogen-limited at a growth-rate of 0.2 hr.$^{-1}$ and pH 6.5 (7).

oxygen tension followed, after a period varying from several minutes to several hours, depending on growth conditions, by a sudden rise in oxygen tension. Typical oscillations in oxygen tension are shown in Figure 2.

That these oscillations were not an artifact of the aeration system was demonstrated by simulating a respiring bacterial culture with a feed of sodium sulphite solution (0.17M) into the culture vessel to give a constant oxygen uptake rate and low steady-state oxygen tensions. The simulated system did not show any tendency to oscillate. The pH and temperature were controlled throughout to ± 0.01 pH unit and ± 0.1°C respectively and no correlation existed between these parameters and the oscillations. Also, there were no fluctuations in the medium feed which could generate such oscillations.

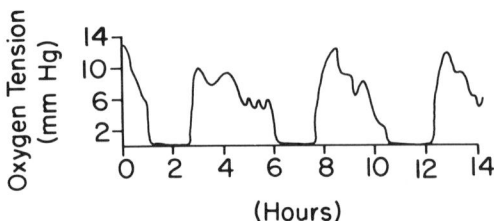

Figure 2. Oscillations in oxygen tension obtained in a continuous culture of K. aerogenes (7). The culture was nitrogen-limited at a growth-rate of 0.2 hr.$^{-1}$

Careful examination of the oscillations showed that the fluctuations in oxygen tension were accompanied by changes in q_{O_2}, CO_2 production and dry weight (Table 1).

TABLE 1

Comparison of respiration rate, glucose utilization and organism yield during oscillations in oxygen tension and aerobic steady-state in a chemostat culture of K. aerogenes. (Growth-rate = 0.2 hr.$^{-1}$, nitrogen-limited).

Condition	dissolved O_2 tension (mm Hg)	q_{O_2} (mmole/g/hr)	$q_{glucose}$ (mmole/g/hr)	Cell dry weight (g/l)
Aerobic Steady-state	>10	5.2 ± 0.1	1.17 ± 0.04	1.09 ± 0.06
Oscillating state:				
i) Maximum O_2 tension	6.13	4.5	1.13	1.13
ii) Minimum O_2 tension	< 0.2	6.4	1.48	0.86

Furthermore, the q_{O_2} values during the minimum oxygen tension phase were significantly higher than that of fully aerobic cultures. The cell dry weight concentration of the culture fell during the minimum oxygen tension phase although glucose utilization was constant. Thus the minimum oxygen tension phase represents a stimulation of respiration accompanied by a lower efficiency of energy utilization for growth, while the maximum oxygen tension phase represents a return to normal aerobic metabolism with an increased efficiency of growth.

The phase of oscillations could be shifted by a sudden change in oxygen tension. When, during a phase of maximum oxygen tension, the oxygen supply to the culture was interrupted for a short while (less than 1 minute) so that the oxygen tension fell below 2 mm Hg, the respiration rate of the culture increased and the culture shifted to the minimum oxygen tension phase and continued to oscillate as before. Similarly, temporarily increasing the oxygen supply to a culture in the minimum oxygen tension phase, so that the oxygen tension rose above 5 mm Hg, caused a sudden decrease in respiration rate and a shift to the maximum phase of oscillations. Undoubtedly, the oscillations resulted from feedback interaction between the respiration rate of the bacteria and the dissolved oxygen.

A mathematical model for these oscillations has been described (9) which shows that such oscillations can result from three conditions. One condition is a region of negative slope in the relationship between respiration rate and oxygen tension in the liquid. Such a relationship has been found for K. aerogenes growing in chemostat culture; the q_{O_2} of the culture being independent of oxygen tension at oxygen tensions above 10 mm Hg but increasing when the oxygen tension falls below 2 mm Hg (Figure 3). A similar phenomenon has recently been found in cultures of E. coli which might therefore also be expected to oscillate at low O_2 tensions (Harrison, unpublished data). The mechanism for the increased respiration rate at low oxygen tensions is not known but it may be a result of partial loss, or uncoupling, of oxidative phosphorylation efficiency (10).

The second condition for oscillations in the chemostat system is a time-dependence of the respiration rate of the culture. The simplest mechanism for time-dependence would be that the respiration rate of the culture depends on the concentration of a substrate which becomes depleted during the high respiration rate and replenished during the low respiration rate. Glucose, the sole carbon source supplied

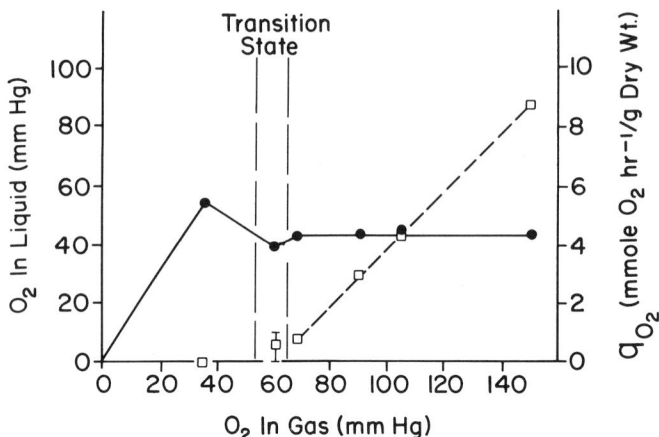

Figure 3. Effect of oxygen tension in the gas phase on q_{O_2} and dissolved oxygen tension in a chemostat culture of K. aerogenes. The culture was nitrogen-limited at a growth-rate of 0.2 hr.$^{-1}$ and pH 6.5 (7). q_{O_2}, •—•—•; dissolved oxygen tension □– –□.

to these cultures could not fulfil this role as it was always present in excess but, possibly, a metabolic intermediate could vary in concentration. Alternatively, time-dependence may be caused by a build-up of an inhibitor of respiration which is removed during the phase of low respiration rate. Another important possibility is that the time dependence be caused by changes in cell concentration: the cell concentration falls during the high respiration phase, and increases during the low respiration phase, as shown in Table 1, and the respiration rate of the culture depends on the number of organisms present.

The third condition required for oscillations is the diffusion limitation of the transport of oxygen from the gas to the liquid. This condition is not related to any physiological or biochemical mechanism in the cells. The resistence to diffusion at the gas-liquid interface is represented by the constant 'K' in equation (1).

The mathematical model shows that when the correct values are given to these three parameters - 1) the negative slope in the respiration/O_2 tension curve, 2) the time-

dependence of respiration rate, and 3) the transfer rate
of oxygen from gas to liquid, - continuing, undamped oscillations may result. Of course, it follows from the model
that if any of the parameters fall outside a certain range
then oscillations will not occur. In fact, it is found
that with increased stirring rates in the culture, which reduce the diffusional barrier to oxygen, stable oscillations
are not obtained. Also the frequency and slope of the oscillation is affected by the growth-rate of the culture
which presumably alters one or both of the first two conditions.

High Frequency Oscillations

In the course of studies on the metabolism of growing K. aerogenes an apparatus was developed to monitor
pyridine nucleotide fluorescence in a chemostat culture.
This technique is the subject of another paper (11). Briefly,
a Johnson Foundation Metabolite Fluorometer (12) was adapted
to record the light emitted at 460 mµ from a chemostat culture
when an incident source of 360 mµ was applied through the
glass wall of the culture vessel.

Damped oscillations were observed in pyridine nucleotide fluorescence after the culture had been subjected to an
anaerobic shock, as shown in Figure 4. The high frequency

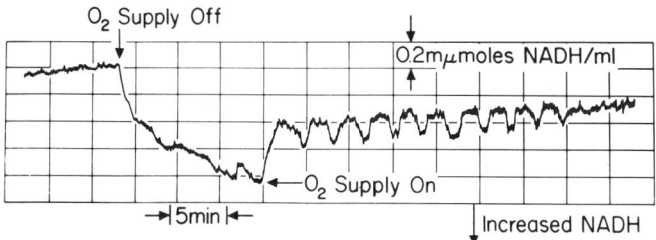

Figure 4. Damped oscillations in pyridine nucleotide
occurring in a chemostat culture of K. aerogenes following
an anaerobic shock. The culture was nitrogen-limited
at a growth-rate of 0.2 hr.$^{-1}$ Glucose was in excess.

of these oscillations and the fact that they were obtained
at oxygen tensions well above 10 mm Hg showed that they were
not connected with the oscillations previously obtained in
oxygen tension discussed earlier in this paper. On repeated

anaerobic shocks to the culture the oscillations became less damped until eventually undamped oscillations (shown in Figure 5) were obtained, which continued for several days if the culture was left undisturbed. Turning off the medium flow stopped the oscillations but they restarted when the medium flow was switched on again. However, if the flow was switched off for a prolonged period, -more than half an hour - then oscillations were not obtained on recommencing the medium flow until another anaerobic shock was applied.

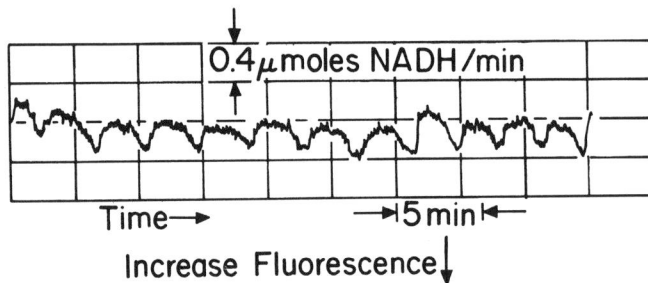

Figure 5. Continuing oscillations in pyridine nucleotide in a chemostat culture of K. aerogenes. The culture was nitrogen-limited at a growth-rate of 0.2 hr.$^{-1}$ Glucose in excess.

Examination of the important control parameters; pH, temperature, medium feed, stirring rate and gas supply revealed nothing which could generate such oscillations. That the oscillations were genuine reflections of metabolic fluctuations, rather than an artifact of the apparatus, was indicated by the fact that they could be started and stopped at will by applying anaerobic shocks and interrupting the medium feed. Further, the frequency of the oscillations could be varied with the growth conditions. Table 2 shows the change of frequency obtained on varying the oxygen supply.

When the oxygen tension of the culture was recorded at high sensitivity using a Mackereth electrode (13), it was found that this also oscillated and with the same frequency as the fluorescence. This is shown in Figure 6. Apparently, the oscillations in pyridine nucleotides were accompanied by oscillations in the respiration rate of the culture, although it was calculated that the amplitude of the oxygen tension oscillations represented only 1% of the

TABLE 2.

Variation of frequency of oscillations in pyridine nucleotide with changes in oxygen availability. Growth conditions: glucose-limited at a growth-rate of 0.2 hr.$^{-1}$

	Period of oscillations mins.
Excess O_2 state	2.5
Limited O_2 state*	3.0 - 5.0
Anaerobic	12.0

* Under limited oxygen conditions the frequency varied with the degree of oxygen deficiency.

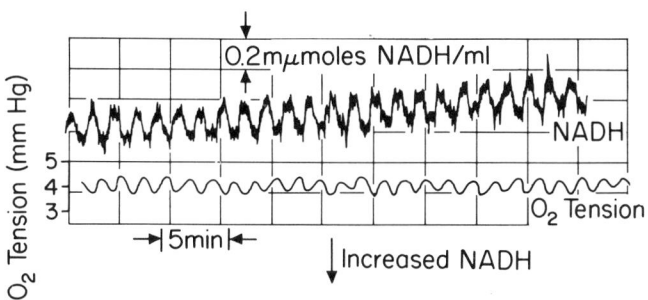

Figure 6. Continuing oscillations of pyridine nucleotide and oxygen tension in a glucose-limited chemostat culture. Growth rate : 0.2 hr.$^{-1}$ (19).

total respiration rate. Oscillations of pyridine nucleotide were also obtained under anaerobic conditions with similar shape and amplitude, but with a much longer period (see Table 2). Apparently respiration is not a prerequisite for the oscillations, and the oscillations in respiration must be a result rather than the cause of the oscillations in pyridine nucleotide.

On four separate occasions oscillations were generated spontaneously in a culture without the aid of anaerobic shocks, demonstrating negative damping. This always occurred within 24 hours after continuing oscillations had been stopped by cutting off the medium supply for a prolonged period. As this was observed only four times the exact events leading to this phenomenon were not ascertained, but the self-generating ability of the oscillations is an important feature.

Continuing oscillations in pyridine nucleotide, of similar frequencies to those reported here, have been obtained in cell-free extracts of yeast (14) and damped oscillations have been demonstrated in resting, intact yeast cells (15,16). These have been shown to be of glycolytic origin (17). The oscillations in pyridine nucleotide in continuous cultures of K. aerogenes were obtained even when the cells were respiring on succinate as the sole carbon source. It would seem that these oscillations are not of glycolytic origin and therefore not homologous with those studied in yeast.

The mechanism of the oscillations in pyridine nucleotide is still obscure; it is difficult to conceive a mechanism based on feedback interaction between cell metabolism and the environment which would give rise to such similar oscillations under such varied growth conditions. The oscillations were quite insensitive to small changes in external parameters such as pH, temperature, O_2 tension and medium feed rate. Possibly, these oscillations arise from intracellular feedback loops similar to those which give rise to the glycolytic oscillations in yeast (16). Conceivably there are oscillations perpetually occurring within individual cells, but these can only be observed when the culture is brought into synchrony by an anaerobic shock. A synchronizing mechanism must exist for oscillations to be maintained for such long periods of time and this implies a form of cell communication. Presumably, repeated anaerobic shocks would serve to strengthen the synchronizing mechanism. The self-generation of oscillations could be explained by the culture being brought into synchrony by interactions between cells, or syntalysis (18).

BIOCHEMICAL OSCILLATORS

Summary

Two examples of oscillations obtained in continuous cultures of K. aerogenes are described which are not simple artifacts of external influences but manifestations of the regulatory properties of the cell. The two examples differ widely in their properties. The first consists of oscillations in respiration rate having a period of several hours. While oscillations in respiration rate may be involved in the second type this is a secondary characteristic and the more important oscillating parameter is the redox state of the pyridine nucleotides. These oscillations have a period of only a few minutes.

The first type of oscillations can be explained by a model based on simple feedback interaction between cell respiration rate and dissolved oxygen tension. There is, as yet, no proposed mechanism to explain the second type of oscillations in pyridine nucleotide, but they probably arise from intracellular feedback interactions.

The observation of oscillations in a chemostat are interesting in that they would not be expected from the simple Monod-type model for bacterial growth (4) and indicate more complex regulatory systems. Also, continuing oscillations in a bacterial culture, which have their origins at the intracellular level, indicate the existence of a synchronizing mechanism which implies that interaction between cells must occur.

Acknowledgements

Some of this work was supported by USPHS GM 12202. Gratitude is due to Dr. Chance for his valuable guidance and providing the opportunity for carrying out this work and to Dr. E. K. Pye for his advice and helpful criticism.

References

1. Malek, I and Fencl, Z. (Editors). "Theoretical and Methodological Basis of Continuous Culture of Microorganisms". Academic Press, N.Y. and London, 1966.
2. Herbert, D., Elsworth, R. and Telling, R.C. J. gen. Microbiol. 14, 601 (1956).

3. Monod, J. Annls. Inst. Pasteur, Paris, 79, 390 (1950).
4. Koga, S. and Humphrey, A.E. Biotechnol. Bioengng. 9, 375 (1967).
5. Gilley, J.W. and Bungay, H.R. Biotechnol. Bioengng. 9, 617 (1966).
6. Günther, H., Knorre, W.A. and Bergter, F. Studia Biophysica, 2, 137 (1966).
7. Harrison, D.E.F. and Pirt, S.J. J. gen. Microbiol. 46, 193 (1967).
8. Harrison, D.E.F. J. Cell Biol. 45, 574 (1970).
9. Degn, H. and Harrison, D.E.F. J. Theoret. Biol. 22, 238 (1969).
10. Harrison, D.E.F. and Maitra, P.K. Biochem. J. 112, 647 (1969).
11. Harrison, D.E.F. and Chance, B. App. Microbiol. 19, 446 (1970).
12. Chance, B. and Legallais, V. I E E E, BME 10, 40 (1963).
13. Mackereth, F.J.H. J. sci. Instrum. 41, 38 (1964).
14. Pye, E.K. and Chance, B. Proc. Natl. Acad. Sci. 55 888 (1966).
15. Ghosh, A. and Chance, B. Biochem. Biophys. Res. Commun. 16, 174 (1964).
16. Pye, E.K. Can. J. Botany. 47, 271 (1969).
17. Betz, A. and Chance, B. Arch. Biochem. Biophys. 108, 36 (1965).
18. Winfree, A.T. J. Theoret. Biol. 16, 15 (1967).
19. Harrison, D.E.F., Maclennan, D.G. and Pirt, S.J. In Fermentation Advances (Ed. D. Perlman) Academic Press, New York, 1969. p. 117.

STABLE SYNCHRONY OSCILLATIONS IN CONTINUOUS CULTURES
OF SACCHAROMYCES CEREVISIAE UNDER GLUCOSE LIMITATION.

H. Kaspar von Meyenburg

Department of Microbiology
Federal Institute of Technology, Zurich, Switzerland

The continuous growth of Saccharomyces cerevisiae under aerobic conditions in a synthetic medium limiting for glucose is characterized by a strong dependence of the type of metabolism on the dilution rate, which equals the specific growth rate, as shown in Figure 1. At low growth rates there is a purely oxidative breakdown of glucose (respiration quotient, RQ = 1.0) resulting in a high yield of dry weight. With increasing specific growth rate - due to increased flow of fresh medium to the cell suspension - an increasingly fermentative type of metabolism arises as evidenced by an RQ > 1.0, ethanol production and decreased yield (1,2).

During various experiments on the growth behavior of Baker's yeast involving studies on catabolite repression (3,4) oscillations of the gas metabolism were often observed. These were detected by automatic analysis of the composition of the air-outflow from the fermenter (5). Such oscillations, shown in Figures 2A, 2B and 2C, appeared only in the range of oxidative growth in continuous culture. There is a strong relationship between the period, T, of these oscillations and the mean generation time, $\bar{g}$, at the corresponding dilution rate, and $T = \bar{g}/2$, $T = \bar{g}/3$, and $T = \bar{g}/1.5$, respectively, in Figures 2A, 2B and 2C, where $\bar{g} = 0.69/D$. This indicates that the oscillations are caused by partial division synchrony. By measuring the population distribution according to the cells in the different states of the budding cycle (1), i.e. counting the single (non-budding) cells and the cells in the budding phase (initial budding cells, late budding cells), the relationship between the oscillations of gas metabolism (V% $CO_2 \sim Q_{CO_2}$) and the budding cycle becomes

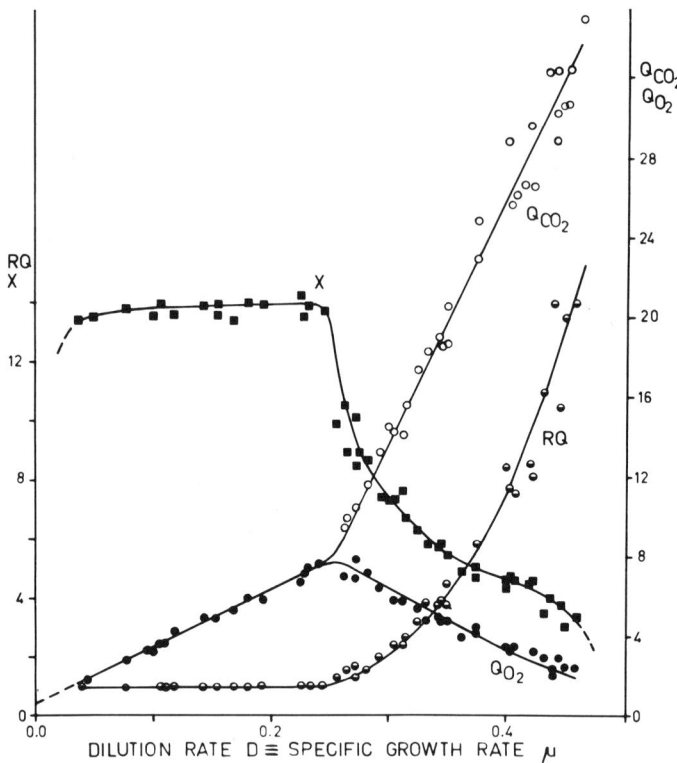

Figure 1. Aerobic growth of <u>Saccharomyces cerevisiae</u> in continuous culture under glucose limitation. Plot of dry weight, X, in the medium (mg/ml), specific oxygen uptake, Q_{O_2}, carbon dioxide production, Q_{CO_2} (mmoles/g dry weight per hour), and respiration quotient RQ (Q_{CO_2}/Q_{O_2}) as a function of the dilution rate = specific growth rate, μ (h^{-1}).

evident as shown in Figures 2A, 2B and 2C. The basic features of the changes of the gas exchange, the course of the change in dry weight, the nitrogen uptake rate, etc., during the budding cycle, at different growth rates, were reported and discussed earlier (2). It was shown that these changes of gas metabolism are due to the storage of reserve carbohydrates during the single cell phase, which is extended at low growth rates - whereas the duration of the

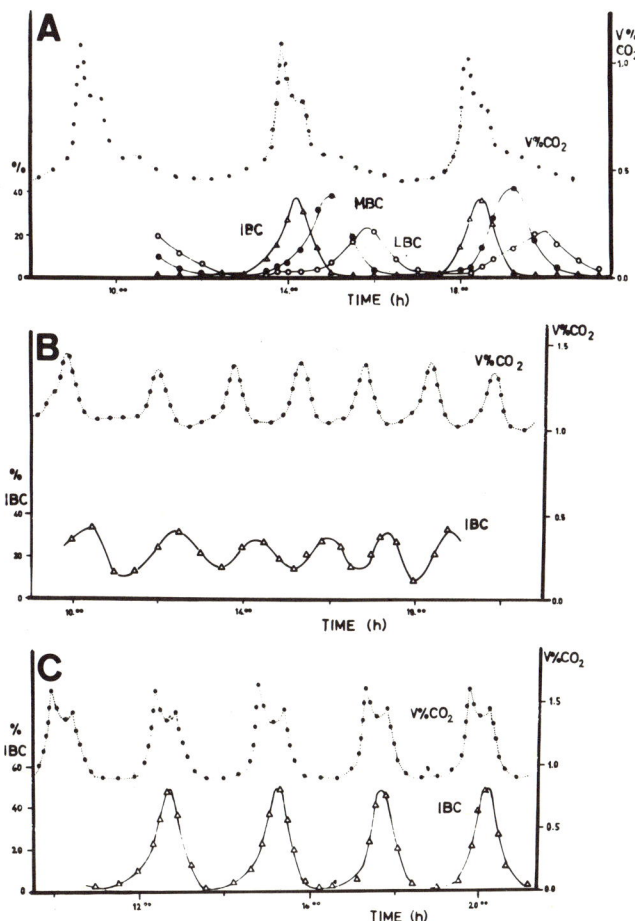

Figure 2. Examples of stable synchrony oscillations during glucose limited aerobic growth of Saccharomyces cerevisiae in continuous culture. Changes of the gas exchange ($Q_{CO_2} \sim V\%CO_2$) and the percentage of initial budding cells (IBC), late budding cells (LBC) and cells from the middle of the budding phase (MBC) as a function of time. A: Dilution rate, $D = 0.076$ h^{-1} → mean generation time, $\bar{g} = 9.1$ h, period of the oscillations, $T = 4.5$ h. B: $D = 0.255$ h^{-1} → $\bar{g} = 3.1$ h, $T = 1.0$–1.1 h. C: $D = 0.185$ h^{-1} → $\bar{g} = 3.7$ h, $T = 2.5$ h.

budding phase is constant, - and the breakdown of this reserve at the onset of and during the budding phase (1,2,6). This sequence of storage and breakdown of reserves enables the cells to complete the important budding phase in a rather constant time independent of the extracellular conditions.

I do not want to discuss here the problem of the intracellular trigger mechanism for the onset of breakdown of reserve carbohydrates at the beginning of bud formation, but I would like to point to another problem which is certainly closely related and which contributes to the discussion of the control of oscillations in biological systems.

TABLE I

Appearance and stability of division synchrony oscillations in the range of oxidative growth in continuous culture of Saccharomyces cerevisiae at different mean generation times following changes in the dilution rate. Growth medium: NL 18, 3% glucose (2).

Change of the dilution rate $D(h^{-1})$	Mean generation time $\bar{g}(h)$	Period T (h)	Number observed oscillations
0.265 → 0.22	3.15	0.9-1.2	>22
0.17 → 0.22	3.15	1.1	5
0.105 → 0.175	3.95	1.2-1.5	>34
0.22 → 0.19	3.65	1.7-2.2	>15
0.0 → 0.17	4.10	2.0-2.4	>14
0.175 → 0.135	5.15	2.4-2.8	>17
0.19 → 0.15	4.65	2.5-3.0	>25
0.26 → 0.10	7.0	7.4	2
0.0 → 0.075	9.25	9.0-12.0	3

BIOCHEMICAL OSCILLATORS

Division synchrony oscillations, shown in Figure 2, often exhibit high stability, as can be seen from the data given in Table I. The appearance of division synchrony in cell populations after a change of the growth conditions is not exceptional and is commonly used in several techniques for the study of synchronized populations (4,7,8,13). However, the high stability or the low damping of these division synchrony waves arising, for example after a change in dilution rate, is not easy to understand because synchrony of cell division normally decays rather fast (9) due to the unequal generation times of individual cells (10). Thus, the question arises as to how these synchrony oscillations are maintained, sometimes over more than 15 generation times, without damping out. Goodwin (11) postulated a self-stabilization of possible biological oscillations in the chemostat by the formation of a product (signal) and its re-utilization (action) in a later stage (in the sense of feedback) leading to the necessary interaction between the cells. He pointed out that repressors might be these couplers which lead to the stabilization of oscillations. In yeast it is most improbable that such macromolecules (proteins) could contribute to the maintenance of the oscillations because they can not penetrate the cell membrane.

I suggest that the concentration of sodium in the cell suspension in the fermenter is of importance for the interaction between the cells and therefore for the maintenance of division synchrony oscillations in yeast populations. When the rate of gas exchange increases because of the sudden breakdown of reserve carbohydrates, the specific uptake rate for nitrogen (NH_4^+) from the medium increases (2). $(NH_4)_2SO_4$ is the nitrogen source in the synthetic medium used. An increase in the uptake of NH_4^+ therefore results in an increased inflow of NaOH solution to the medium for the control of the pH at 5.5. This can be seen from the recording of the pH controller in Figure 3. The frequency of the small changes in the pH indicates the rate of NaOH inflow. When the gas exchanges decrease again the NaOH inflow decreases too. These changes lead to changes in the Na^+ concentration in the broth which are in close relation to the synchrony oscillations. They may therefore be the cause of a constant timing or reinforcement of the waves of division synchrony. Finn and Wilson (12) observed oscillations of the pH and the optical density of continuously growing populations of Baker's yeast (at relatively low growth rates) in a

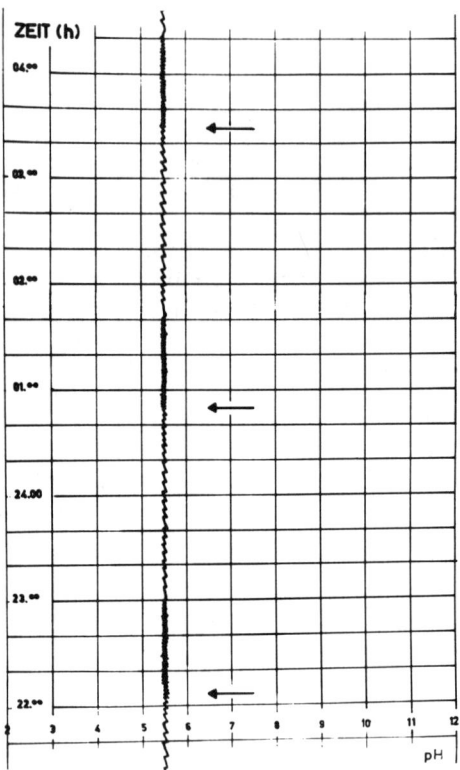

Figure 3. Recording of the pH controller unit during stable synchrony oscillations in continuous culture of Saccharomyces cerevisiae. The frequency of the pH changes is proportional to the inflow of NaOH solution to the cell suspension. Arrows point to increases of the gas exchange at the onset of budding. $\bar{g} = 4.1$ h, $T = 2.7$ h.

cultivation system without pH control. These may be interpreted on the basis of our study as effects of partially synchronous budding, also.

Another possibility for the stabilization of the oscillations is changes of the concentration of the limiting substrate (glucose), or a possible product (ethanol), during the cycles. Some preliminary measurements seem to rule out this possibility, the changes of the glucose concentration, for example, being rather weak.

If one was able to stabilize synchrony oscillations in continuous culture of yeast, or other microorganisms, at a period equal to the mean generation time, one could take great advantage of them for the detailed study of the division cycle under various growth conditions.

REFERENCES

1. Meyenburg, H.K.von. Path. Microbiol., 31, 117 (1968).

2. Meyenburg, H.K. von. In Proceedings of the 4th International Symposium on Continuous Cultivation of Microorganisms, Prague, 1968. p.129.

3. Beck, C. and Meyenburg, H.K. von. J. Bacteriol. 96, 479 (1968).

4. Meyenburg, H.K. von. Dissertation ETH, Zürich, 1968.

5. Fiechter, A. and Meyenburg, H.K. von. Biotechn. Bioeng., 10, 535 (1968).

6. Küenzi, M.T. and Fiechter, A. Arch. Mikrobiol., 64, 396 (1969).

7. Dawson, P.S.S. Can. J. Microbiol., 11, 893 (1965).

8. Mueller, J. and Dawson, P.S.S. Can J. Microbiol., 14, 1115 (1968).

9. Hirsch, H.R. and Engelberg, J. Bull. Math. Biophys., 8, 391 (1966).

10. Powell, E.O. J. Gen. Microbiol., 15, 492 (1956).

11. Goodwin, B.C. Nature, 209, 479 (1966).

12. Finn, R.K. and Wilson, R.E. J. Agric. Food Chem., 2, 66 (1956).

13. James, T.W. In Cell Synchrony, (Ed. J.L. Cameron and G.M. Padilla) Academic Press, New York, 1966, p.1.

PHYSIOLOGICAL RHYTHMS IN SACCHAROMYCES CEREVISIAE POPULATIONS

G. Kraepelin

Botanisches Institut der Technischen Universität,
Braunschweig, Germany.

Yeast cells synchronized with regard to visible budding need not also be synchronous for other physiological characteristics not directly coupled to the process of cell division. With the culture conditions usually used to synchronize cell multiplication cells are accumulated in the state of "maximal ripeness" by selectively delaying the onset of budding (minimum predivisional lag). On the other hand cells of a population can be "equalized" to a high degree by starvation. The fundamental difference from the former situation is that here individual differences are cancelled by accumulating the cells in the opposite state within the cell cycle, i.e. in a state of "maximal distance from ripeness" (maximum predivisional lag). From this point of view it is not surprising that in some instances synchronized cell division may even disturb or exclude synchrony of other reactions and vice versa.

This paper briefly describes rhythmic phenomena observed in various yeast populations synchronized merely by nutrient starvation. Though our experimental results in this field are still of a preliminary nature they may show that starved populations can be very suitable for the study of oscillating cell reactions of a generally unknown type. The procedure used to produce this kind of synchrony is very simple: Yeast cells from an aerated stationary phase culture (glucose-containing medium with yeast extract and peptone) were washed and starved in deionized water at $28°C$ for at least 2-4 h without agitation. The suspension density of this stock suspension was adjusted to approximately $2-4 \times 10^8$ cells/ml. Under these conditions synchrony developed progressively with time, the optimum depending on the preculture conditions and the strain used. In some cases starvation was maintained up to 48 h before cell death became prominent. It is noteworthy that among the yeast strains tested so far the best results were always obtained with haploids of the rough and flocculent type mostly derived from ascospore isolates of a normal Saccharomyces cerevisiae.

The mechanism responsible for the synchronizing effect of starvation is still unclear but it seems that a particular kind of excretion product from the cells plays an important role by mediating regulatory cell-cell interactions. Following a simplified model this would mean that deficiency of exogenous nutrient supply induces within a population of unicellular organisms such as yeast (or bacteria) a differentiated state where the sum of autonomous individuals becomes progressively transformed into a kind of loosely coupled multi-cellular system, almost comparable with a tissue-like body. Cellular interactions needed to coordinate the reactions within such a complex of cells could be mediated by exchanging actively-excreted "informational molecules".
Since this state of a population develops without any addition of exogenous nutrients the presumed regulatory interactions may in some way increase the chance of survival under stress conditions.

I. Time dependent excretion of material absorbing at 260mμ.

The first set of experiments concerns the time course of the excretion into the medium of some material from the cell having an absorption maximum around 260 mμ. To demonstrate this phenomenon yeast populations previously synchronized by starvation were transferred into a simple salts medium containing 5% glucose (final suspension density 2-4 x 10^7 cells/ml; constant temperature 26°C). A series of samples (7 ml) were removed automatically from this vigorously-stirred suspension at a constant time interval - usually 15 sec - and directly transferred into ice-cold tubes. After centrifugation in the cold samples were filtered through membrane filters attached to a syringe and the supernatants tested at 260 mμ in a spectrophotometer. In several cases the experimental procedure was modified as follows: a) After removal each sample was postincubated, without cold shock for exactly 2 min at 26°C and then "stopped" in the ice bath; the further treatment was the same as previously described. b) Samples were directly taken from the starved stock suspension and poured into a series of tubes containing nutrient medium adjusted to 26°C. After postincubation for exactly 2 min each sample was "stopped" by cooling and treated as before. c) To stop reaction in the samples more rapidly perchloric acid (final conc. 0.7 M) was combined with the cold shock. It may be added that the concentration of nutrients and the suspension density in the tubes was always the same.

Though the absolute values of the E_{260} varied between identical as well as modified experiments the curves obtained were of the same type: Within each series of samples the E_{260} in the supernatants increased and decreased with time in a more or less regular fashion following in general the pattern of an oscillation. The period of the individual cycles was mostly about 45-60 sec but sometimes 90 sec cycles were also found. The "amplitude" varied from 0.01 to 0.3 extinction units depending to a great extent on the strain and the method used to "stop" the samples. From our results the release of 260 mμ absorbing material into the medium can be interpreted as an endogenously-controlled cyclic process for which the term "leakage" seems inadequate. It is further suggested that the rhythm is already present in the starved population; it is not induced but can merely be intensified by the addition of nutrients. Experiments, like that shown in Fig.1, indicate that the oscillating behavior is a rather stable property since it is only transiently disturbed by changing the conditions of sampling. On the other hand it is surprising that the rhythmic process could be stopped and transient states "fixed" even by rather harmless shocks. It is also not clear at present whether 260 mμ material is really excreted into the medium and then reabsorbed by the cells during one cycle, or if the cyclic variation of E_{260} found in the samples must be regarded as a secondary reaction, i.e. the shock itself induces excretion with the intensity depending on the time the cells were submitted to the shock. In both cases an oscillating system has to be present which controls reversible excretion or permeability at the level of the cell membrane.

II. Time-dependent lag phase.

The second phenomenon concerns a time-dependent variation of the ability of starved cell populations to initiate growth and multiplication when transferred into nutrient medium. One would expect that samples taken from a suspension

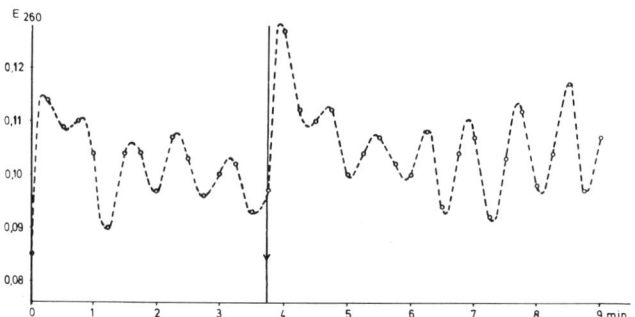

Figure 1. Rhythmic excretion of 260 mµ absorbing material at 26°C. The sampling interval was 15 sec. Starved cells (S. cerevisiae, haploid T 11a/R-form) were sampled into tubes containing glucose medium, incubated for exactly 2 min at 26°C and stopped by cooling. At the arrow an adequate volume of glucose medium was added to the stock suspension and the following samples stopped immediately by cooling.

of starved cells at the very short time interval of 15 sec following the addition of nutrient medium would all start to grow after an equal lag period. What we found was a cyclic variation very similar to that described previously. To produce measurable values samples were postincubated at 26°C for 5-7 h and the optical density read at 546 mµ. The result of such a series is shown in Fig. 2. One can see that optical density increased more or less according to the time sequence of sampling. It may be added that the same result was obtained when cell multiplication was determined by direct counting in a THOMA chamber following an 18 h incubation under sterile conditions. (RÜGER 1967). The cycle length in both types of experiments was about 45-70 sec. We conclude from these results that in starved cells the ability or readiness to initiate growth and multiplication is under the control of an oscillating mechanism. As a consequence cells adapt to growth conditions more or less readily within a series of samples taken during different phases of the endogenous rhythm.

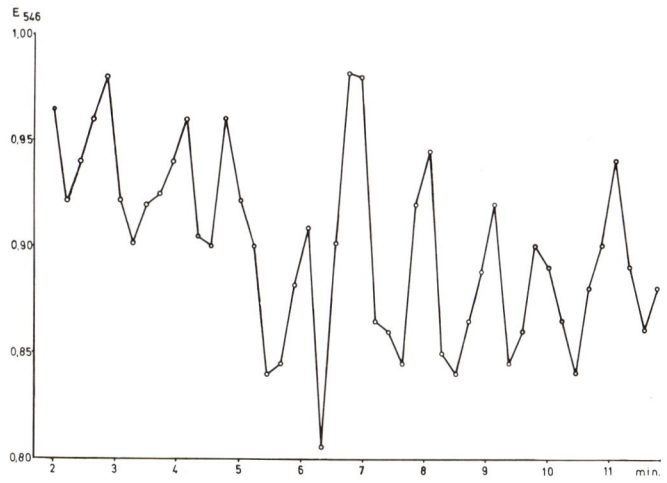

Figure 2. Variation of the optical density at 546 mµ in a series of samples collected at 15 sec intervals from a suspension of starved cells (haploid T 11a/R), following the addition of glucose medium (zero time). Samples were further incubated in tubes at 26°C for 7 h and "stopped" by cooling before readings were made.

Since no shock was used in these experiments the only "fixing agent" could be the introduction of anaerobiosis (transfer of 7 ml samples from a vigorously stirred suspension into test tubes incubated without agitation at 26°C for several hours). We are presently testing this possibility.

III. Time-dependent RD-lability.

It is already known that the tendency to pass into a stable respiratory-defective state (RD state or petite mutation) is a strain dependent property in yeasts. Moreover our results suggest that this species specific RD lability is not constant since cyclic variations can be found if determinations of RD cells are repeated within short time intervals. Samples collected at 15 sec intervals from a suspension of starved cells following addition of nutrient medium were immediately stopped by dilution in 200 ml of ice-cold water. From these dilutions five-fold platings were made on nutrient agar (complex medium with glucose, peptone, yeast extract) and the number of pure RD colonies determined after 5 days with the TTC-overlay technique. The most regular cycles in the production of spontaneously-produced RD cells (percentage as well as absolute values) were found with unstable heploids of the rough type, i.e. the same strains used to demonstrate the other rhythms mentioned previously. Though the "amplitude" in the experiment represented in Fig. 3 is rather low the similarity of the curve to the former one may be seen. Oscillations were previously found when RD cells were determined at 2.5 min intervals, but only about 1 cycle/10 min was registered (KRAEPELIN 1968, 1969). We can not

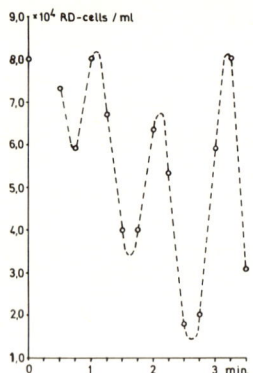

Figure 3. Rhythmic changes in the number of cells producing RD colonies in a suspension of starved cells (haploid T 11a/R) following the addition of glucose medium. Samples collected at 15 sec intervals were immediately "stopped" by dilution in ice-cold water and plated five times on complete medium.

exclude the possibility that the real phase length is even shorter than the 45-60 sec suggested from the 15 sec sampling experiments. Apart from this question it seems to us a matter of fact that RD transitions are also under the control of an oscillating system. It is worthwhile to test further the possibility that the striking similarity of the curves for spontaneous RD lability and the other rhythmic phenomena observed under the same conditions is more than merely coincidental.

IV. Time-dependent glucose uptake.

As known from MAITRA (1966) yeast cells aerated for one hour in buffer take up glucose stepwise with the duration of one full cycle being about 75 sec. Determinations of glucose uptake made in our laboratory (GÄB, unpublished data) with yeast populations synchronized by starvation revealed even more pronounced oscillations. After the addition of nutrient medium containing glucose (1% instead of 5%) samples were taken at 15 sec intervals, "stopped" by cooling, centrifuged and filtered as before. Glucose was assayed in the cell-free supernatants with the O-toluidine method (Merckotest), as well as enzymatically with hexokinase (Biochemica Test-Combination). Both methods gave essentially the same results concerning the oscillatory shape of the curves. The regular sequence of decreasing and increasing values in the supernatants suggests that glucose uptake (or binding?) is coupled to a partial release (transitional binding?). In this process uptake always prevails over longer periods of time but seems to be fully reversible if single cycles (45-60 sec cycle length) are considered, particularly during the first minutes after addition of the glucose medium. Since these results will be published in full detail elsewhere they are only mentioned here as an additional example of the oscillatory behavior of yeast populations synchronized by starvation.

Concluding remarks.

As already pointed out the rhythmic phenomena described here should be regarded as a preliminary basis from which further analysis may start. Nevertheless some remarks may be added about their possible meaning and interrelation. At first sight excretion of 260 mµ absorbing material, growth initiation, RD lability or glucose uptake seem to have nothing in common but the experimental procedure used to measure their oscillatory character. This may be true but it should also be considered whether these different physiological activities could not have in common an oscillating primary mechanism controlling them all at the same time. For instance there is a remarkable inverse proportionality over a wide range between the intensity of 260 mµ excretion and suspension density. If in a simple nutrient medium, containing glucose,

suspension density was lowered below a critical value the
extent of excretion/cell became lethal and cells died within a few hours. Of those things tested so far only the addition of yeast extract and peptone prevented this "dilution
effect". On the other hand a certain degree of excretion
(or concentration of excreted products in the medium) seems
to be necessary or perhaps useful to initiate the series of
reactions leading to adaptation and growth in a fresh medium.
Apparently 260 mµ excretion, as well as a kind of predisposition to initiate growth, are both controlled by an endogenous
oscillatory mechanism. The most simple interrelation could
be that during the transitional phases of increased excretion
the lag phase is more pronounced and growth is delayed. One
question which is difficult to explain is the fact that the
initial situation, from which each of the collected samples
started, led to long lasting measurable differences in growth
and multiplication (see RÜGER for generation time). This indicates that the "time fixing agents", though seemingly harmless, still produce an unexpected degree of permanence in
the otherwise transitional states within the endogenous
rhythm.

If this assumption proves correct then one has only to
proceed one step further to explain the cyclic variation in
RD lability or RD mutability. If, during rhythmically repeating states, yeast cells are transitionally marked by a
specific or even unspecific physiological lability, then
adaptational stress produced by dilution may also increase
"spontaneous" RD transitions during this time period. It is
very difficult to explain the oscillatory behavior of RD lability on the basis of accidental loss mutations within mitochondrial DNA. One possibility could be that the mitochondrial "rho factor", which is apparently involved in petite
mutation, is itself under oscillatory control or subjected
to a kind of turnover cycle with regular phases of increased
lability.

Concerning the question of rhythmic glucose uptake
our curves could be interpreted as the result of a partly
reversible binding process combined with a pulsating transport into the cells. It is tempting to look for an oscillating control system at or near the cell membrane. The
question then arises as to whether uptake as well as the release of substances from the cells into their surrounding
medium may be controlled by the same mechanism. This means

possibly that the cell membrane itself (i.e. rhythmic changes of its conformation and permeability) could be the responsible factor. This hypothesis of course includes a primary or secondary function in the exchange reactions at the external cell barrier. However the cell membrane, or the more generally described "reacting cell surface", is one of the central and most complex problems within cell physiology. It is therefore not our intention to further increase the mass of speculations and hypotheses in this field.

REFERENCES

Maitra, P.K. Biochem. Biophys. Res. Commun., 25, 462 (1966).

Kraepelin, G. Paper presented at the 5th FEBS meeting, Prague, 1968.

Kraepelin, G. Ant. van Leeuwenhoek, 35, Suppl.: Yeast Sympos. (1969).

Rüger, H.-J. Gärungsabhängige Vermehrungsrhythmen bei verschiedenen Hefetypen. Untersuchungen an fraktionierten Populationen. Thesis : Technische Universität, Braunschweig, 1967.

ACKNOWLEDGEMENTS

These investigations were supported in part by Forschungsmittel des Landes Nidersachsen. I thank Miss U. Schneider and Miss U. Dittloff for their careful assistance and Dr. Rüger and Mr. Gäb for their permission to mention results in advance of publication.

LONG- AND SHORT-PERIOD OSCILLATIONS IN A MYXOMYCETE WITH
SYNCHRONOUS NUCLEAR DIVISIONS*

W. Sachsenmaier and K. Hansen

Institut für Experimentelle Krebsforschung, Deutsches
Krebsforschungszentrum, Heidelberg, and Zoologisches
Institut der Universität Heidelberg

The slime mold or myxomycete Physarum polycephalum represents a naturally synchronous system which has become a promising tool for biochemical studies of periodic events related to the mitotic cycle. This organism has been well known to biologists for a long time because of its rapid cytoplasmic shuttle streaming. This paper is concerned with recent studies referring both to the long term periodicity of synchronous nuclear division and to the short period oscillation of the streaming cytoplasm.

Plasmodia of Physarum polycephalum are multinuclear and may be cultivated axenically according to Daniel and Rusch (1) on a defined liquid medium. A single disc-shaped macroplasmodium with a diameter of about 6 cm contains more than 100 million nuclei which all divide simultaneously every 8 to 10 hours within a period of 10 minutes or less. This organism thus may be regarded as a giant single cell which is particularly suited for studies on the biochemistry of the mitotic cycle.

One possible approach to the molecular level of growth control is to look for biochemical events which exhibit periodicity in relation to the division cycle. So far, the best known periodic process in dividing cells is the replication of chromosomal deoxyribonucleic acid (DNA) which usually is restricted to a particular segment of the cell cycle, the so-called S-period. In Physarum polycephalum

---

* Supported by the "Deutsche Forschungsgemeinschaft".

all nuclei synthesize their DNA synchronously during the first three hours of interphase (2,3). Our earlier studies with various inhibitors of macromolecular biosynthesis support the idea that the mitotic cycle follows a sequence of events which is programmed in the nuclear genome (3-8). Replication of the DNA appears to be a rate-limiting step in this sequence to the extent that further steps cannot be initiated unless all or most of the nuclear genome has replicated. These further steps probably involve the sequential induction of proteins which must be newly synthesized prior to each mitosis.

Periodic synthesis of enzymes in relation to the division cycle has been observed in various systems. In Physarum polycephalum the enzyme thymidine kinase is induced periodically (6) about 1 hour prior to the onset of prophase, as shown in Figure 1. The increase of enzyme activity probably represents a gene-dependent stimulation of new enzyme synthesis since it is sensitive to actinomycin and actidione treatment (8).

Treatment of synchronous plasmodia with UV-light or X-rays, which causes a delay of the onset of the next mitosis (7,9,10), equally delays the timing of enzyme induction (Figure 2). This suggests a close correlation between the mechanisms controlling the periodic synthesis of thymidine kinase and the onset of nuclear division. On the other hand, these mechanisms probably are not completely identical since the periodic behavior of thymidine kinase may be altered independently of the mitotic cycle. Exposure of synchronous plasmodia to the anti-metabolite 5-fluoro-2'-deoxyuridine (FUDR) during the G_2-period largely inhibits the induction of thymidine kinase without affecting the schedule of the first mitosis (Figure 3). Later on the enzyme increases more or less steadily while at the same time it is repressed in the control cultures. FUDR blocks the de novo synthesis of thymidine triphosphate (TTP) by inhibiting the enzyme thymidylate sythetase (11). Replication of nuclear DNA following the first (uninfluenced) mitosis therefore is blocked in the presence of FUDR which in turn prevents the onset of the second mitosis (4).

The effect of FUDR on thymidine kinase induction suggests that one or more components of the deoxynucleotide pool, perhaps TTP or one of its intermediates, are involved

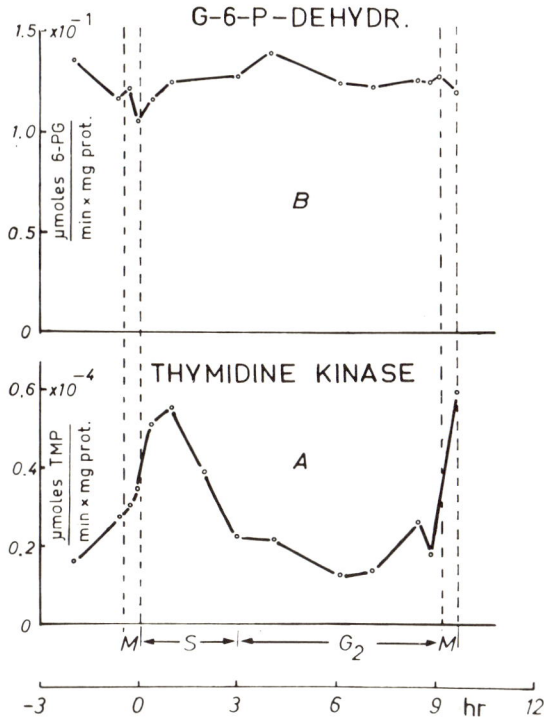

Figure 1. Specific activities of thymidine kinase (ATP: thymidine 5'-phosphotransferase, EC 2.7.1.21) and glucose-6-phosphate-dehydrogenase (EC 1.1.1.49) at various stages of the mitotic cycle of Physarum polycephalum. Enzyme activities and protein were measured in the supernatant of centrifuged (30 min, 120,000 g) plasmodial homogenates (8). M = mitosis, S = period of nuclear DNA replication, G_2 = premitotic rest-period.

in the regulation of this enzyme. One possibility would be that TTP accumulates at the end of the G_2-period and functions as a derepressor of the thymidine kinase gene. Normally, messenger synthesis would continue until the onset of DNA replication and cease as soon as the DNA precursor pool is emptied. The periodicity of thymidine kinase thus would reflect periodic fluctuations of the DNA precursor pool which in turn are linked to the periodicity of DNA replication.

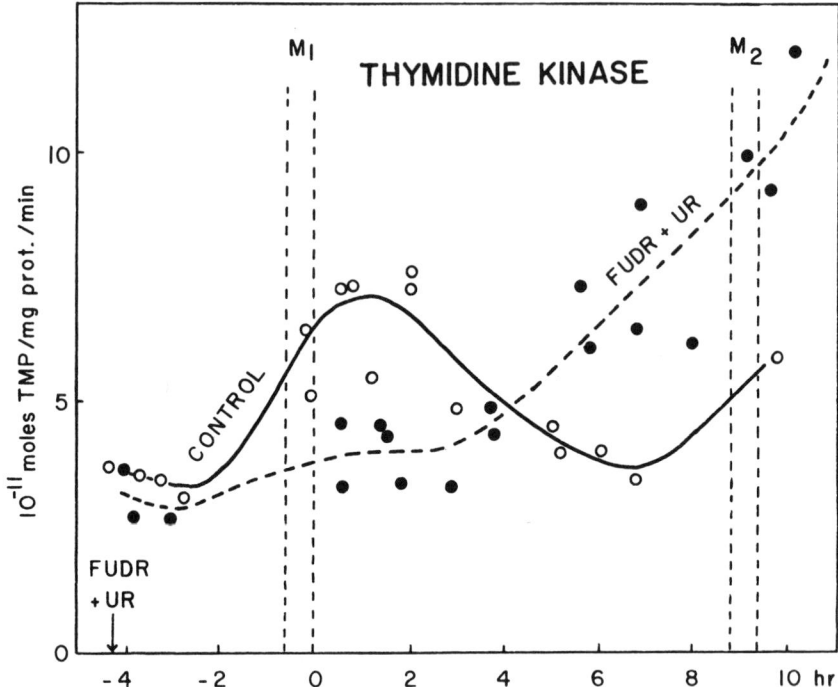

Figure 2. Specific activity of thymidine kinase in controls (-o-) and in plasmodia treated with 2×10^{-5}M 5-fluoro-2'-deoxyuridine plus 4×10^{-4}M uridine (-●-). Treatment started 4 hrs prior to mitosis. M_1, M_2 = synchronous nuclear divisions in controls. The first mitosis (M_1) following the addition of FUDR + UR occurred at the normal time in treated plasmodia; the next mitosis (M_2), however, did not appear until the end of the experiment.

We have not yet investigated possible changes of the level of free deoxynucleotides in Physarum polycephalum. However, recent results obtained from the analysis of the ribonucleotide pattern strongly suggest that nucleic acid precursors in fact do accumulate at the end of interphase and during mitosis (12). Table I shows the average level of four nucleoside triphosphates at three different stages of the division cycle: interphase, mitosis and reconstruction period (first hour immediately following telophase

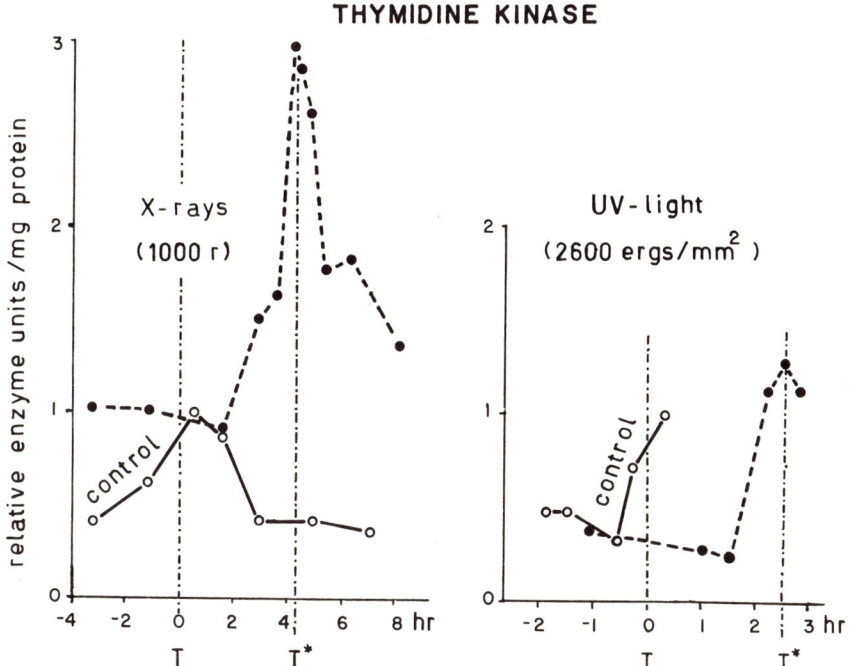

Figure 3. Specific activity of thymidine kinase in controls (-o-) and in plasmodia treated with X-rays or UV-light (-•-). T = synchronous nuclear division (telophase) in controls. T* = delayed mitosis in treated plasmodia. Irradiation with X-rays occurred during the preceding mitosis (T-1) with a Siemens Dermopan irradiator (29 kV, 25 mA, 0.3 mm Aluminum filter, dose rate 870 r/min, total dose 1000 r). UV-light was applied with a germicidal lamp five hours prior to mitosis (T) (λ max 258 mμ, dose rate 140 ergs/mm^2/sec., total dose 2600 ergs/mm^2).

during which the nucleolus reappears). The table summarizes the analytic data of 20 plasmodia and subjected to anion exchange chromatography with the automatic nucleotide analyzer of Schnitger et al (13,14). The contents of guanosine-, uridine-, and cytidine-5'-triphosphate in the nucleotide pool are significantly higher around the time of mitosis than at any other stage of the cycle. The individual values for ATP varied considerably between different plasmodia so that the differences of the mean ATP contents are not significant.

The increase of the triphosphate nucleotides prior to and during mitosis very likely reflects an accumulation of RNA precursors due to a decreased rate of RNA synthesis. As shown previously (3,5) RNA synthesis drops during the last 2 to 3 hours prior to nuclear division and this alone could fully account for the observed changes of the triphosphate level.

TABLE I.

Contents of nucleoside-5'-triphosphates at different stages of the synchronous mitotic cycle in Physarum polycephalum.

Nucleo-tide	I			M			R		
	$\bar{x}$	s	n	$\bar{x}$	s	n	$\bar{x}$	s	n
ATP	7.48	± 0.67	10	8.48	± 1.52	6	7.68	± 2.70	4
GTP	2.10	± 0.15	9	3.23	± 0.32	6	2.65	± 0.37	4
UTP	1.51	± 0.18	9	2.42	± 0.39	6	2.20	± 0.18	4
CTP	1.29	± 0.15	10	2.08	± 0.27	6	1.86	± 0.13	4

Four to ten small pieces (~30 mg wet weight) were cut from synchronous macroplasmodia during interphase (I), mitosis (M), and reconstruction period (R) (i.e. first hour following telophase), extracted with cold perchloric acid and the acid soluble fraction was chromatographed on Dowex (1 - X8) with an automated nucleotide analyzer (13,14). ATP = adenosine-5'-triphosphate, GTP = guanosine-5'-triphosphate, UTP = uridine-5'-triphosphate, CTP = cytidine-5'-triphosphate. $\bar{x}$ = mean values (μmoles/g protein), s = standard error of the mean, n = number of samples. The differences of the mean values between stage M and stage I or R are significant for GTP ($\alpha \leqslant 0.01$), UTP ($\alpha \leqslant 0.05$), and CTP ($\alpha \leqslant 0.05$), but not significant for ATP, as tested by variance analysis (12).

The periodicity of all long-period oscillatory functions discussed so far (DNA and RNA synthesis, nucleotide accumulation, enzyme induction) appears to be the result rather than the cause of a specific stimulus which ultimately controls the onset of synchronous mitosis in Physarum polycephalum. The chemical nature of this triggering event is still unknown. The sharp synchrony of nuclear divisions in this syncytial organism suggests that mitosis is controlled by an extranuclear factor which periodically accumulates in the cytoplasm up to a threshold level and triggers the onset of prophase simultaneously in all nuclei. This concept is supported by experiments in which two plasmodia, representing different stages of the division cycle, are fused (Figure 4). In this way a new mixed plasmodium is formed, containing two sets of nuclei which have originated from the last mitosis at different times. Nevertheless, all nuclei enter the next mitosis simultaneously about half way in between the mitotic times of unfused control plasmodia.

Although the hypothetical trigger finally appears to act upon the nuclei from outside, presumably from the cytoplasm, its formation or function very likely is controlled by the nuclei themselves. Plasmodium (A) which normally speeds up mitosis of nuclei from plasmodium (B) in a fusion experiment loses its accelerating ability when it is treated with FUDR prior to the fusion process (lower part of Figure 5). FUDR inhibits DNA replication during S-period but protein synthesis proceeds at an almost normal rate for several hours as shown previously (4). The formation or function of the cytoplasmic trigger therefore depends on the normal replication of nuclear DNA. Perhaps a late replicating gene or operon which governs the production of this factor is activated only after it has replicated at the end of the S-period.

The participation of DNA in the mechanism which controls the onset of mitosis is further suggested by a fusion experiment in which one moiety of the mixed plasmodium was treated with UV-light (Figure 5). Total irradiation of a plasmodium with UV-light during the first half of the G_2-period delays the initiation of the next mitosis (7,9). The primary target very likely is nuclear DNA, since incorporation of 5-bromouracil into DNA stimulates the antimitotic effect of UV-light (7). In a fused plasmodium, containing about

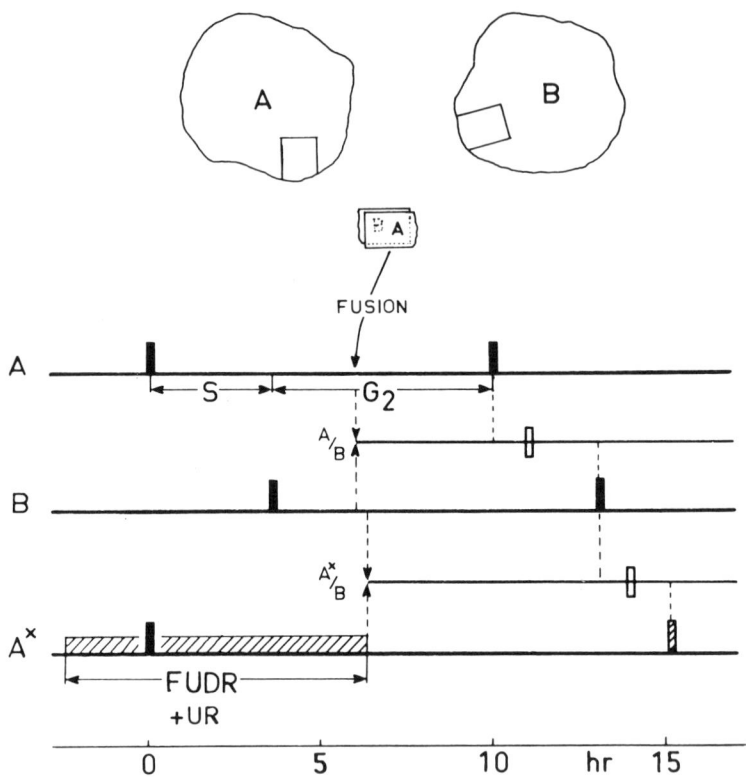

Figure 4. Nuclear mitosis in mixed plasmodia. Two macroplasmodia (A) and (B) were prepared with mitotic schedules shifted from each other by about 3 hrs. At the time indicated by dotted arrows rectangular pieces (~2 cm^2) were cut from each plasmodium and fused by "sandwiching" within one hour in the absence of nutrient medium. At the same time, two homologous pieces from each parent culture were fused with each other (controls). A third macroplasmodium (A$^+$) with the same mitotic schedule as in (A) was treated with 3×10^{-5}M 5-fluoro-2'-deoxyuridine plus 4×10^{-4}M uridine during the period marked with a horizontal shaded bar. The first mitosis was not altered, the second mitosis, however, appeared with a delay of 5 hrs. A mixed plasmodium was prepared by fusion of pieces from (A$^+$) and (B), analogous to the procedure applied to untreated pieces (A) and (B).
▮ = mitosis in untreated controls, ▨ = mitosis in FUDR treated control, ☐ = mitosis in mixed plasmodia.

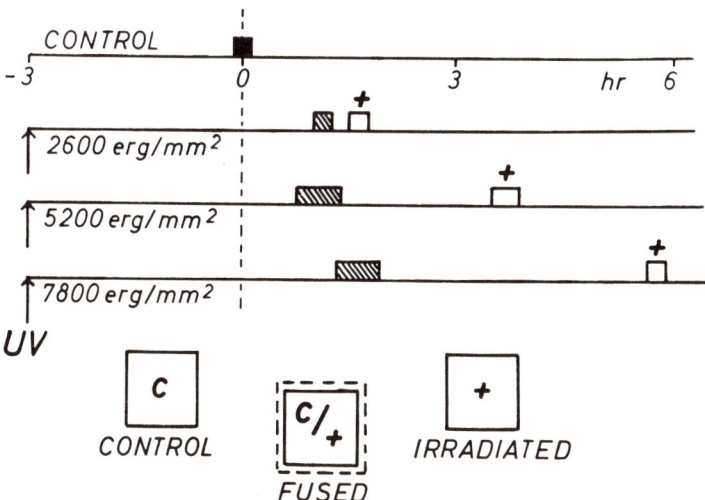

Figure 5. Nuclear mitosis in mixed plasmodia containing normal and UV-treated nuclei. Four replicate macroplasmodia with the same mitotic schedule were prepared. At the time indicated by the arrow, three cultures were irradiated with different doses of UV-light from a germicidal lamp (λ_{max} = 258 mµ). Immediately thereafter small pieces (~2 cm^2) were cut from each plasmodium and fused with an untreated control piece as shown in Figure 4. ■ = mitosis in the control, ☐ = mitosis in irradiated unfused plasmodia, ▨ = mitosis in mixed plasmodia.

equal numbers of irradiated and normal nuclei, only a single synchronous mitotic wave is observed. This again strongly suggests the ultimately extranuclear localization of the trigger factor although nuclear DNA is probably involved in the production or function of this factor.

Let us now turn briefly to some new observations related to the interesting phenomenon of short-period oscillations of cytoplasmic streaming in <u>Physarum polycephalum</u>. The protoplasm of a syncytial macroplasmodium moves rapidly back and forth through a system of fan-shaped channels with a period of 1 to 2 minutes at room temperature (15). This reciprocal movement probably results from rhythmic local

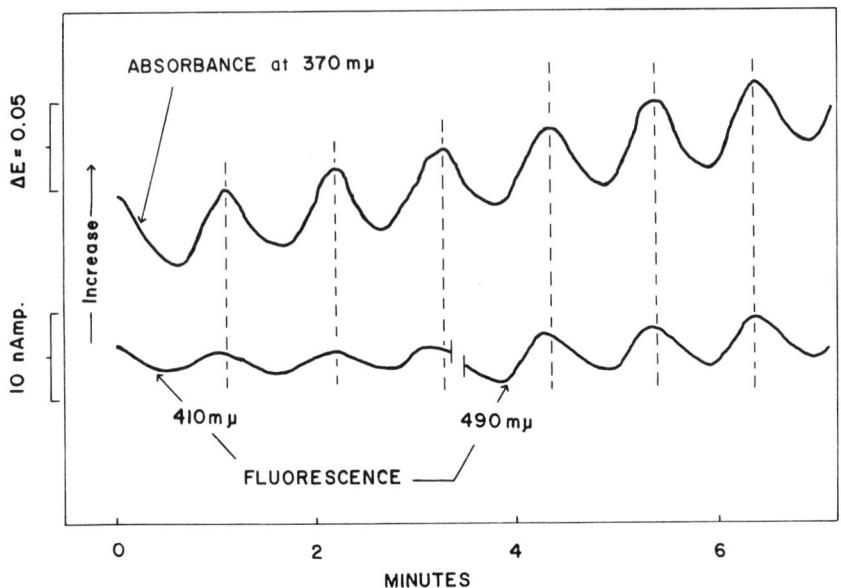

Figure 6. Simultaneous recording of light absorption and fluorescence of a living plasmodium with the Ferrand fluorometer. A rectangular plasmodium (1.5 x 2.5 cm) was placed on a microscopic slide (0.1 mm thick) and oriented at an angle of 70° toward the light entrance slit of the fluorometer. The intensity of fluorescent light (410 and 490 mμ) emitted at 90° as well as the energy of the incident light (370 mμ) transmitted directly through the plasmodium ($\sphericalangle$ 0°) were recorded simultaneously using two separate photomultipliers and two recording units.

contractions (16) of submicroscopic fibres (17) located inside the protoplasm. The channels represent transient areas of different viscosity rather than rigid permanent structures. The branches eventually extend into a sponge-like mass which swells and shrinks according to the rhythm of the cytoplasmic flow.

Oscillatory changes of light absorption and fluorescence with a period of 1 to 2 minutes are observed when a living plasmodium is placed into the light path of a photometer or a fluorometer (Figure 6). Maxima and minima of

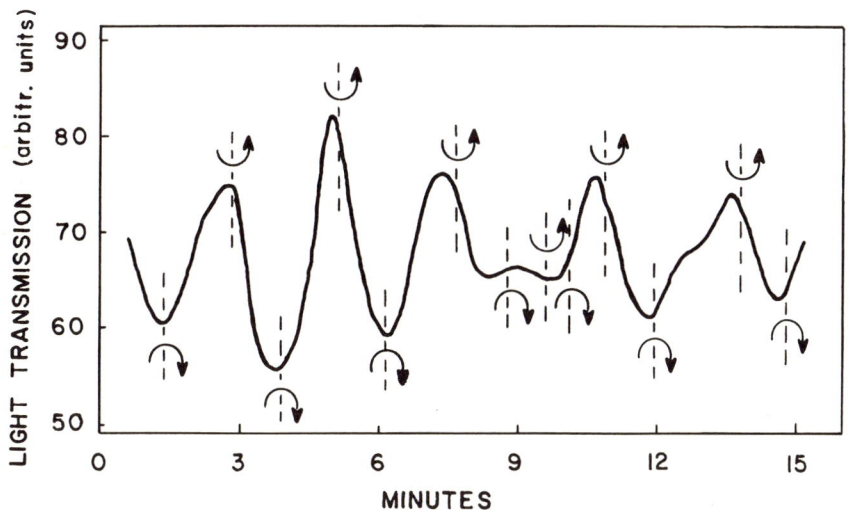

Figure 7. Correlation of cytoplasmic streaming with rhythmic variations of light transmission. A plasmodium was mounted on an agar-coated microscopic slide and observed through a Zeiss-Photomicroscope. The intensity of the unfiltered white light penetrating through the peripheral region of the plasmodium was recorded with a photometer-head, attached to the phototube of the microscope. Cytoplasmic streaming along the plasmodial channels was observed simultaneously through the eye piece of the phototube. Curved arrows indicate turning points of flow direction: ↶ toward, ↷ away from the plasmodial periphery.

light absorption coincide with maxima and minima of fluorescence and are correlated with turning points of the cytoplasmic flow (Figure 7). These optical oscillations occur with nearly equal relative amplitudes (10% to 20%) over a wide range of the absorption and the emission spectrum. They are recorded only, however, if the plasmodium extends beyond the limits of the illuminated area. Distant portions of a plasmodium usually oscillate independently, suggesting that a macroplasmodium represents a polyrhythmic system. From these observations we suspect that optical oscillations result mainly from rhythmic local alterations of the plasmodial geometry. A closer examination of the absorption spectrum of

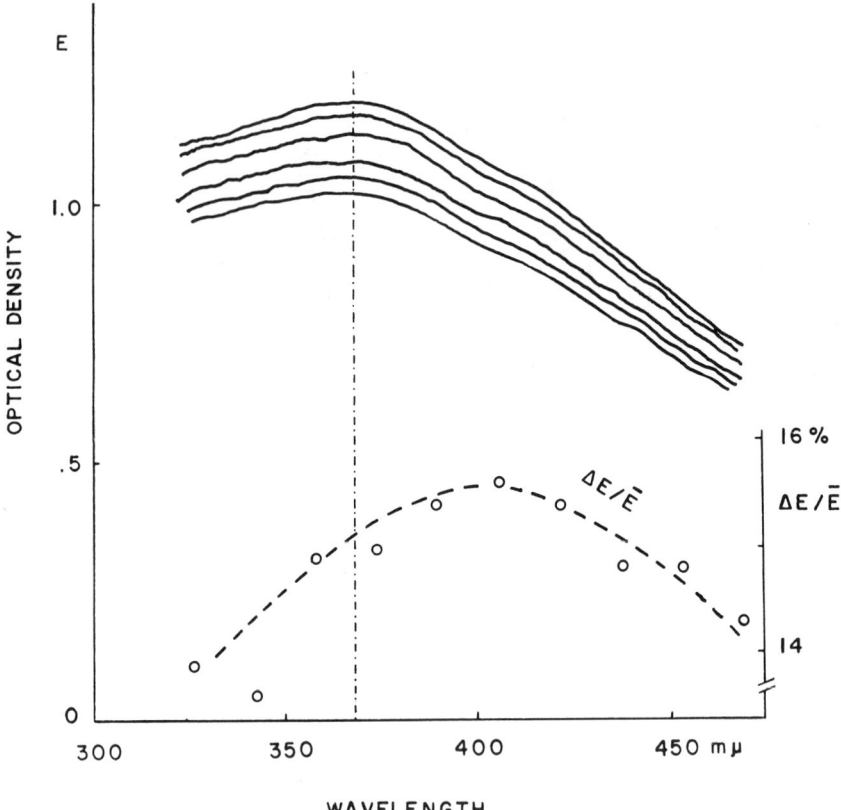

Figure 8. Oscillation amplitude of light absorption within the spectral region of 310 to 470 mµ. A rectangular plasmodium (1x2 cm) growing on agar was placed in a quartz cuvette and the absorption spectrum was recorded every 10 seconds with the Howaldt-Rapid spectrometer (20) from minimum to maximum of one half oscillation period. Scanning time: 1/40 second. -o- = relative amplitude ($\Delta E/\bar{E}$) calculated for various points of the spectrum. ΔE = maximum minus minimum of optical density, $\bar{E}$ = average optical density at a particular wavelength. (This experiment was carried out in collaboration with B. Brauser and Th. Bücher at the Institut für Physiologische Chemie der Universität München).

living plasmodia, however, reveals a slightly increased relative amplitude of the oscillation around 400 mµ compared to adjacent regions of the spectrum. This follows from an experiment in which the spectrum of a living plasmodium was recorded every 10 seconds during one half oscillation period with the Howaldt-Rapid-spectrometer (recording time 1/40 sec.) (Figure 8). The plasmodium shows an absorption maximum at 365 mµ which may be ascribed to a group of yellow pigments (19,20) which are located in cytoplasmic granules. The largest relative change of light absorption ($\Delta E/\bar{E}$) occurs at about 410 mµ which corresponds to a shoulder in the absorption spectrum of the isolated pigment. This shoulder appears only at acidic pH and indicates a dissociable group with a pK value of 4.3 (21).

It is not clear yet whether the increased oscillations at 400 mµ merely reflect local changes of the composition of the heterogeneous protoplasm, due to the in- and outflow of cytoplasmic constituents, or true absorption changes of an oscillating metabolite. Perhaps a component of the yellow pigment participates in the oscillating metabolic system which controls the periodic movement of the cytoplasm. The main component of this system is a contractile protein (22, 23,24) which is arranged in bundles of submicroscopic fibres (17) and exhibits a high ATPase activity (25). Glycolytic ATP serves as the energy source (15, 26). A strong influence of the intracellular ATP level on the periodicity of plasmodial pulsations is suggested by the experiment shown in Figure 9. Removal of oxygen slows down the frequency of oscillations while slightly increasing the amplitude. The ATP content of the plasmodium drops by about 40% under these conditions (Figure 10).

Interestingly, cytoplasmic oscillations decay during synchronous nuclear division (Figure 11). The reason for this effect is not yet clear. Perhaps a drop of the ATP level due to an increased energy demand during mitosis may be responsible for this transient breakdown of cytoplasmic oscillations. From the respiration rate (~35 µl O_2/mg protein/hr) an approximate turnover time of 5 sec may be estimated for the ATP pool (12). A slight increase of ATP consumption by only 1 percent would empty the ATP pool within 10 min, provided that the rate of oxidative phosphorylation does not change. Based on this assumption, the additional amount of energy required during mitosis would be equivalent

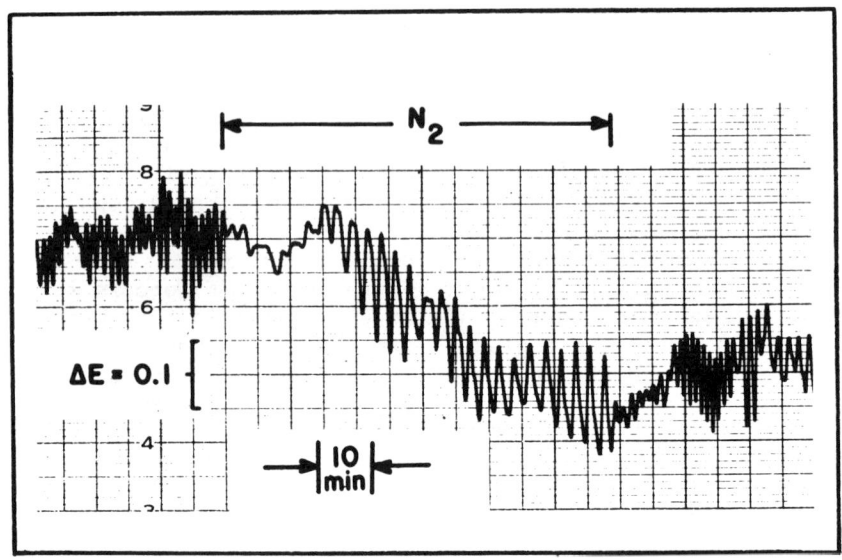

Figure 9. Influence of anoxia on plasmodial oscillations. A rectangular plasmodium (1.5x5 cm) was mounted on an agar-coated slide and placed in a plexiglass chamber (5.5 x 7.5 x 3 cm) perpendicular to the light beam of a Zeiss-photometer (PMQ II). Light absorption at 380 mµ was recorded while moistened air or nitrogen (99.96% N_2) was passed through the chamber.

to ~10^9 molecules ATP per nucleus. Although in some experiments a definite drop of the ATP level in mitosis has been observed (12), this was not as reproducible as the effect on cytoplasmic oscillation. Alternative explanations for this effect may, therefore, be discussed as, for example, a change of the concentration of free calcium ions in the cytoplasm during mitosis.

Calcium greatly stimulates the ATPase activity of plasmodial myosin B (25) and might well function as the control variable in this particular oscillating system. Plasmodial myosin B appears to be the most likely candidate of the oscillophore which may alter its affinity to ATP or to Ca^{++}, depending on its mechano-chemical state. Oscillations thus could arise by a similar mechanism as discussed for the high frequency oscillations of insect flight muscles (29). In any case, it would be most interesting to learn more about

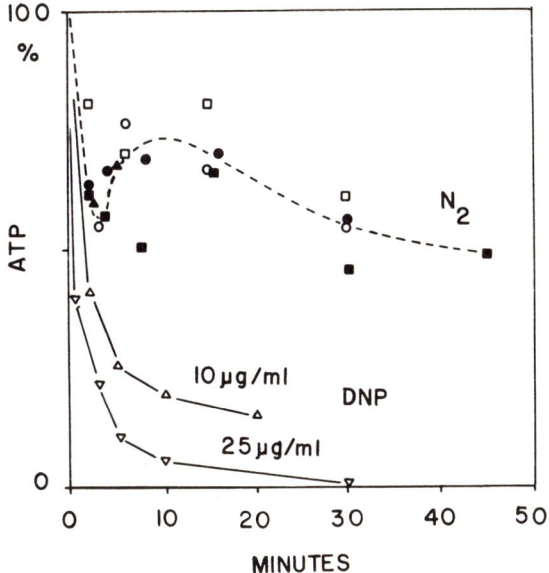

Figure 10. Change of the ATP level during anoxia. Macroplasmodia with a diameter of about 6 cm were cut into 4 to 6 segments (4 to 6 cm^2 each) and placed into a cylindrical glass chamber. Moistened air or nitrogen (99.6% N_2) was passed through the chamber. Prior to and different times after filling the chamber with nitrogen, single segments were removed and immediately immersed into 5 ml of preheated (97°C) distilled water in a 15 ml centrifuge tube. After 2 min, the samples were chilled in ice, centrifuged and the ATP content of the supernatant measured with the fire-fly method (27). The sediment was used for protein determinations (28). Different symbols indicate the results of 5 separate experiments. For comparison the effect of dinitrophenol (DNP) on the ATP level of an aerobic suspension culture of microplasmodia is shown by the lower curves.

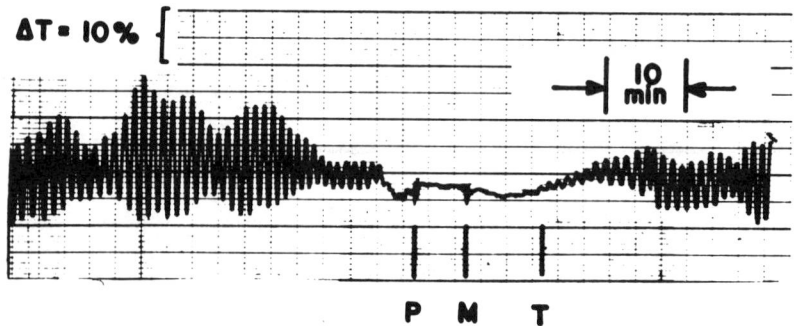

Figure 11. Decay of plasmodial oscillation during synchronous nuclear division. Light transmission of an aerobic plasmodium was recorded at 500 mµ as described in Figure 9. Mitotic stages were observed in a sister culture incubated outside of the photometer under equivalent conditions. P = prophase, M = metaphase, T = telophase.

the biochemistry of plasmodial pulsation and its breakdown during mitosis. The latter effect obviously indicates a yet unknown specific event related to nuclear mitosis.

It is tempting to speculate that the rhythmic movement of the plasmodial cytoplasm may be geared to metabolic oscillations of a glycolytic system. The techniques used so far in our studies did not permit detection of rhythmic changes of NADH fluorescence correlated with plasmodial pulsations. Any variation of fluorescence due to metabolic changes of the NADH/NAD ratio would have been obscured by oscillations of the intensity of fluorescent light resulting from rhythmic local changes of the plasmodial geometry (Figure 6). Moreover, no significant changes of the fluorescence emission spectrum were observed during aerobic - anaerobic transitions (18), suggesting the presence of a metabolically stable fluorochrome other than pyridine nucleotides. It appears difficult, therefore, to demonstrate glycolytic oscillations in Physarum by conventional fluorometric techniques which have been used so successfully in experiments with yeast (30-36). Although glycolytic oscillations have been observed even in single cells (37), there is still no good evidence for a functional role of these oscillations. If the phenomenon of reciprocal plasma streaming in Physarum should prove to be correlated with glycolytic oscillations, this would be an interesting example for such a functional role of an oscillating enzyme system. The rapid movement of the cytoplasm in the giant syncytial plasmodium apparently serves to facilitate the exchange of matter (nutrients, oxygen, catabolites, etc.) which otherwise would be greatly hampered by the low ratio of cell surface to volume compared to uninucleated cells.

Summary

Multinuclear plasmodia of the myxomycete Physarum polycephalum exhibit naturally synchronous mitoses every 8 to 10 hours. DNA and RNA synthesis as well as the production of certain enzymes (i.e. thymidine kinase) occur periodically in close correlation with the synchronous mitotic cycle. Nucleosidetriphosphates accumulate prior to and during mitosis. These "long-period" oscillations appear to be controlled by a triggering event which initiates the onset of nuclear division. Fusion experiments suggest that the "trigger factor" is formed during the G_2-period under the control of nuclear DNA and accumulates in the cytoplasm up to a threshold level.

The plasmodial cytoplasm moves rhythmically through a system of branched channels with a period of 1 to 2 minutes. Oscillations of light absorption and fluorescence of the living plasmodium are observed with a similar periodicity. These oscillations largely reflect changes of the plasmodial geometry due to rhythmic local contractions and dilatations of the protoplasm. Light transmission near a shoulder in the absorption spectrum of the particle-bound yellow pigment (~400 mµ) oscillates with a slightly increased amplitude compared to adjacent regions of the spectrum. This suggests that the pigment or a component with similar absorption characteristics participates in an oscillating metabolic system which controls reciprocal cytoplasmic streaming. Plasmodial pulsations persist in the absence of oxygen with an extended period (2 to 4 minutes) and with a markedly increased amplitude. Light absorption as well as cytoplasmic streaming almost cease to oscillate during the time of a synchronous nuclear mitosis. The possible role of oscillating enzyme systems in the mechanism which governs cytoplasmic shuttle streaming is discussed.

REFERENCES

1. Daniel, J.W. and Rusch, H.P., J. Gen. Microbiol., 25, 47 (1961).
2. Nygaard, O.F., Guttes, S. and Rusch, H.P., Biochem. Biophys. Acta, 38, 298 (1960).
3. Sachsenmaier, W., Biochem. Z., 340, 541 (1964).
4. Sachsenmaier, W. and Rusch, H.P., Exp. Cell Res., 36, 124 (1964).
5. Sachsenmaier, W. and Becker, J.E., Mschr. f. Chem., 96, 754 (1965).
6. Sachsenmaier, W. and Ives, D.H., Biochem. Z., 343, 399 (1965).
7. Sachsenmaier, W. In Problems of Biological Reduplication, (Ed. P. Sitte), Springer Verlag, Berlin-Heidelberg-New York, 1966, p. 139.
8. Sachsenmaier, W., v.Fournier, D. and Guertler, K.F., Biochem. Biophys. Res. Commun., 27, 655 (1967).
9. Clausnizer, B., Dönges, K.H., Remy, U. and Sachsenmaier, W., Hoppe-Seyler's Z.f.Physiol. Chem., 349, 1242 (1968).
10. Nygaard, O.F. and Guttes, S., Int. J. Rad. Biol., 5, 33 (1962).
11. Harbers, E., Chauduri, N.K. and Heidelberger, C., J. Biol. Chem., 234, 1255 (1959).
12. Sachsenmaier, W., Immich, H., Grunst, J., Scholz, R. and Bücher, Th., Europ. J. Biochem., 8, 557 (1969).
13. Schnitger, H., Papenberg, K., Ganse, E., Czok, R., Bücher, Th. and Adam, H., Biochem. Z., 332, 167 (1959).
14. Schnitger, H., Bücher, Th., Grunst, J., Patat, U., Schnitger, St., and Scholz, R. In preparation.
15. Kamyia, N. In Die Zelle, Struktur und Funktion (Ed. H. Metzner), Wissenschaftliche Verlagsgesellschaft, Stuttgart, 1966, p.329.
16. Allen, R.D., Pitts, W.R., Speir, D. and Brault, J., Science, 142, 1485 (1963).

17. Wohlfahrt-Bottermann, K.E. In *Primitive Motile Systems in Cell Biology*, (Ed. R.D. Allen and N. Kamyia), Academic Press Inc., New York, 1964, p.79.

18. Hansen, K. and Sachsenmaier, W. Unpublished observations.

19. Brewer, E.N. Doctoral Thesis, University of Wisconsin, 1965.

20. Lübbers, D. and Niesel, N., Pflügers Arch. Ges. Physiol., 268, 286 (1959).

21. Sachsenmaier, W. In preparation.

22. Loewy, A.G., J. Cell. Comp. Physiol., 40, 127 (1952).

23. Ts'o, P.O.P., Eggman, L. and Vinograd. J., J. Gen. Physiol., 39, 801 (1956).

24. Nakajima, H., Protoplasma, 52, 413 (1960).

25. Hatano, S. and Tazawa, M., Biochem. Biophys. Acta, 154, 507 (1968).

26. Hatano, S. and Takeuchi, I., Protoplasma, 52, 169 (1960).

27. Addanki, S., Sotos, J.F. and Rearick, P.D., Analyt. Biochem., 14, 261, (1966).

28. Lowry, O.H., Rosebrough, N.J., Farr, A.L. and Randall, R.J., J. Biol. Chem., 193, 265, (1951).

29. Rüegg, J.C. 5th FEBS-Meeting, Prague, 1968, Abstr. No.448.

30. Chance, B., Estabrook, R.W. and Ghosh, A., Proc. Nat. Acad. Sci. (USA), 51, 1244 (1964).

31. Chance, B., Schoener, B. and Elsaesser, S., J. Biol Chem., 240, 3170 (1965).

32. Hess, B., Chance, B. and Betz, A., Ber. d. Bunsenger. f. Phys. Chem., 68, 768 (1964).

33. Hess, B., Brand, K. and Pye, K., Biochem. Biophys. Res. Commun., 23, 102 (1966).

34. Hess, B. and Boiteux, A., Biochem. Biophys. Acta, Library, 11, 148 (1968).

35. Betz. A and Chance, B., Arch. Biochem. Biophys., 109, 579, (1965).

36. Pye, K. and Chance, B., Proc. Nat. Acad. Sci. (USA), 55, 888 (1966).

37. Chance, B. and Williamson, G., 5th FEBS Meeting, Prague, 1968, Abstr.948.

OSCILLATIONS IN THE EPIGENETIC SYSTEM: BIOPHYSICAL
MODEL OF THE β-GALACTOSIDASE CONTROL SYSTEM

W. A. Knorre

Department of Biophysics
Institute for Microbiology and Experimental Therapy,
German Academy of Sciences Berlin, 69 Jena, GDR

The Model

Many bacterial enzymes are synthesized periodically rather than continuously. Such oscillations in synthesis of macromolecules, *i.e.* enzymes and corresponding mRNA, are called oscillations in the epigenetic system. Using the operon model, Goodwin (1) has made suitable simplifications and approximations and proposed a simplified mathematical model for oscillations in the epigenetic system. The basic element of Goodwin's theory is a closed feedback repression circuit which is described by two differential equations for enzyme and mRNA synthesis. One of the major dynamic consequences of the negative feedback in this circuit is the possibility of oscillations in the synthesis of both mRNA and enzyme. It is possible to combine such elements into a regulatory network.

Based upon Goodwin's model, we have proposed a model for the synthesis of β-galactosidase in <u>Escherichia coli.</u> The scheme in Figure 1 represents the β-galactosidase system. It contains two closed feedback circuits. The function of SG_1 and SG_2 is to produce a messenger $x_{1,2}$ for both β-galactosidase and galactoside-permease, which, after complexing with ribosomes, P, controls the synthesis of β-galactosidase, y_1, and galactoside-permease, y_2, respectively. The active transport of substrate (lactose) into the cell is designated m_1 and the splitting products of the β-galactosidase, m_2 and m_3 for glucose and galactose, respectively. Furthermore, the β-galactosidase produces the inducer, m_1', by a transgalactosidation. For simplicity m_2

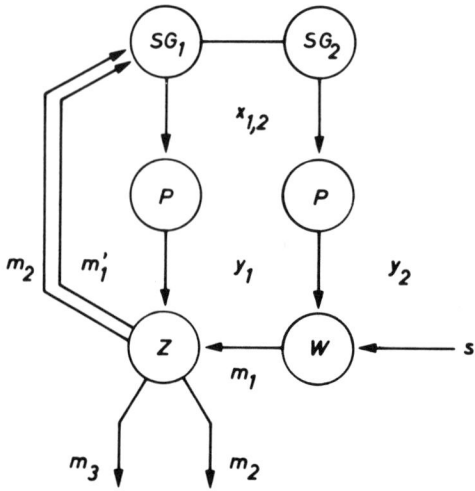

Figure 1. Simplified scheme of the β-galactosidase system.

and m_1' can be regarded as controlling factors for the rate of mRNA synthesis. The inducer, m_1', increases and the catabolite, m_2, decreases by catabolic repression the rate of mRNA synthesis. Consequently, the β-galactosidase control system contains a closed feedback induction loop and a closed feedback repression loop. The model is described by the following set of differential equations.

$$\frac{dy_1}{dt} = \alpha_1 x - \beta_1 \quad ; \quad \frac{dy_2}{dt} = \alpha_2 x - \beta_2 \quad ;$$

$$\frac{dx}{dt} = \frac{am_1'}{1+c_3 m_2} - b \quad ; \quad \frac{dx}{dt} \leq \left[\frac{dx}{dt}\right]_{max} \quad ;$$

$$\frac{dm_1}{dt} = c_1 s y_2 - c_2 m_1 y_1 \quad ; \quad \frac{dm_2}{dt} = c_2 m_1 y_1 - c_4 m_2 \quad ;$$

$$m_1' = c_5 m_1 \quad ;$$

$$\alpha_1, \beta_1, \alpha_2, \beta_2, a, b, c_1, c_2, c_3, c_4, c_5 = \text{const.}$$

Analog computer techniques are used to study the effect of variable parameters of the system and to record the dynamical behavior of all functional entities (2,3). For example the response of the system to a step in the substrate concentration from zero to a constant value is shown in Figure 2. The same figure also shows the sequence of events when the system becomes operational at the time indicated by the arrow. After a lag phase, the β-galactosidase concentration increases and shows an overshoot, followed by continuous oscillations of the mRNA and the enzyme concentrations. The interactions of induction and catabolic repression are responsible for such behavior. Whereas the mRNA concentration is changing between zero and a maximum, the enzyme concentration shows undamped oscillations about a steady state value. However, a relative constancy in the concentrations of metabolites is seen. Further, in the transient phase the concentration of substrate in the cell undergoes damped oscillations with a comparatively small period. These oscillations can be regarded as metabolic oscillations.

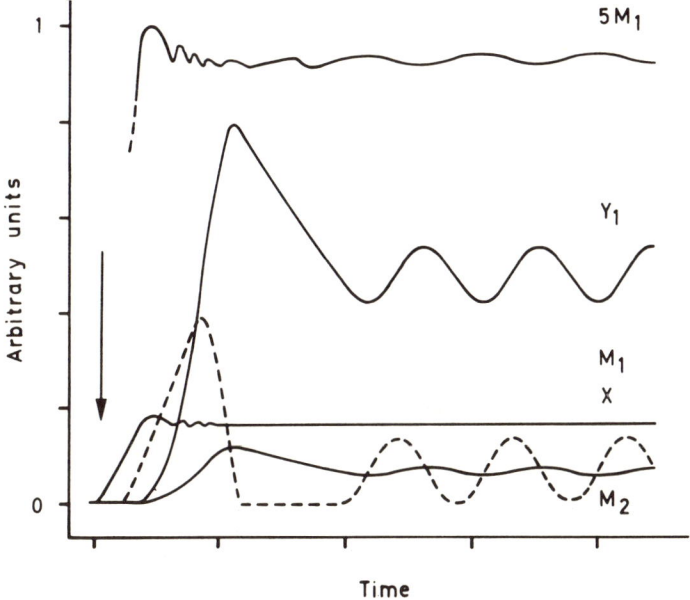

Figure 2. Time course of concentrations of β-galactosidase (Y_1), mRNA (X), lactose in the cell (M_1), and galactose (M_2) in the model.

In a second example, Figure 3, the effect of small differences between the parameters of three "cells" is demonstrated. The enzyme concentrations oscillate all approximately with the same periods and amplitudes but of differing phases about their steady state values. Consequently, the oscillations vanish by averaging as shown in the upper curve. Such behavior may be generally expected when we measure the specific β-galactosidase concentration in an experiment.

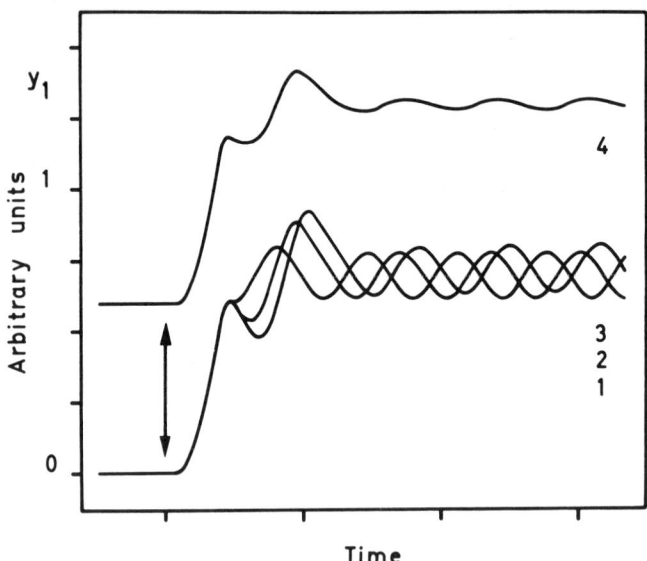

Figure 3. Effect of small differences between three model cells on the time course of β-galactosidase.

This simplified model was extended to reduplication of the genes and exponential growth of the cells. Some results of the investigations of models for β-galactosidase are the following:

1. The β-galactosidase system exhibits properties similar to those of a technical control system, i.e. the concentrations of the metabolites are relatively constant and the concentrations of the enzymes are a function of the growth rate of the cell.

BIOCHEMICAL OSCILLATORS

2. The concentrations of enzymes and mRNA undergo continuous oscillations with a period of about the doubling time of the cell.

Results

A number of different experiments were performed to demonstrate the occurrence of oscillations in the β-galactosidase system. The substrate, lactose, can be regarded as an input signal for the control system and the rate of enzyme synthesis as an output. Therefore, we measure the specific enzyme activity in cultures growing exponentially in a synthetic medium after a change of the carbon source from glucose to lactose. The specific rate of enzyme synthesis, designated $f(t)$, was evaluated by a digital computer from data of specific activity, Y, and growth rate μ, by the equation $f(t) = dY/dt + \mu(t)Y$ (4,5). It can be assumed that the specific rate of synthesis of β-galactosidase, $f(t)$, is proportional to the concentration of corresponding mRNA. Consequently, the curves of specific synthesis rate in Figures 4,5, and 6 also show the changes in the level of mRNA.

The time course of the specific rate of β-galactosidase synthesis after a change from glucose to lactose for the inducible strain ML30 and the constitutive strain ML 308 is shown in Figure 4.

The specific rate of the inducible strain periodically increases from 0.2 (steady-state value on glucose) to 100 (steady-state value on lactose), whereas the rate of the constitutive strain periodically decreases from 200 (steady-state value on glucose) to 100 (steady-state value on lactose). The period of oscillations is about 50 minutes. This is comparable with the doubling time of the cells of 49 and 45 min. for the inducible strain and the constitutive strain, respectively. It can be seen that in both cases the period is about the same, but there is a phase shift of 180° between the oscillations in the constitutive and the inducible strain. Furthermore, the shape of the oscillations exhibits a fine structure.

This fine structure is more clearly shown in Figure 5 in another experiment with the inducible strain, ML30. It is possible to explain this shape of the curve by super-

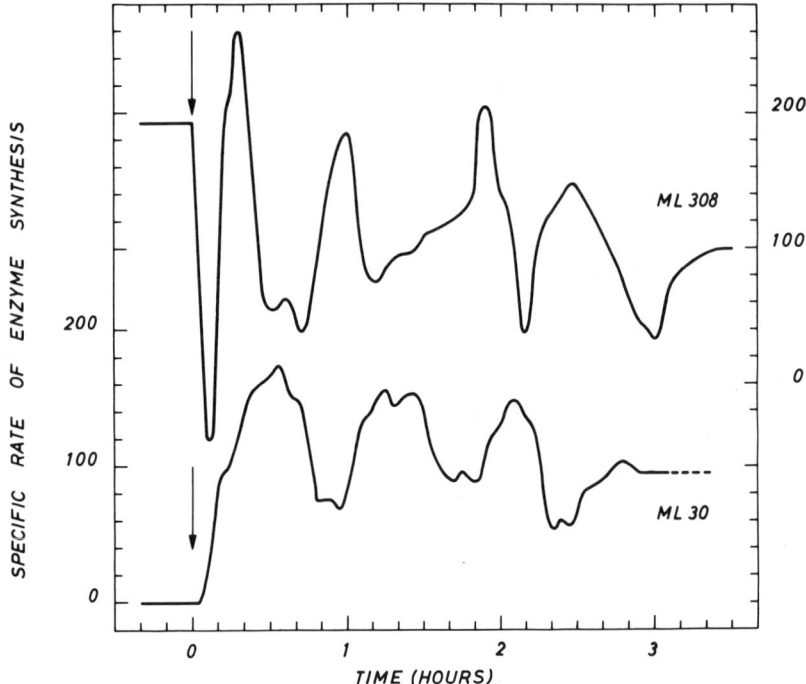

Figure 4. Time course of the specific rate of β-galactosidase synthesis in an inducible strain (ML30) and a constitutive strain (ML308).

position of the oscillation of β-galactosidase synthesis with another enzyme system oscillating in a like manner. The rate of β-galactosidase synthesis is temporarily decreased when the amplitude of the second oscillating system achieves its maximum. Because there is only a small lag between the maximum of β-galactosidase and the other enzyme, our tentative explanation is that this is the expression of the sequential induction of the galactose enzymes which are induced by the released galactose. This hypothesis is supported by the relationship of growth rate, $\mu(t)$, and specific rate of enzyme synthesis, $f(t)$. The growth rate increases stepwise and the second step to the steady-state value is beginning, even when the galactose enzymes have achieved their maximum rate and the galactose can be metabolized.

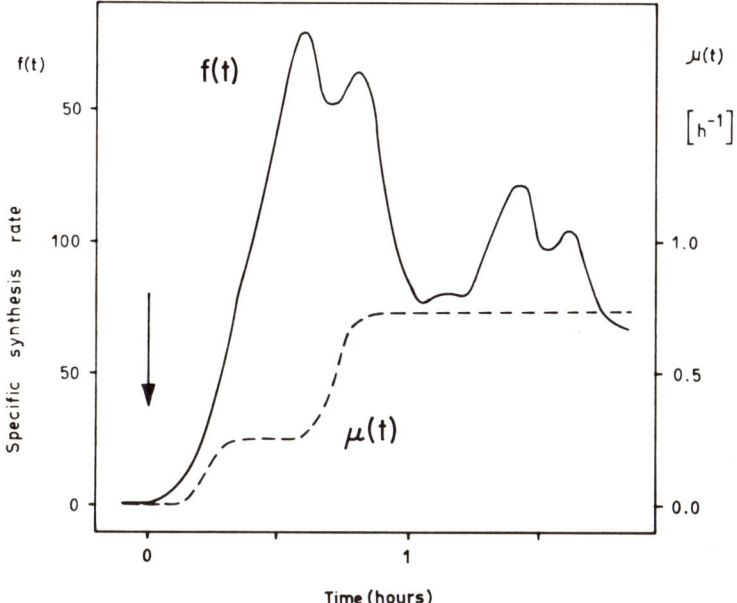

Figure 5. Relationship between the growth rate µ(t) of the cells and the specific rate of β-galactosidase synthesis f(t) in the inducible strain (ML30).

One of the predictions of the model was the dependence of the period of oscillations on the growth rate of the cells. In a chemostat we have the possibility of adjusting the growth rate in a suitable range. The results of two experiments in a chemostat with the dilution rates of 0.7 h^{-1} and 0.4 h^{-1} are shown in Figure 6. In both cases the time of occurrence of the first minimum in the rate of enzyme synthesis depends on the dilution rate and is in good agreement with the generation times of about 60 min and 105 min, respectively. Similar results were found with other growth rates.

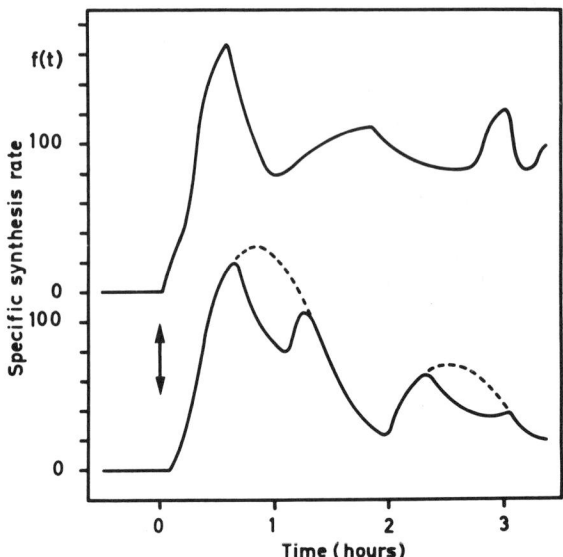

Figure 6. Time course of the specific rate of β-galactosidase synthesis f(t) in two experiments in a chemostat with dilution rates of 0.7 h^{-1} (upper curve) and 0.4 h^{-1}.

Discussion

The results presented reinforce our earlier findings that the synthesis of β-galactosidase in E.coli is periodical (6) and the period of oscillations is dependent on the generation time of the bacteria. Oscillations were not detectable when a gratuitous inducer instead of lactose was used. Therefore, it may be assumed that the oscillations of the β-galactosidase synthesis are intimately connected with the interaction of induction and catabolic repression. An extensive discussion of these problems is given by Knorre (3).

We have already noted that the oscillations of β-galactosidase synthesis in the model are undamped, but we have found only damped oscillations in the experiments. This leads to two possible interpretations. The first emphasizes that the parameters of the system do not allow continuous oscillations; the observed behavior is the transient response of the system to the input signal. In a second interpretation, mentioned above, the loss of synchrony between the single system is considered to be responsible for the appearance of damped oscillations, whereas the variables in the single systems still undergo continuous oscillations. Furthermore, an effect of synchrony of cell division can be excluded. Before and after the step from glucose to lactose, a synchrony of cell division was not found. However, from studies of β-galactosidase synthesis in synchronously dividing cultures, the second interpretation becomes plausible.

REFERENCES

1. Goodwin, B.C., Temporal Organisation in Cells, Academic Press, New York, 1963.
2. Knorre, W.A., Dissertation, Leipzig, 1967.
3. Knorre, W.A., Studia Biophysica, 6, 1 (1968).
4. Knorre, W.A., Z. Allg. Mikrobiol., in press.
5. Knorre, W.A., Biochem. Biophys. Res. Commun., 31, 812 (1968).
6. Knorre, W.A. and Bergter, F., Mber. Dt. Akad. Wiss., 8, 127 (1966).

VI
CIRCADIAN OSCILLATIONS

THE INVESTIGATION OF OSCILLATORY PROCESSES
BY PERTURBATION EXPERIMENTS
I. THE DYNAMICAL INTERPRETATION OF PHASE SHIFTS*

Arthur T. Winfree

Biology Department, Princeton University
Princeton, New Jersey 08540

The analysis of phase shifts resulting from discrete perturbation of biological rhythms was developed and exploited by Perkel and co-workers (1,2), and independently by Rawson, Pittendrigh and co-workers (3-9). By these methods, they achieved important insights into the entrainment behavior of neural and circadian rhythms, respectively.

In the present paper, I will attempt to extend this method to the investigation of the internal dynamics of certain kinds of oscillatory processes of unknown mechanism and unknown complexity. The second paper, "A Singular State in the Circadian Clock-Oscillation of Drosophila pseudoobscura" (hereafter referred to as "SS"), gives experimental results.

Though evolved in the context of circadian behavioral rhythms, the analysis here illustrated should be applicable to any oscillatory systems in which:

1. If not directly observable, the hypothetical oscillator at least projects a reliably periodic rhythm into some observable quantity.

2. We have experimental access to a quantity affecting the oscillator. By manipulation of this quantity we can perturb the oscillator.

3. Following perturbation, the observable rhythm is generally deranged, but eventually returns to its normal

*This work was done while enjoying the support of an NSF Regular Fellowship.

+ Present Address: Department of Biological Sciences,
Purdue University,
West Lafayette, Indiana 47907

period. There remains an asymptotically constant residual phase shift with respect to a control not so perturbed. For example, limit-cycle oscillators, relaxation oscillators, and linear oscillators would behave in this fashion. This residual phase shift, and not the intervening "transients" will be our primary object of study following the suggestion of Pittendrigh (5). Phase information is reliably transmitted after transients subside, at least as regards certain qualitative features of its dependence on parameters of perturbation. We shall base our conclusions only on such features (22).

Phase of Observable Rhythm

The term "phase", describing the observable rhythm at any given instant, is used in this way: Let the rhythm have period TAU. We employ dimensionless time, scaling TAU to unity by $t'=t/TAU$. An arbitrary reference event is chosen to mark phase $\phi=0$; for example, a maximum or a centroid on the observable rhythm. Phase is then marked off in proportion to time in the unit interval terminated by the next reference event (see Figure 1).

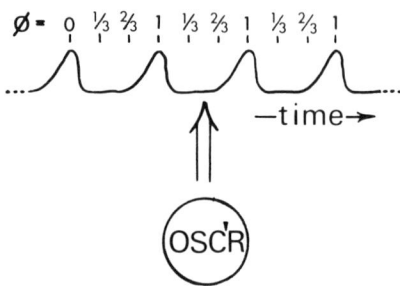

Figure 1. "Phase" indicates what fraction of a cycle has elapsed since last reference event, when rhythm projected by the underlying oscillator is strictly periodic.

BIOCHEMICAL OSCILLATORS

The State of the Driving Oscillator

We need also to speak of the "state" of the underlying oscillator. This is the set of N independent measurements (state variables, or degrees of freedom) required to compare unambiguously the conditions of two identical oscillators: if all the state variables are equal in both cases, then the subsequent behavior of the two oscillators will be the same in the same environment.

Because in any "simple clock" (20) there is only one important state variable, it therefore bears a 1:1 correspondence to the "phase" of the oscillation: the amount of water accumulated on the lip of a dripping faucet is sufficient to identify the state of the process at any instant. Condenser voltage plays the same role in a neon discharge oscillator. In a simple pendulum, by contrast, the bob can have any velocity at a given point in the arc, depending on initial conditions. There are two degrees of freedom, or two state variables (position and momentum) and therefore two measurements are required to reproducibly determine the instantaneous state (and therefore future behavior) of the pendulum. The same is true (N=2) of most of Higgins' biochemical oscillators, the state variables being concentrations of two chemical species (10). In the Hodgkin-Huxley description of nerve membrane behavior, four measurements are required (N=4) to specify local membrane state following arbitrary perturbations (17).

Degrees of Freedom of the Perturbation

Except where specifically noted, I believe but cannot prove that the analysis here presented in terms of a state <u>plane</u> is generally valid for N>2, as long as the perturbations by which the dynamics is explored vary in only two ways, e.g. duration of exposure to the perturbing agent and the time at which the perturbation is applied. In this case only a two-dimensional subspace is actually accessible to observation, given that observations are limited to the measurement of a single scalar quantity.

In fact, the experiments here proposed (and illustrated in "SS") employ only a two-parameter family of perturbations: the oscillation is initiated and allowed to free-run in a standard dark environment (DD); the perturbation consists of switching at time T from dark (DD) to light (LL) (at standard illumination), and then back to dark after a duration, S. We then observe the residual phase shift, $\Delta\phi$, as in Figure 2.

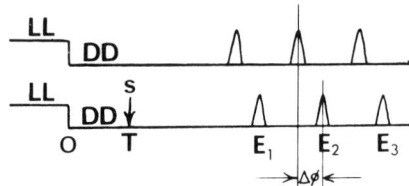

Figure 2. Switch of environments (LL/DD) initiates oscillation. Top: control; bottom: rhythm reset by one of a two-parameter family of perturbations.

Cophase

It will be convenient to measure the phase shift in terms of the time, E, of the phase reference event measured (modulo 1) after any transients subside. We are interested in the time to E from the end of perturbation at T+S. We call this interval θ, the cophase of the state $\underline{x}$ reached by applying the perturbation (T,S).

$\theta(\underline{x}[T,S])$ = asymptotic value of (E-S-T) mod 1.

The N-dimensional state space of an oscillator with N state variables thus seems to be stratified into N-1 dimensional manifolds (isochrons) parametrized on $0 \leq \theta < 1$. Each such contour (if N=2) or surface (if N=3) or hyper surface (if N>3) has constant cophase: it connects together all states which correspond to the same phase of the observable rhythm.

The loci of constant cophase in the dynamical plane of $\underline{x}$, and in the plane of possible perturbations, TxS, will all be called isochrons. The isochrons have no particular physical significance, but serve as a convenient conceptual tool, like

BIOCHEMICAL OSCILLATORS

a bookkeeping trick in the accounting of phase shifts.

Geometrical Description of Dynamics in Standard Dark Environment

Let us consider the structure of the isochrons and of the dynamical flow, $\dot{\underline{x}}(\underline{x})$ in standard dark conditions. All we know about $\dot{\underline{x}}$ is that its trajectories are roughly closed loops, presumably smooth and continuous unless experimental evidence proves otherwise. A necessary topological feature (16) of such bounded quasi-periodic flow is that there is an isolated singular state, $\underline{x}^*$, such that $\dot{\underline{x}}(\underline{x}^*)=\underline{0}$. See Figure 3.

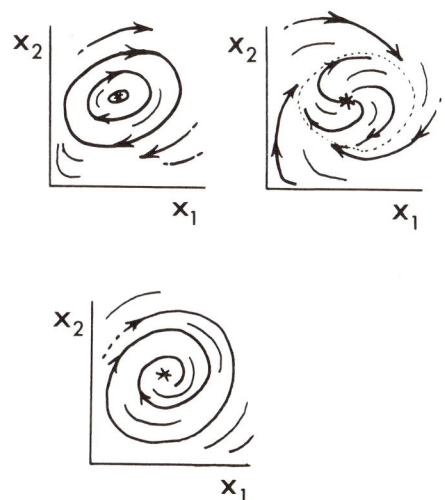

<u>Figure 3</u>. Three possible forms of oscillatory interaction between state variables X_1, X_2; left to right: nonlinear conservative, limit-cycle; lower: damped. *marks the singularity.

Since every value of ϕ and therefore of θ is passed through in each cycle of oscillation, each isochron must cut across every loop of every dynamical trajectory surrounding $\underline{x}^*$, i.e. the isochron contours converge to $\underline{x}^*$ (but see "Testable Consequences", section 2), stratifying the dynamical plane in a pinwheel-like fashion. See Figure 4.

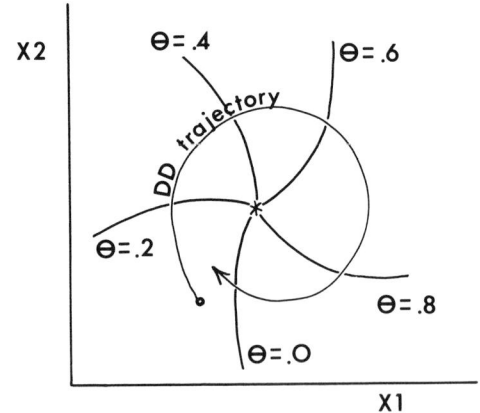

Figure 4. In the same plane as Figure 3; general appearance of contours of constant cophase.

Modified Dynamics During Perturbation

We know that the dynamical behavior of the clock in the presence of light differs from its behavior in the dark, from the fact that a sojourn in the light phase-shifts the clock with respect to a control in the dark. We insist upon a sufficiently comprehensive description of the state of the oscillating system (including, for example, any rapid adaptation of sensitivity to light) so that the dynamics, $\underline{\dot{x}}(\underline{x})$, can be regarded as <u>fixed</u> in a fixed environment. We will employ only two environments: dark and light (both at $20.0^\circ C$).

BIOCHEMICAL OSCILLATORS

In general, any physical parameter affecting the investigated oscillator might be reset at time T to a new (fixed) value to provide altered dynamical flow for a duration S. In a chemical system, the perturbation might consist of a fixed rate of input of one reactant, or operation at elevated temperature. In the case of the D.pseudoobscura clock it is convenient simply to increase the blue light intensity from zero to a standard value. The light-altered dynamical flow, $\dot{\underline{x}}'(x)$, and the state at which it vanishes, $\underline{x}^{*\prime}$ such that $\dot{\underline{x}}'(\underline{x}^{*\prime})=0$, will be indicated by primes as shown. It is possible to make some simple inferences about the shape of $\dot{\underline{x}}'$ in our particular experimental situation. But for present purposes let us note only that the state $\underline{x}^*$ which is singular in the dark is generally not so in the light: $\underline{x}^* \neq \underline{x}^{*\prime}$, and $\dot{\underline{x}}'(\underline{x})$ is smooth and uncomplicated parallel flow in the neighborhood of $\underline{x}^*$.

Resetting Maps, In abstracto

Without attempting to specify further the altered dynamical trajectories, we consider some qualitative aspects of phase-shifting experiments as interpreted in this intuitive geometrical language. The qualitative conclusions are "robust" in the sense of Levins (11) i.e., that considerable quantitative changes in the assumptions do not affect the conclusions.

First of all, we consider the formal structure of the dependence of cophase, θ, on T for a given duration, S, of perturbation. If the process perturbed is nearly periodic and was initiated from a standard state, then we can replace T by φ as an identification of the state of the clock when it enters into the altered dynamics. θ(φ) then represents the "new phase" vs "old phase" relationship mediated by perturbation, S. There is no operational distinction between θ and θ±n, nor between φ and φ±m, where n,m are integers, i.e. multiples of TAU. The function θ(φ) is therefore most appropriately plotted on biperiodic graph paper: on the surface of a toroid. This function, which forms a smooth closed loop on the toroid, is called the "resetting map" since it describes the resetting of phase accomplished by a standard perturbation administered at any phase of the cycle (see Figure 5).

Resetting maps fall into completely separate types, without possible intermediates, due to the biperiodicity:

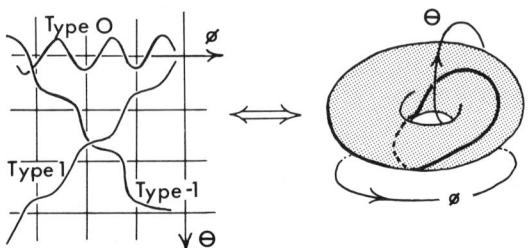

Figure 5. Left: Multiple ("wallpaper") plot of three of the topological types of resetting maps. Right: Equivalent plot of a single unit square out of "wallpaper," with opposite edges identified. Only the Type 1 curve is shown.

they may loop through the hole in the toroid in this direction, or that, or not at all, any integer number of times.* (Let us name these types by the number of times the resetting map passes clockwise through the hole.) Does this mean that all possible oscillators having the properties given earlier in this paper are apriori discretely categorized, like the eigenstates of a wave function?

Resetting Maps, In Fact

Pittendrigh, in 1957, measured a resetting map on the circadian clock of the fruitfly D. pseudoobscura (4; data presented in response-curve form on rectangular coordinates). It is of type 0, i.e. not threading the hole. In 1966 Quinn (personal communication) measured the resetting map of the species D. robusta, using the same experimental arrangements. It is of type 1, threading the hole once clockwise. Can two

---

* The same is, of course, true of the response-curve of Pittendrigh, Perkel, etc., with only this difference, that response-curves plot $\Delta\phi$ vs ϕ. The subtraction and the use of cartesian coordinates can introduce some artifactual discontinuities.

flies of the same genus have evolved fundamentally different clock processes? More perplexingly, Minis and Horne (12), using the moth Pectinophora, measured resetting maps of types 1 and 0, using short and long perturbations, respectively, of identical intensities. So "oscillator type" is not even species-specific! Apparently the topological types of resetting maps do not correspond to different kinds of oscillators. So what is the significance of resetting map types?

A search of the literature of circadian rhythms reveals two empirical generalizations about resetting maps. To map the response curves, $\Delta\phi$ vs ϕ, perturbation experiments of the sort here studied have long been used on the physiology and behavior of diverse plant, animal, and unicellular species. In the twenty cases known to me in which the $\Delta\phi$ is measured after transients have subsided, the resetting map can be constructed.

1. In all cases it is either type 0 or type 1. No type -1, type 2, or other possible biperiodic relations between θ and ϕ are found.

2. In several instances two different types of resetting maps are obtained on the same species by simply changing the duration of perturbation; the type 0 always corresponds to the more prolonged illumination.

Interpretation of Empirical Generalizations

Are these regularities a kind of adaptation? Or are they a clue to the mechanism of circadian rhythms? Or are they possibly only a general feature of a very broad class of oscillatory control processes? Possibly all three, but with the emphasis on the latter. For these apparently universal features find a simple and natural interpretation in terms of isochrons and the modification of standard (dark) dynamics during perturbation.

During perturbation, oscillators initially on a nearly closed loop trajectory cease to follow the common trajectory, and move off into other parts of the state space, following now $\underline{x}'$ dynamics, e.g. as in Figure 6. After a short interlude of the altered dynamics, these oscillators would be found at states indicated in Figure 7. Returned to the standard dark environment, they resume motion along $\underline{x}$ trajectories.

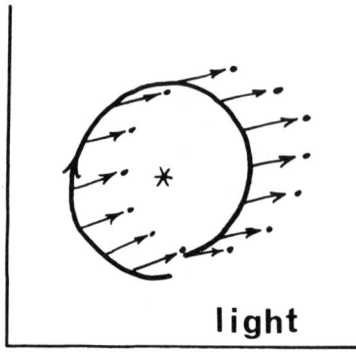

Figure 6. Heavy curve: a segment of dark trajectory (one cycle). Thin curves: light trajectories starting from each phase of the dark oscillation. Sketched to resemble Figure 9A, but B or C would serve the same purpose.

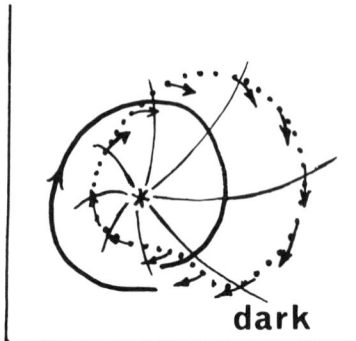

Figure 7. Oscillators displaced in Figure 6 continue along dark trajectories. Perturbation left an oscillator on every isochron.

This mapping of the ring of initial states to a new deformed ring leaves some oscillators on every isochron, since the confluence of isochrons is inside the new ring as, for example, in the limiting case S=0, when the two rings are the same. Therefore, if in a series of perturbation

BIOCHEMICAL OSCILLATORS

experiments varying ϕ at fixed small S, we measure $\theta(\phi)$, we find θ scanning a full cycle (clockwise) as ϕ scans a full cycle (clockwise). The resetting map threads the $\theta \times \phi$ toroid once, clockwise (type 1).

To obtain a type 0 resetting map, using a more prolonged sojourn in the light, it is necessary to imagine obtaining the situation of Figure 8, in which the new θ values all fall within a limited range. The confluence of isochrons is external to the new ring of states. Each θ is represented twice on the new ring. As ϕ scans the cycle, θ rises and falls within this limited range, and the closed loop, $\theta(\phi)$, does not thread the toroid (type 0).

In such a two-variable system the mapping of before-perturbation states to after-perturbation states either encloses $\underline{x}^*$, and $\theta(\phi)$ is type 1, or it does not enclose $\underline{x}^*$, and $\theta(\phi)$ is type 0. In the latter cases, there must exist a trajectory in $\underline{\dot{x}}'$ leading from the ring of initial states across the $\underline{x}^*$ to a point on the new ring. And, if such a trajectory exists then the type 0 case is obtained. Such a situation would prevail if, for example, the $\underline{\dot{x}}'$ dynamical flow resembled Figure 9C or A or B; that is, approaching an equilibrium, both components activated, or (as in the model

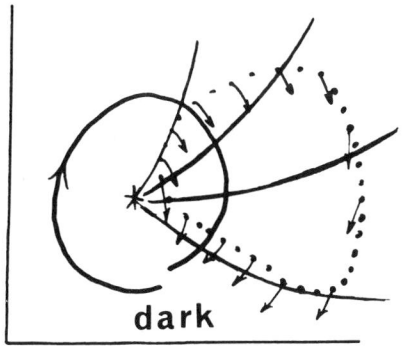

Figure 8. Same as in Figure 7, but more prolonged perturbation. At the ends of the light trajectories, oscillators from every initial phase lie within a limited range of cophases.

of Pavlidis) one component destroyed. Under continuous dynamical flow in two degrees of freedom, the mapping of the initial ring into a new ring is progressive and preserves the topology of the ring; self-crossings such as in Figure 10 are not possible.

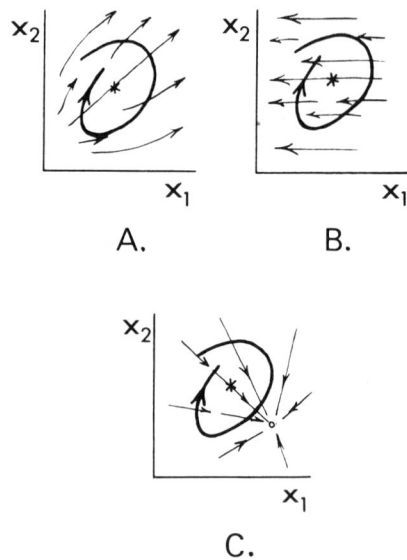

Figure 9. Heavy curve: A segment of dark trajectory. Thin curves: three possible forms of altered (light) dynamics in which a trajectory passes through the singularity. * Marks the dark singularity.

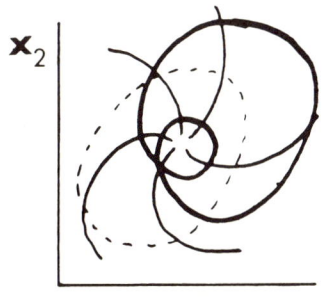

Figure 10. Analogous to Figures 7 and 8 showing a transformation of the ring of initial states which is forbidden by requirements of continuity and uniqueness of flow.
Dotted = original states
Solid = new states

Testable Consequences

If this is a correct interpretation of empirical generalizations 1 and 2, then four further consequences must be observed:

1. If a Type 0 resetting map is measurable using a duration S, then there may be some T* from which an $\underline{\dot{x}}'$ trajectory runs straight across $\underline{x}^*$, as in Figure 9. Therefore, there is some S*<S such that at the end of the perturbation begun at T* the oscillator is left in its "phaseless" state at the intersection of isochrons, at $\underline{x}^*$. This critical annihilating perturbation will immediately suppress the periodic responsiveness of the oscillator to perturbations, and the overt rhythmicity will die out with the time constant characteristic of transients. Since stimuli of duration S>S* have no adverse effect on the clock, but simply reset its phase, it seems at first strange to predict that a more delicate tap (a shorter duration) will completely inhibit the oscillation -- particularly since as far as I know, such a phenomenon has never been reported in the literature of circadian rhythms. (see ‡ on the next page). Yet, if these effects are not observed, then our conjecture is exploded; no trajectory of the altered dynamics runs from the initial ring to the confluence of isochrons and, therefore, either:

a. The oscillator shows only a single degree of freedom, so that the steady-state of the overt rhythm corresponds unambiguously to the state of the underlying oscillator.

or b. The geometrical dynamics are specialized in some peculiar way, perhaps involving discontinuities, multiple singularities, or large N.

or c. The clock process may not be describable in terms of any finite number of continuously-variable quantities.* As far as I can tell, such processes would not be expected to exhibit the behavior here anticipated for biochemical-like oscillatory dynamics.

2. The intersection of isochrons is an (N-2)-dimensional locus in the N-dimensional state-space of the oscillator. This manifold <u>contains</u> the singularity, but only in case N=2 is it <u>identical</u> to the singularity. Only in that case is (T*,S*), defined by the abolition of all circadian rhythmicity, necessarily identical to the <u>singularity</u>, and, therefore, a strictly time-invariant state.

---

‡ Pavlidis' model (7) is of the general type here considered, restricted by five special conjectures not required by, but consistent with, the data available at the time. In (19) he explicitly recognizes that a weak perturbation might, at least temporarily, stop the clock, if it brought the oscillator to its singular state.

M. Rosenzweig also recognized the probable existence of an annihilating pulse on the basis of his "predator-prey" clock model (personal communication).

Engelmann (23) obtained a similar phenomenon experimentally, in <u>Kalanchoe</u>, but apparently for a different reason. He used <u>two</u> pulses.

— — — — — — —

* Example: The finite-state oscillations recently proposed by Kauffman for gene-interaction dynamics in the cell cycle (13). Example: The spatially-distributed oscillations revealed in the remarkable experiments of Zhabotinsky.

BIOCHEMICAL OSCILLATORS

From such a "frozen" state, any subsequent perturbation will simply <u>continue</u> the critical perturbation as though there had been no interruption - we may hope, therefore, to show that exposure to S_2 seconds of light at any time after (T^*,S^*) reinitiates the rhythm at the same phase as would the more familiar resetting pulse, (T^*,S^*+S_2).

Such a finding would contribute circumstantial evidence that N is only 2.

3. By varying T and S as indicated in Figure 11 and measuring the subsequent θ, we should be able to construct an image of the isochrons in the TxS plane. They will be seen to converge to a center which is the same as (T^*,S^*) in point 1, above. The surface θ(T,S) corresponding to such a pinwheel-shaped contour map is a helicoid or corkscrew.

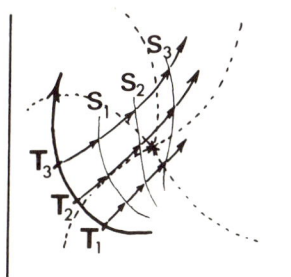

Figure 11. Heavy curve: a segment of a dark trajectory showing three instants $T_1<T_2<T_3$. Thin arrows: light trajectories prolonged to durations $S_1<S_2<S_3$. Dotted lines: isochrons, showing the cophases of states reached by (T_i,S_j). The TxS grid picks up an image of the isochron pinwheel and singularity.

4. The dependence of θ on T at fixed S is a resetting map. The isochron map may, therefore, be regarded as a one-parameter family of resetting maps. On account of the anticipated pinwheel shape of θ, this family is expected to have a number of distinctive pecularities. A variety of otherwise perfectly plausible resetting map or response curve shapes should never be found, e.g. sinusoids with peak-to-peak amplitude 3/4 cycle.

In "SS" these predictions seem to be confirmed using the circadian rhythm of the fruit fly Drosophila pseudo-obscura.

Acknowledgments

I am grateful to Jay Mittenthal of the Johns Hopkins University Biophysics Department who introduced me to the idea of a dynamical space, and to the many workers in the Princeton "clocks lab" who shared with me their unpublished response-curve data. I am particularly grateful to Britton Chance, Joseph Higgins, and Kendall Pye of the Johnson Research Foundation for their consistent encouragement and thoughtful criticism, and to Theodosios Pavlidis of the Princeton Electrical Engineering Department for stimulating discussions. Through his essays, "Strong Inference" and "The Art of Creative Thinking," I am indebted to John R. Platt for the inspiration of this whole project.

REFERENCES

1. Perkel, D.H., Schulman, J.H., Bullock, T.H., Moore, G.P., Segundo, J.P., Science, 145, 61 (1964).

2. Moore, C.P., Segundo, J.P., Perkel, D.H. In Proceedings of San Diego Symposium for Biomedical Engineering, 1963.

3. Rawson, K.S. Ph.D. Thesis, Harvard University, 1956.

4. Pittendrigh, C.S. In Circadian Clocks, (Ed. J. Aschoff) North-Holland Publishing Co., Amsterdam. 1965, p.277.

5. Pittendrigh, C.S., Proc. Nat. Acad. Sci., 58, 1762 (1967).

6. Ottesen, E. (1965) Bachelor's Thesis, Biology Department, Princeton University.

7. Pavlidis, T., Bull. Mathe. Biophys., 29, 291 (1967).

8. Aschoff, J. In *Circadian Clocks*, (Ed. J. Aschoff), North-Holland Publishing Co., Amsterdam, 1965, p.95.

9. Pittendrigh, C.S. and Bruce, V.G. In *Rhythmic and Synthetic Processes in Growth*, (Ed. D. Rudnick), Princeton University Press, Princeton, 1957, p.75.

10. Higgins, J., Indus. Eng. Chem., $\underline{59}$, 19 (1967).

11. Levins, R., Amer. Sci., $\underline{54}$, 451 (1966).

12. Minis, D. In *Circadian Clocks*, (Ed. J. Aschoff), North-Holland Publishing Co., Amsterdam, 1965, p.333.

    Horne, D. Bachelor's Thesis, Biology Department, Princeton University, 1967.

13. Kauffman, S., J. Theor. Biol., $\underline{22}$, 437 (1969).

14. Zhabotinsky, A. This volume.

15. Winfree, A.T., J. Theor. Biol., $\underline{16}$, 15 (1967).

16. Wei, J., J. Chem. Physics, $\underline{36}$, 1578 (1962).

17. Fitzhugh, R., Biophys. J., $\underline{1}$, 445 (1961).

18. Platt, J.R., Science, $\underline{146}$, 347 (1966).

    Platt, J.R., *The Excitment of Science*, Houghton-Miffling, Boston, 1962.

19. Pavlidis, T. In *Lectures on Mathematics in the Life Sciences*, vol. 1, (Ed. M. Gerstenhaber), American Math. Soc., Providence, R.I., 1968, p.88.

20. Campbell, A. In *Synchrony in Cell Division and Growth*, (Ed. E. Zeuthen), Interscience, 1964, p.469.

21. Kalmus, A. and Wigglesworth, L.A. In *Cold Spring Harbor Symposium on Quantitative Biology*, (Ed. L. Frisch), $\underline{25}$, 1969, p.211.

22. Winfree, A.T. In *Lectures on Mathematics in the Life Sciences*, vol. 2 (Ed. M. Gerstenhaber) Amer. Math. Soc., Providence, R.I. (1970) p. 109.

23. Engelmann, W. and Honegger, H.W., Z. Naturforschg. $\underline{22b}$, 200, (1967).

THE INVESTIGATION OF OSCILLATORY PROCESSES BY
PERTURBATION EXPERIMENTS
II. A SINGULAR STATE IN THE CLOCK-OSCILLATION
OF Drosophila pseudoobscura*

Arthur T. Winfree

Biology Department, Princeton University
Princeton, New Jersey 08540 +

1. Introduction

Much of the interest in circadian rhythms has centered about their physiology and ecological involvements. Some have also inquired into the ultimate source of the rhythmicity, the basic "driving" oscillator. Very little is known of its concrete mechanism (presumably biochemical or hormonal). Even within the province of formal dynamics, several fundamental questions have remained unanswered, for example:
1. Is the clock a limit-cycle oscillator?
2. Is it self-exciting?
3. Does it have one, two, or more important state variables?
4. Does it have a stationary state (a singularity)? One, or more than one?
5. If so, is that state stable or not?

These are questions of phenomenology, rather than of mechanism. If the mechanism should prove to be complex, as many physiological control processes are, such abstract distillates of phenomenology may assume vital roles in our understanding. Even as few as three dynamical variables, nonlinearly connected in a causal loop (a common situation

---

* These investigations were supported by an NSF Regular Fellowship and Grants ONR 1858(28) and NASr-223.

+ Present Address: Department of Biological Sciences,
Purdue University,
West Lafayette, Indiana 47907

in biochemical regulation) suffice to embarrass our ability to analytically derive behavior from a fundamental mechanism.

In the previous paper, "The Dynamical Interpretation of Phase Shifts", (page 461, referred to below as "DI"), a perturbation method was suggested for the experimental investigation of complex nonlinear oscillatory processes. The most important observable quantity, following Pittendrigh, was the asymptotic residual phase shift, $\Delta\phi$, measured on any overt rhythm driven by the oscillator. An equivalent measure is the complement of the new, reset phase:
θ = constant $- (\phi - \Delta\phi + S)$, where ϕ is the phase of the rhythm at the beginning of a perturbation of duration S. This θ is the "cophase" of the previous paper, "DI". We attempt to design simple crucial experiments utilizing θ measurements to exclude broad classes of models as suggested in 1 through 5 above. Topological criteria are employed. Their application is illustrated using the circadian clock of the fruit fly Drosophila pseudoobscura. This clock regulates the time of emergence of the fly from its pupal case. Those facts and inferences about this system which are crucial to the interpretation of experiments reported below are summarized in the Appendix.

The purpose of this paper is not so much to support particular hypotheses about the Drosophila clock as to illustrate an approach to the study of oscillatory dynamical processes. Consequently, references are omitted to many excellent experimental and theoretical investigations of circadian rhythms, even on Drosophila. My own experiments are not reported here in complete detail; they will be published later (1,2).

2. Perturbation and Phase Shifts

The D. pseudoobscura clock process is perturbable (perhaps indirectly) by irradiation with blue light (3). One of the intriguing properties of many circadian rhythms, including the one investigated here in an all-female* strain of D. pseudoobscura is that they return to the normal period rather soon after perturbation - in this case within 72 hours.

---

*The pupal population is obtained using an X-chromosome ("sex-ratio") which eliminates Y-bearing sperm in the fathers.

In the experiments reported, pupae are irradiated with less than three minutes of approximately 10 μwatt/cm² blue light. (These signals are two to three orders of magnitude weaker than those used in (4,5,6). The pupae are at this time 2-5 days into the 9-day metamorphosis. Transients have died out before emergence is monitored and the latter appears with a constant phase shift relative to the control. We record the centroid time, E, of the emergence peak measured from initiation of the clock oscillation by transfer of pupae from continuous light into the darkroom. Cophase, θ, will be defined as:

$$\theta(\underline{x}) = (E-S-T) \text{ modulo } 1$$

where $\underline{x}$ is the state of the oscillator reached at the end of a perturbation of duration S, administered at time T. Time is measured in units of TAU, the period of the oscillator. That is, the quantity θ measures some function of the state of the oscillator immediately after the perturbation (T,S). The nature of this cophase function and the influence of the perturbation on the oscillator's state are our basic subjects of investigation.

3. The Basic Oscillator "State"

First of all, the scalar function of state θ turns out to have a vector argument; i.e. the "state" of the clock process requires for its unambiguous description at least two simultaneous measurements (of two state variables; see "DI", section 3). It can vary in a continuum of at least two degrees of freedom. Though the scalar "phase" may adequately describe the state of the underlying oscillator in particular kinds of experiments (e.g. 4,5,6), it seems not to serve as an adequate description in general, in the following senses:
a. It is very difficult (strictly speaking, impossible) to imagine how a negative resetting map* slope can be measured on a continuous mechanism whose state is determined by a single scalar (2). Section 10C below indicates that the clock mechanism is continuous. Yet signals with S greater than 60 seconds always produce resetting maps with a region of negative slope in the experiments to be described.

---

* The dependence of θ on T at fixed S. See section 8 in "DI".

b. There is a clock state associated with no phase (see sections 11 and 12 below).

c. It is possible to alter the shape of the resetting map to subsequent perturbations by prior application of a perturbation which causes no phase shift ((2) and unpublished experiments).

d. On the 1-degree-of-freedom hypothesis, if the resetting map is periodic measured with $S=S_1$, then so must it be with any other $S=S_2$. In D. pseudoobscura, this is not the case, with S_1 = 120 sec, S_2 = 30 sec. This behavior seems to require postulation of a second internal degree of freedom (unpublished experiments).

4. Postulates

In "DI" a geometrical language proved convenient for analysis of phase shift experiments on hypothetical dynamical processes having at least two degrees of freedom. By induction from the known phenomenology of circadian rhythms, in particular of D. pseudoobscura, three hypotheses were hazarded regarding the geometry of dynamical flow in the D. pseudoobscura clock:

a. The dynamical space is at least two dimensional.

b. The dynamical flow in the dark differs from that in the light — though by definition the space of states is the same in both cases, and

c. There seems at this time no evidence compelling recognition of any geometrical complexity in the flow, such as discontinuities, multiple singularities, separatrices, extreme local curvature or convolutedness of flow, more than two degrees of freedom. Thus, we sketch smoothly rotatory flow in the dark, unspecified flow in the light, and in each case a unique singular point.

5. Predictions

On this basis, the existence of the Type 0 resetting map (data of (6)) tells us that a sufficient prolongation of light sends the oscillator state across the region interior

BIOCHEMICAL OSCILLATORS

to the normal trajectory initiated by LL/DD (see section 10 of "DI"). We can therefore, use the TxS plane as in "DI" to map $\theta(\underline{x})$ near the singularity and to thus verify the existence of the singularity, $\underline{x}^*$, and measure its stability - or, failing this, to exclude the whole class of models represented by the conjunction of hypotheses a, b, and c in Section 4.

If these hypotheses are realistic, and if the analysis of "DI" is essentially correct, then we expect to find constant-cophase contours ($\theta(\underline{x})$ = const.) radiating from a unique (T^*,S^*). At (T^*,S^*) the overt rhythmicity must begin to decay with the time-constant of transients, and the circadian periodicity of responsiveness to the subsequent perturbations must be immediately annihilated. (T^*,S^*) is expected to lie near that T at which the phase inversion occurs and at S shorter than any which measure a Type 0 resetting map. So the acceptability of this view of the clock process hangs on the paradoxical prediction that the clock oscillation will be completely abolished by a delicate tap at exactly the right moment - even though a more powerful perturbation merely resets the clock without causing any such catastrophe.*

6. Experimental Format

To test this implication of the proposed hypotheses, a set of experiments was executed in the following format:

a. We induce coherent[+] clock oscillations in a population of pupae by LL/DD transition, the larvae having been reared in constant LL, as in (4), but using developmentally heterogeneous populations.

b. After permitting $0 < T < 24$ hours of dark dynamics, we interpose $15 \leq S \leq 180$ seconds of light dynamics (by irradiation with the standard blue light) and return to the dark (DD), see Figure 1.

c. Beginning 72 hours after perturbation (after subsidence of transients), we record the emergence peaks of adult flies hatching from their pupal cases.

* As mentioned in "DI" (fn., p.474), Pavlidis and Rosenzweig also independently, and on somewhat different grounds, anticipated this phenomenon.

[+] See Appendix B1 c, and Section 10 B below.

483

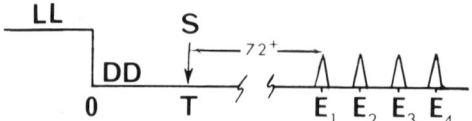

Figure 1. Format of the basic experiment: DD after LL is punctuated by a single exposure to light for S seconds at hour T. The triangles represent daily emergence peaks, occurring in the dark.

Figure 2 shows the distribution of experiments over the TxS plane. This plane divides into three zones as regards results. In zones b and c the normal 24 hour rhythmicity with normal 5.7 hour peak width reappears in the first 72 hours after perturbation. In zone c, the perturbation has

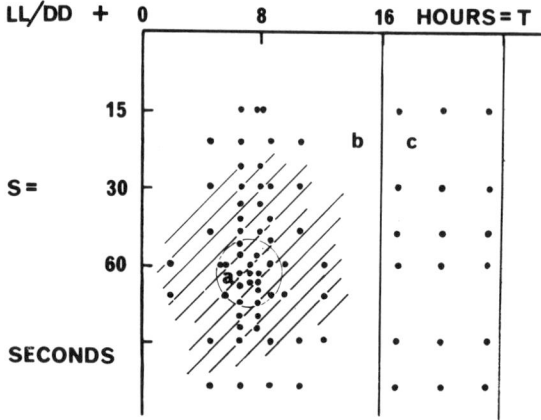

Figure 2. Black dots mark the values of T and S for which emergence data were obtained.

little or no effect; in zone b there is a phase shift. In zone a, the period recovers but peaks remain substantially broadened. In this report, we will be concerned exclusively with the cross-hatched region, the environs of the singularity.

7. Controls

In Figure 3, a random sampling of control experiments is presented, vertically overlapped so that only the bottom part of each emergence peak is visible. Each x represents one emerging fly. The horizontal scale is in hours measured from LL/DD, and the large dots mark the centroids of daily emergence peaks. Their scatter about the vertical measures

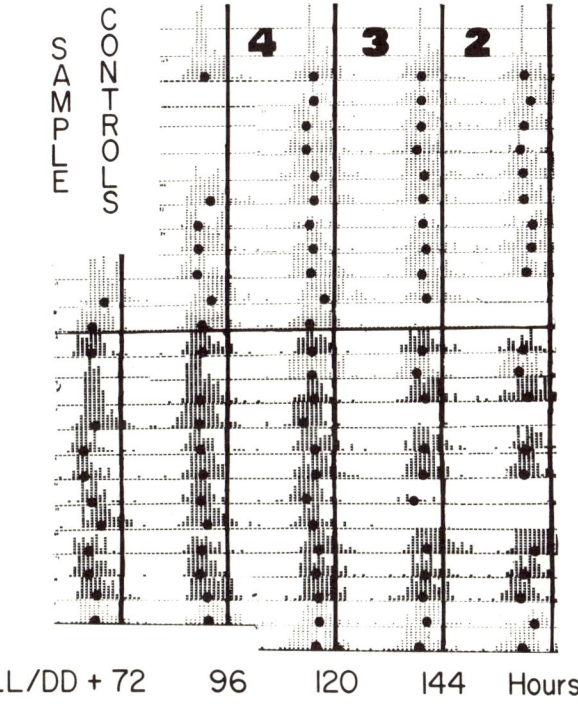

Figure 3. Twenty four control experiments (S=0): histograms of hourly emergence activity for 5 days. Numerals "4,3,2" indicate developmental age of hatching pupae when LL/DD was given. (Age is 9 days at emergence.)

the basic "noise level" of phase measurements, not including any additional sources of variance associated with the perturbation. These LL/DD transitions occurred under diverse circumstances, at all times of day during an eight week interval.

8. The Helicoid of Cophase

To display phase information alone, a three-dimensional graph of perturbation experiments was constructed: for each experiment the centroids of emergence peaks were marked above the TxS plane at an altitude proportional to E-S-T, that is, to $\theta(\underline{x}\#) \pm 24n$, where $\underline{x}\#$ is the state immediately after perturbation. Emergence peaks occurring 72 to 132 hours (during 2-1/2 days) after the perturbation were plotted representing pupae who saw LL/DD and (T,S) in the 5th to 2nd days of pupal life. This cloud of about 100 dots forms a

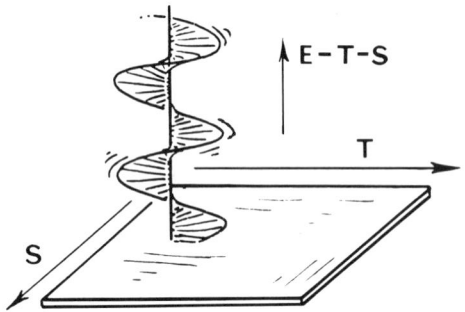

Figure 4. Cloud of data points in central portion of $\theta(T,S)$ is sketched as a continuous spiral ramp. Two and one half days' data = 2 1/2 turns. Helical boundary of sketch does not really exist; the full θ surface terminates only at T=0 and S=0.

distinct spiral ramp, a corkscrew winding upward from
E = T+S + 72 for 2-1/2 turns above the TxS plane as predicted in "DI" Section 11.3. This helicoid is the three-dimensional structure corresponding to a radial contour map of θ (see sketch in Figure 4). Its symmetry axis lies in zone $\underline{a}$ at (T=7.3 hours, S=60 sec), which seems to correspond to the singularity.#

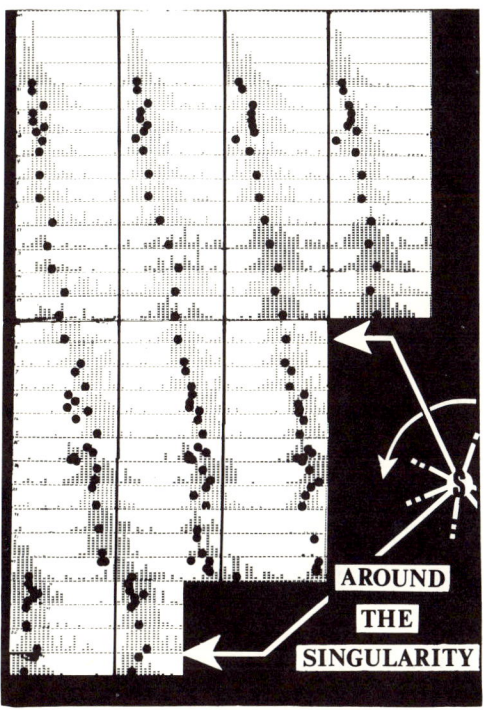

Figure 5. Emergence histograms as in Figure 3 but with perturbations (T,S≠0), stacked in vertical order of angular position on TxS plane relative to symmetry axis of helicoid. Time is measured from T+72 at left-hand edge. Dots represent centroids of peaks for the histograms plotted, and for others for which there was no room on the graph. Zone a of Figure 2 is deleted because "peaks" are indistinct and centroids capriciously positioned.

Note added in proof: a better estimate, based on additional experiments of 1969, is T=6.8, S=50.

9. Isochron Contours

Lacking stereo images of the three-dimensional graph, the radial structure of the isochrons on the TxS plane (the loci of fixed cophase) is illustrated in Figure 5: here a sampling of experiments in the cross-hatched zone have been stacked as in Figure 3, but not randomly. They are vertically in order of arctan (T-7.3/S-60). That is, at each level in the stack the emergence rhythm is displayed following perturbations whose (T,S) coordinates lie in the same radial direction from (7.3, 60) irrespective of radial distance. It is thus seen that each such direction corresponds to a particular resetting of the rhythm. As the perturbation passes around (7.3, 60) ascending vertically in Figure 5, the reset rhythm also scans one full cycle horizontally; i.e. the isochrons map into the expected approximately radial contours, at least in the cross-hatched part of Figure 2, minus zone a.

10. What This Experiment Excludes

A) It seems to me that this precludes any interpretation of the clock mechanism involving only a single important state variable, (like the volume of water accumulated on the lip of a dripping faucet) e.g. the position of the polymerase molecule on a single chronon (7). This is so regardless of how one imagines the perturbation to affect the clock; whether in proportion to the total energy, or only at the moment of turning on and/or off, or with rapidly lessening effect due to adaptation of sensitivity, etc.

One argument runs as follows: the points along an isochron measured in the TxS plane either 1) represent different ways of getting to the same state, each cophase corresponding to a unique state, or 2) represent different clock states having the same cophase. On alternative 1, the point of intersection* of the isochrons must represent the same state as each isochron represents. Therefore, conversely, all isochrons represent the same state, i.e. _any_ perturbation (T,S) sends the clock to this unique state. This is plainly nonsense, particularly in the case S = 0. Therefore the clock mechanism must enjoy at least a two-dimensional continuum of states.

---

* Technically, this "intersection" is not in any isochron, but is the closure of their union.

A second argument is an experiment (unpublished) in which states lying in the same isochron are found to react differently to a common subsequent perturbation.

B) It is now possible to support the crucial points B.1,2 of the Appendix: that the individual pupae of the population have the same phase and sensitivity following LL/DD. Outside zone a of Figure 2 the peak widths of reset emergences are indistinguishable from the normal 5.7 hours. But many of these experiments were done in regions of the TxS plane where the observed $\frac{\partial \theta}{\partial T}$ is substantially different from unity, e.g. 3 or 0. If the 5.7 hours represented a distribution of clock phases following LL/DD then application of these perturbations would respectively triple or eliminate the incoherence (peak width). But no measurable difference is found. Population variance in clock phases could not therefore account for more than a fraction of normal variance in emergence times. The population is homogeneous in respect to phase. These same data suggest that pupae are not very diversely sensitive to or diversely exposed to the light. Since the pupae do not interact and since all have the same clock phase within narrow limits following LL/DD and are equally sensitive to the perturbation, we are entitled to regard the population experiment as many simultaneous replicas of the same experiment on single pupae (of diverse developmental ages).

C) In view of the smoothness of the response surface $\theta(T,S)$, clock models involving discontinuities (e.g. threshold-relaxation processes) seem implausible.

11. The Singularity: Criterion One

In zone a of Figure 3, however, the peak widths increase as the radial images of isochrons continue toward the center. The symmetry axis of the helicoid is not represented by dots in the three-dimensional graph because flies emerged continuously, not in peaks.

In Figure 6 this effect is illustrated by stacking experiments in vertical order of distance to (7.3, 60), irrespective of angular position of the perturbation on the TxS plane. Data have all been shifted horizontally as much

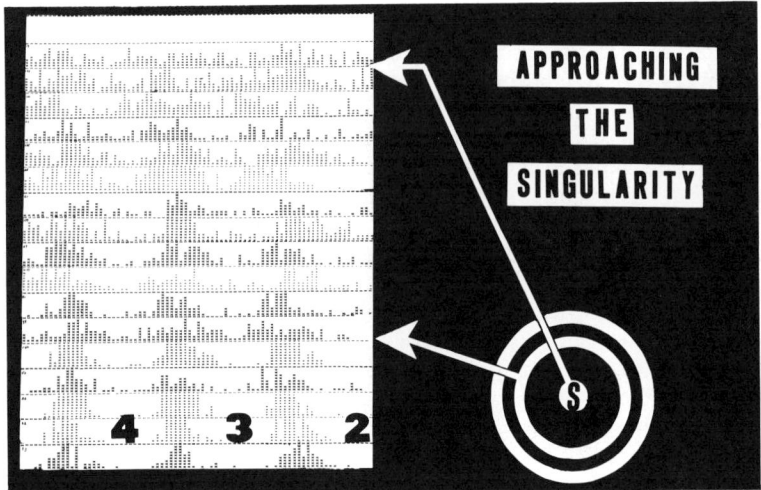

Figure 6. As in Fig. 5, but stacked in vertical order of radial distance from symmetry axis. Time axes are individually shifted horizontally to put peaks in vertical columns for clarity. Large numerals have the same meaning as in Fig.3.

as ±12 hours to put peaks in vertical columns for clarity. All data shown begin at least 72 hours after the perturbation, so transients have decayed away. Due to overlapping of records, peaks are truncated in the lower, more sharply rhythmic records. The top record shows the effect of hitting a population near hour 7.3 with 60 seconds of standard blue light. This overt persistent arrhythmicity is the first criterion by which a singularity is demonstrated. Thus far it would appear that $(7.3, 60) = (T^*, S^*)$.

12. The Singularity: Criterion Two

A second criterion is that following LL/DD and (T^*, S^*), the response to a second perturbation should be time-invariant: the rhythmicity of emergence should be re-initiated at a particular phase, the same no matter when the reinitiating perturbation is applied (at least if $N = 2$). Specifically, since upon a second exposure to light, the system simply continues upon the trajectory which led to the singularity of the dark dynamics during the first perturbation, therefore the new phase should be the same as is obtained by giving the full perturbation, 60 + S_2, all in one dose at hour 7.3. And that new phase is marked by vertical bars in Figure 7.

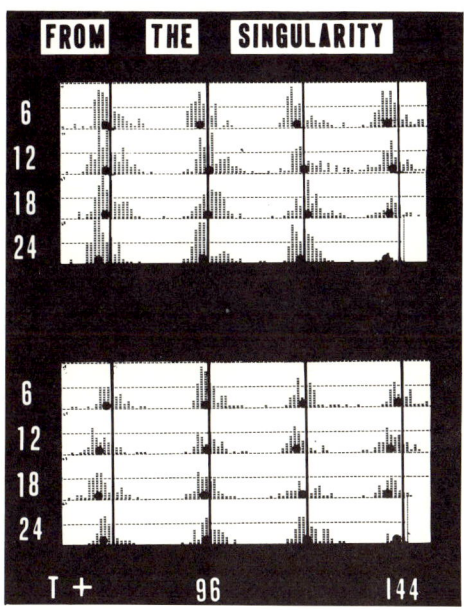

Figure 7. Histograms of emergence activity following reinitiation from singularity; two experiments each of four parts. Time measured from reinitiating pulse.

In Figure 7 this experiment is illustrated by applying LL/DD followed by (7.1, 65) in the top panel and (7.5, 55) in the bottom panel, in each case followed by S_2 = 120 seconds after 6, or 12, or 18, or 24 hours. In all eight cases the rhythm IS reinitiated near the predicted phase. Apparently the first perturbation left the clock process in a "limbo" at the singularity, a relatively time independent state from which the perturbation can be continued at any later time as though never interrupted.

It would appear by both criteria therefore that (7.3, 60) is close to the predicted isolated critical perturbation which annihilates the clock oscillation.

This result also suggests that the clock process in

D. pseudoobscura may be quite simple in the sense of involving ONLY two important variables in geometrically uncomplicated dynamical interaction.* Only in that case should we confidently expect the "phaseless" aperiodic state to also be time-independent in its reaction to subsequent perturbations - as it seems to be.

13. The Question of : Stability versus Instability
Dead versus Scattered Clocks
Initiation versus Synchronization

However, there is an alternative interpretation of sections 11 and 12. We are operating on and observing a population of oscillators. Both the overt arrhythmicity of emergence and the time-independence of responsiveness could be due not to the annihilation of the oscillations but to a random scattering of their phases ... i.e. the progressive broadening of peaks as (T,S) approaches the confluence of isochrons might represent progressive loss of coherence. The nearly infinite grad θ near the singularity might amplify the inevitable small population variance of initial states, and of sensitivity and/or exposure to the light. We must discriminate between:

Hypothesis A: In terms of dynamics: it was tacitly assumed that the singularity would be stable or at least not violently unstable, so that oscillators moved near to this state would stay near it. This reduced amplitude (to zero in the limit as (T,S) approaches (T*,S*)) of periodic state fluctuations would certainly not fortify the periodic modulating influence on emergence behavior, and would certainly weaken it as amplitude approaches zero. The clocks on this hypothesis

---

* If this were so it would not necessarily mean that there are only two genetic loci or only two reaction pathways determining the biochemical components of the clock. There might be 500. But they apparently fall into two groups interacting with a time constant of many hours, but within each of which equilibrium is always maintained or is restored after perturbation too quickly to be of significance on the time scale of circadian phenomena.

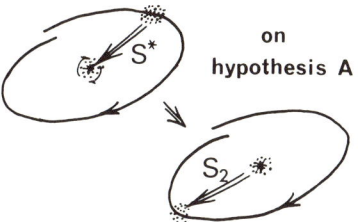

Figure 8. Theoretical effect of (T*,S*) and subsequent S_2 (as in Fig.7) on a population of oscillators if the singularity is stable.

are all DEAD following (T*,S*). The second perturbation then has an effect analogous to striking a bell whose ringing had been previously quenched. See Figure 8.

and

Hypothesis B: Alternatively, a scattering of phases would be expected if the singularity were unstable ... and if it is, then we are dealing with a relaxation oscillator or with a strongly self-excitatory limit-cycle oscillator. In that case the (T*,S*) identified by the disappearance of population rhythmicity would be that which so places the little cloud of oscillators near the singularity and confluence of isochrons that they spiral back out to the normal self-sustaining oscillations at uniformly randomized phases. Such a scattered population would be time-independently responsive to a second perturbation, the effect of which would be NOT the initiation of oscillations in equilibrated clocks, but partial synchronization among asynchronously running clocks. See Figure 9.

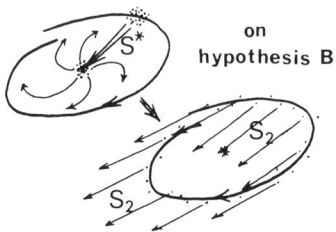

Figure 9. Same as Figure 8, but if the singularity is unstable, oscillators return toward limit cycle between S* and S_2.

14. A Criterion Distinguishing between Hypotheses A and B

The alternatives of initiation (corresponding to dead clocks and stable singularity) or partial synchronization (corresponding to normally-running, but scattered clocks, unstable singularity, and limit-cycle oscillators) are distinguishable. The first (Hypothesis A) results in a narrow unimodal distribution of phases, $\rho(\Phi)$.* But in the second case (Hypothesis B) the clocks are initially at quite different states and react to perturbation in different ways. The resulting distribution of phases depends upon the duration of perturbation used. Selection of a perturbation which measures a resetting map of Type 0 topology (e.g. 120 seconds) gives us a wide two-horned distribution of final phases as illustrated in Figure 10.

To measure $\rho(\Phi)$ we must "see through" the superposition of 5.7-hour-wide emergence peaks typical even of

---

* "Phase" is used to indicate the fraction elapsed in the oscillator's "cycle": ϕ for initial phase, and Φ for reset phase.

BIOCHEMICAL OSCILLATORS

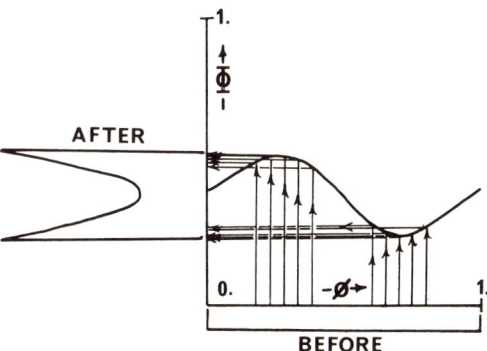

Figure 10. Redistribution of uniformly scattered phases ("BEFORE") by Type 0 resetting map to "AFTER" distribution.

highly coherent populations. This is done by numerically inverting the convolution, using as input the normal 5.7-hour-wide basic peak and the observed emergence peak.

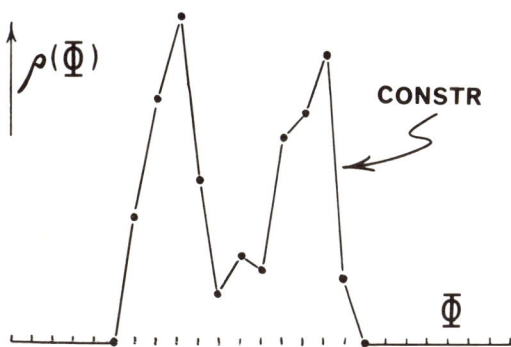

Figure 11. Control experiment: distribution of phases resulting from experiment of Fig.10. Ordinate is fraction of population assigned by computer program to coherent subpopulation at phase Φ. Abscissa is Φ in units of 1/24.

To check the validity of these procedures, the situation of Hypothesis B was artificially constructed by mixing 12 populations stepped down from LL to DD at two hour intervals throughout a day. This population of scattered running clocks was arrhythmic until shocked with 120 seconds of standard blue light. After 72 hours, the measured distribution of phases was as in Figure 11.

On the other hand the same procedure applied to the data of Figure 7, produced the narrow unimodal "T*,S*" of Figure 12.

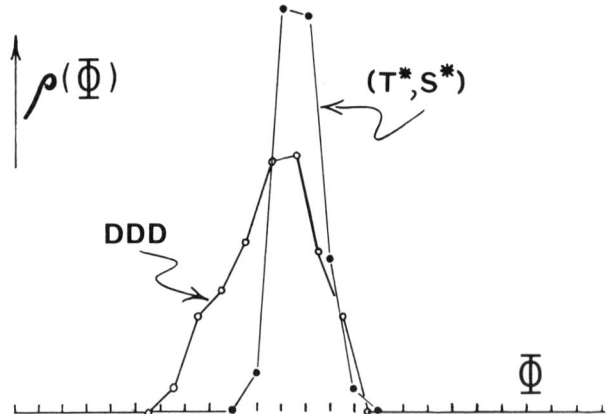

Figure 12. Distribution of phases, computed by program used in Fig. 11, resulting from re-initiation (of Fig.7, T*,S*) and initiation (DDD) experiments.

This experiment seems to exclude Hypothesis B* and to favor A. Even if the theory were wrong it is unmistakably clear that the arrhythmic population resulting from LL/DD and (T*.S*) differs in some essential way from an arrhythmic population of normally running clocks with scattered phases.

A further check is obtained by examining the well known (3,8) arrhythmicity of populations of pupae reared as

* The possibility is NOT excluded that each individual pupa's clock consists of a population of self-exciting cellular oscillators which are only scattered in phase by (T*,S*).

larvae in constant dark. If the singularity is unstable, i.e. the clock oscillator self-exciting (and therefore a limit-cycle or relaxation oscillator), then such populations are equivalent to the artificially constructed population of running clocks. Subjecting such a population to a 120 second inducing pulse, the unimodal distribution of phases, "DDD" in Figure 12 was obtained, centered near the same phase as expected on Hypothesis A. This seems to confirm the exclusion of Hypothesis B*.

A plausible remaining interpretation would be that clocks in this developmentally heterogeneous population are created (in epigenesis) near to or at the singularity, and remain there incoherently oscillating at diverse small amplitudes until synchronously knocked farther from the singularity by the 120 second pulse.

15. Limit Cycles

I mention for the reader's amusement that these experiments (Spring, '68) were undertaken in order to prove that the circadian rhythms are limit-cycle processes (10,12) at least in the case of D. pseudoobscura. It was necessary only to demonstrate a violently unstable singular state (11). Though the singularity turned out to be stable, more is required to rigorously exclude the limit-cycle hypothesis; a further extension of these perturbation methods may provide the needed experimental criteria.

Though the resetting map shapes in circadian rhythms of diverse other organisms resemble sections of a helicoid and therefore suggest continuous dynamics in two or more degrees of freedom, they say nothing about the stability of the singularity. Indeed the persistence of mammalian activity rhythms for hundreds of cycles in a constant environment is suggestive of a limit-cycle process.

16. Conclusions

A light pulse $(T^*, S^*) \cong$ (7.3 hrs., 60 sec.) sends

* By a different line of reasoning and an elegantly simple experiment, Dr. William Zimmerman has also concluded that temperature-shock induces rhythmicity in "DDD" populations by initiation rather than by synchronization (9).

each pupa's photosensitive circadian oscillator near to a time-independent stationary state. (T*,S*) is isolated and seems unique; very likely, so is the conjugate state (singularity) in the dynamical plane. The isochron contours converge to the singularity in the way expected if and only if the circadian oscillator derives from continuous mutual interaction among two or more dynamical variables. The time-independence of state in the manifold reached by (T*,S*) (at the confluence of isochrons) suggests but does not prove that <u>only</u> two variables are importantly involved. The apparent stability of the singularity appears to exclude self-excitatory processes, including simple limit-cycles and relaxation oscillators, unless the time constants are so long as to have little interest for experiments in the nine days of pupal metamorphosis.

More extensive data from this series of experiments show that the resetting map shapes of both topologies Type 0 and Type 1 depend upon S in the way expected if the action of light is direct, continuous, and simple, as e.g. in the model of Pavlidis without adaptation(12). These shapes are sagittal sections of the helicoid partly described in Section 8. This is a very definitely constrained family of shapes, and it lends additional support to the two empirical generalizations of Section 10 in "DI". The resetting maps of all other circadian rhythms for which I was able to find data fall into or near the same family. "Verboten" shapes - e.g. sinusoids of peak-to-peak amplitude near 3/4 - are conspicuously absent. Where the data are available these resetting maps change shape in the expected way when perturbation duration, S, is increased (2,11).

There may be some value, beyond demonstration of its existence, in the possession of a recipe for sending an oscillatory process of generally unknown and nonlinear complexity to its stationary state, especially if that state is stable (in the sense that trajectories through nearby states do not rapidly leave the neighborhood). For example:

a) If the critical perturbation can be applied quickly, as in the case illustrated, then one can immediately inhibit all change in the state variables of the basic oscillator, and in the same environment which previously permitted oscillations. Any quantity subsequently observed to change is therefore excluded as a potential component of the basic clock process or even as a factor affecting that process.

b) It becomes possible to examine the dynamics of clock-influenced quantities (e.g. pupal emergence probability) in the absence of the modulating or driving clock oscillations, but without changing anything else.

c) Factors which influence clock dynamics only very weakly (e.g. most metabolic inhibitors, unless used in pathologically large doses) cause only small phase shifts in pulse-chase experiments. However, the same small displacement of state applied when the oscillator is at the singularity would initiate a distinct rhythm. Moreover, all factors impinging on the clock in the same way will initiate from the singularity to the same phase. The number and nature of such classes may suggest the number and nature of the discrete processes or variables affected.

APPENDIX

A. Summary of Relevant Details Re: D.pseudoobscura clock (based on (4))

Metamorphosis in the pupa of Drosophila pseudoobscura nominally terminates with the emergence of the fly from the puparium. In a population of females under constant illumination (LL) at 20.0°C, these events occur in an approximately normal distribution $2\sigma = 11$ hours wide, centered 208 hours after the larva forms its puparium.

If, however, the illumination is terminated at hour $0 < t < 160$, then the subsequent distribution of emergences in the dark (DD) consists of one or two discrete peaks less than 24 hours apart, each about $2\sigma = 3$ hours wide. The centers of these peaks are displaced from 208 towards hours e (emergence) = $k + t + 24n$, where $0 < k < 24$ is a constant characteristic of the fly strain and culture temperature (5) and $3 < n < 9$. It is inferred that some periodic or asymptotically-periodic time measuring process (clock), endogenous to the individual pupa, is in some sense induced by the LL/DD transition. This oscillation, together with other uncontrolled factors, determines the time of emergence.

In a developmentally homogeneous population, the actual center of the emergence peak deviates from e by one to several hours, somewhat in proportion to /e − 208/. These deviations being roughly symmetric, in a developmentally heterogeneous population, emergences are distributed approximately symmetrically about e with a somewhat greater peak width: 2σ = 5.7 hours in the developmentally heterogeneous populations used in these experiments.

B.   Population Aspects

The amplitude of an emergence peak indicates only the number of pupae aged 195-225 hours on that day. Information regarding the "clock" is derived from the location of the peak (its centroid time), and its shape, especially width. It is possible to show that:

1. The normal variance of an emergence peak, $(5.7/2)^2$, does not represent a dispersion of phases among the clock-oscillators in a population of pupae. On the contrary, such incoherence in the first 24 hours after LL/DD accounts for at most two hours out of 5.7 (10B above).

2. The population of pupae is also fairly homogeneously sensitive to light (10B above).

3. The distribution of periods in the population is narrower than 4 hours/208 hours = 2% (from the data of 4).

4. To a good first approximation the parameters of the oscillator's dynamics (the "Structural Control Variables" of (13)) are essentially invariant with developmental age from days 2 through 5 after prepuparium formation.

C.   The Effect of LL/DD Transition:

Since the pupal clocks are found to be in the same state after LL/DD, it is tempting to conjecture that they were all in this same state immediately before LL/DD, too, and that the termination of light only releases them onto a dark trajectory from this state. This state, reached after long illumination ($S \to \infty$), would then be the stable singularity of the altered (light) dynamics as in Figure 9 of "DI", B or C.

ACKNOWLEDGEMENTS

I thank my thesis advisor, Colin Pittendrigh, who made it possible for me to pursue these ideas and experiments in his labs at Princeton. To Wolfgang Engelmann I am indebted for a myriad of invaluable helps and hints on experimental technique. Most of the experimental apparatus was built by Russell Mycock and Roman Charydczak. The flies were constructed from mutant chromosomes kindly provided by Ronald Quinn and Theodosius Dobzhansky. Finally, it is a pleasure to thank my lab technician, secretary, and wife, Trisha, for her cheerful labors throughout and patient criticism of the many drafts of these papers.

REFERENCES

1. Winfree, A. T. Ph.D. Thesis, Princeton University (1970).
2. Winfree, A.T. "The Temporal Morphology of a Biological Clock" in <u>Lectures on Mathematics in the Life Sciences</u> $\underline{2}$, (Ed. M. Gerstenhaber) Amer. Math. Soc. Providence, R.I. (1970).
3. Honegger, H.W., Zeit für Vergl., $\underline{57}$, 244 (1967).
4. Skopik, S.D. and Pittendrigh, C.S. Proc. Nat. Acad. Sci., $\underline{58}$, 1862. (1967).
5. Zimmerman, W.F., Pittendrigh, C.S., and Pavlidis, T. J. Insect Physiol., $\underline{14}$, 669. (1968).
6. Pittendrigh, C.S., Proc. Nat. Acad. Sci., $\underline{58}$, 1762. (1967).
7. Ehret, C. and Trucco, E., J. Theor. Biol., $\underline{15}$, 240. (1966).
8. Pittendrigh, C.S., Proc. Nat. Acad. Sci., $\underline{40}$, 1018 (1954).
9. Zimmerman, W.F. Biol. Bull. $\underline{136}$, 494. (1969).
10. Winfree, A.T. J. Theor. Biol., $\underline{16}$, 15 (1967).
11. Winfree, A.T. "Puzzles and Paradoxes ..." (mimeo circullated in limited number). 1967.
12. Pavlidis, T., Bull. Mathe. Biophys., $\underline{29}$, 781. (1967).
13. Higgins, J., Indus. Eng. Chem., $\underline{59}$, 19 (1967).

THE CIRCADIAN OSCILLATION:
AN INTEGRAL AND UNDISSOCIABLE PROPERTY OF
EUKARYOTIC GENE-ACTION SYSTEMS

C. F. Ehret, J. J. Wille and E. Trucco*

Division of Biological and Medical Research
Argonne National Laboratory
Argonne, Illinois

One is tempted to identify the high frequency ticks of the short-period biochemical oscillations emphasized at this colloquium with the escapements for low-frequency circadian clocks. However, theoretical considerations and experimental observations of the macromolecular properties of circadian systems persuade us to reject such a view. Of course, we accept as self-evident the cell's reliance upon its own molecular milieu for energetic and material inputs, but we believe that concomitant high frequency oscillations are not closely related to circadian phenomena. Thus in terms of temporality, the circadian outputs of a cell may be regarded as being nearly as independent of the characteristic periods of its high frequency molecular oscillators as the timekeeping capacity of a mechanical clock equipped with a pendulum escapement is relatively independent of its energy source (even if, for example, the latter energy source consists of a 60 r.p.m. synchronous motor). The basic problem remains of how and for what reason one should obtain an over-all period of approximately 24 hours from whatever mechanism.

As we have discussed elsewhere (1, 2) the universality of circadian capacities amongst eukaryotes, taken together with the small temperature dependence of the circadian clock make any single mechanism to explain circadian outputs

---

* The mathematician-biologist, Ernesto Trucco, coauthor of this paper and of the chronon theory, and friend and colleague for many years, died on 25 August 1970 at the age of 48. - C.F. Ehret.

from all eukaryotes by means of high frequency metabolic
oscillators merely gratuitous and extremely unlikely. At
the same time we showed that the genetic algorithm -- linear
sequential replication, and especially linear sequential
transcription, cistron by cistron, of a very long polycistronic unit termed the chronon -- provides a cell with a remarkably simple interval-measuring and event enumerating
device (or "clock", as clocks are so defined). In monorep-
liconic prokaryotes and in multirepliconic eukaryotes grown
in a rapid exponential mode of growth (the ultradian mode),
(3), these clocks are extremely temperature dependent (viz,
Fig. 1), they have no evident circadian outputs, and unless
the temperature is constant they are poor timekeepers;
nevertheless the property of temporal ordering is maintained
(discussed for Tetrahymena below in Figure 5) and under constant conditions one can for example distinguish, even in
the bacteriophage T-4, temporally unique species of RNA less
than twenty minutes apart from one another (8). According
to chronon theory (1) the multirepliconic nature of eukaryotic chromosomes has permitted in all eukaryotes the preservation of naturally selected primitive polycistronic sequences acting as escapement mechanisms for the circadian period.
Under the slow mode of growth, the infradian mode, (2,3,4),
in all eukaryotes natural selection has pruned the chronon
to precisely the number of cistrons to provide the circadian
period, i.e. $\tau \approx 24$ hours; thus one or several chronons per
cell will rate-limit, with great precision, the cycle length
of a cell in replication and/or transcription. The relative
temperature independence of circadian outputs in the infradian
mode of growth may be accounted for by a drastic switchover
in cellular regulatory mechanisms from mass action limitations
on net reaction velocities during ultradian, to diffusion
limitation during the more nutritionally stringent conditions
of infradian growth (1,2). We have reached the conclusion
that a cell in switching its state from the temperature dependent ultradian mode to the infradian mode is then capable
of light-synchronizable circadian outputs in the latter mode.
Because this effect is best demonstrated in Tetrahymena
(Figures 1 and 2) and in Gongaulax and Euglena, we have
referred to it as the G.E.T. effect (2,4).

Occasional circadian rhythms that are observed in the
ultradian mode of growth may be accounted for using the
model of multiple fork replicons proposed by Helmstetter and
Cooper, without changing the basic premises of the chronon

theory (2). If, as postulated by the theory, linear sequential transcription is the principal regulatory mechanism that controls the low-frequency-oscillatory and temporally-characteristic properties of eukaryotic cells then one should be able to isolate distinct classes of RNA that correspond to temporally unique (e.g., "early" and "late") cistrons in the cell cycle, whether the cells be those grown in the fast (ultradian) or slow (circadian-infradian) modes of growth.

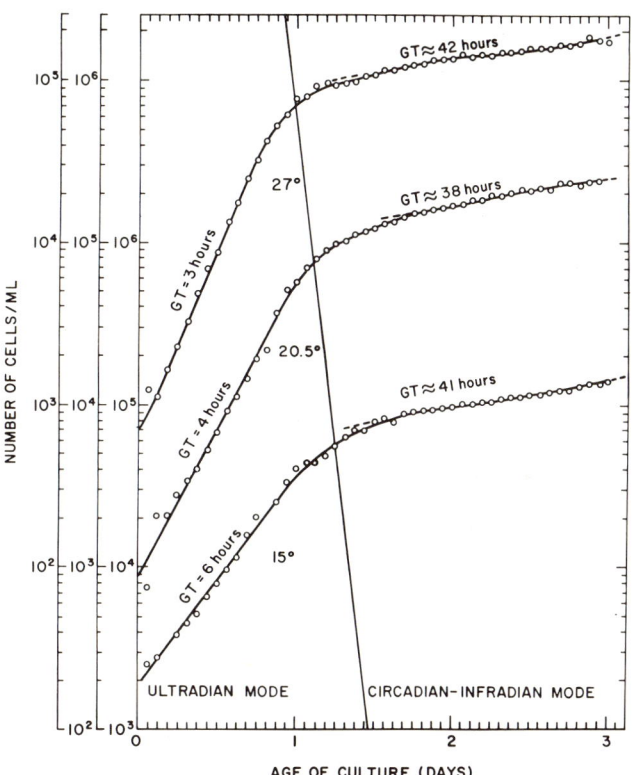

Figure 1. Growth curves of asynchronous batch cultures of Tetrahymena pyriformis (W) at three different temperatures. The outermost ordinate is for the growth curve at 27°C, the middle ordinate is for the 20.5° growth curve, and the innermost ordinate pertains to the 15°C growth curve. Note well the conspicuous temperature dependence during the ultradian mode. After Szyszko et al., 1968 (3).

505

Furthermore, if the transcription algorithm is essentially similar in either mode, ultradian cells and circadian-infradian cells should synthesize temporally unique classes of RNA common to both modes. The experiments reported below (5) affirm each set of expectations by the method of molecular hybridization between DNA and RNA macromolecules in the protozoan Tetrahymena pyriformis (W).

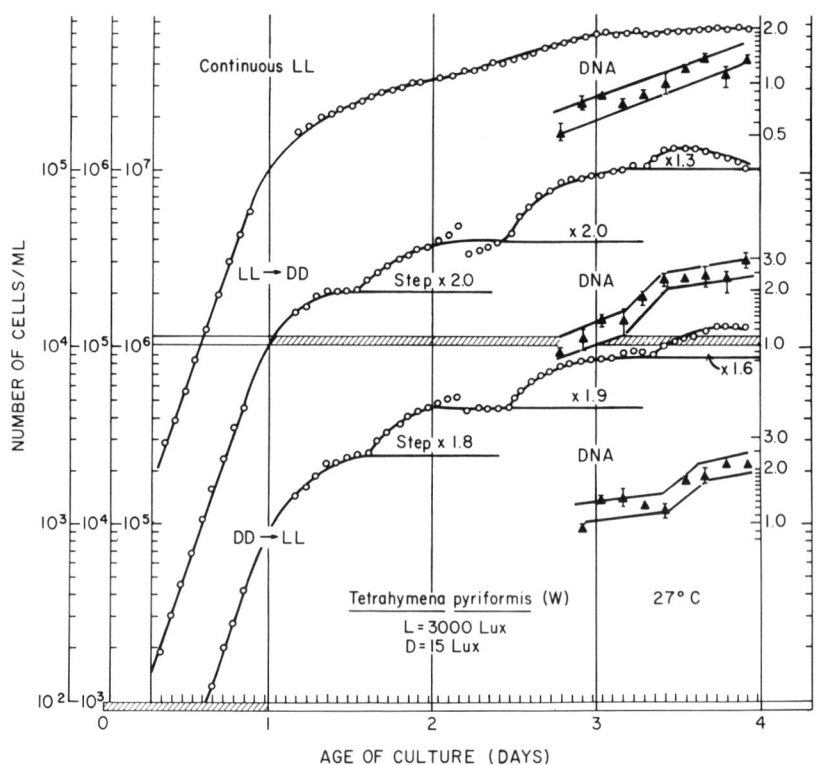

Figure 2. An example of the light synchronization of eukaryotic cells in the infradian growth mode. At about the time that an asynchronous population of Tetrahymena cells changes in batch culture from the ultradian (GT ≈ 3 hrs) to the infradian (GT ≈ 40 hrs) mode of growth (around 1 day of age in this experiment) either a "switch-down" (LL → DD) or a "switch-up" (DD → LL) in illumination induces not only synchronization of the cells, but also circadian ($\tau \approx 21$ hrs) steps. Not all cells in the infradian population divide, but those that do, do so on circadian time. From Wille and Ehret, 1968 (4).

RNA suitable for hybridization was obtained by a low-temperature extraction method followed by density gradient sedimentation in sucrose (5-20%)(6). Clearly discernible in the absorption spectrum of the sucrose layers are 25S, 17S and 4S fractions, and by its high specific activity, a fraction probably representing early RNA in the 45S region (Figure 3). This RNA was prepared from an unsynchronized late ultradian (GT ≥ 6 hrs) population of cells whose "turned-on" genes were presumably randomly distributed at the time of harvest. Useful as a somewhat arbitrary reference point in the studies that follow, it is referred to herein as "Steady State RNA" (SS-RNA). Temporally unique RNA was obtained by inducing synchronization in either ultradian (U)

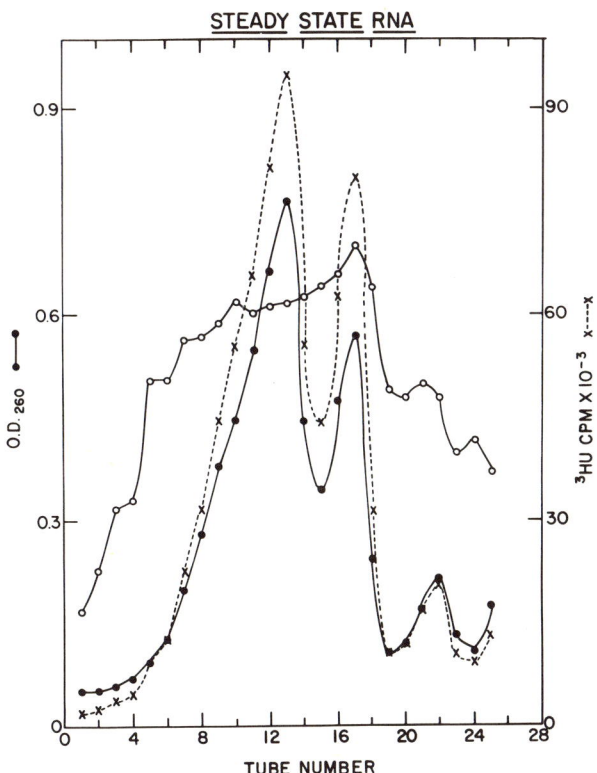

Figure 3. Sucrose gradient of tritium-uridine labeled RNA. Major peaks (O.D. at 260 nm) occur at 25S, 17S and 4S; high specific activity (open circles) is also evident in the very heavy region. Wille and Ehret (5).

or in circadian-infradian (C) modes of cell growth. In the former case (U), the method of successive heat-shocks (7) was employed resulting in a highly synchronized population with a generation time (GT) of about 2-1/2 hours. In the latter case (C), the technique of light-induced (switch-down) synchronization (4) was employed, on a program-fed modification of the continuous culture apparatus earlier described (3,4). Figure 4 shows the relation between the actual cell count (middle curve) and the steps and levels that the growing population of cells would have shown (top curve) had it not been for the washout resulting from feeding the culture 96 times a day (an aliquot every 15 minutes). In a number of experiments RNA was pulse labeled either with ^{32}P or with ^{3}H labeled uridine for either 2 hours (in C) or for 1 hour (in U), but always at a well-defined and temporally unique point in the cycle (viz. "C-3" or "U-3").

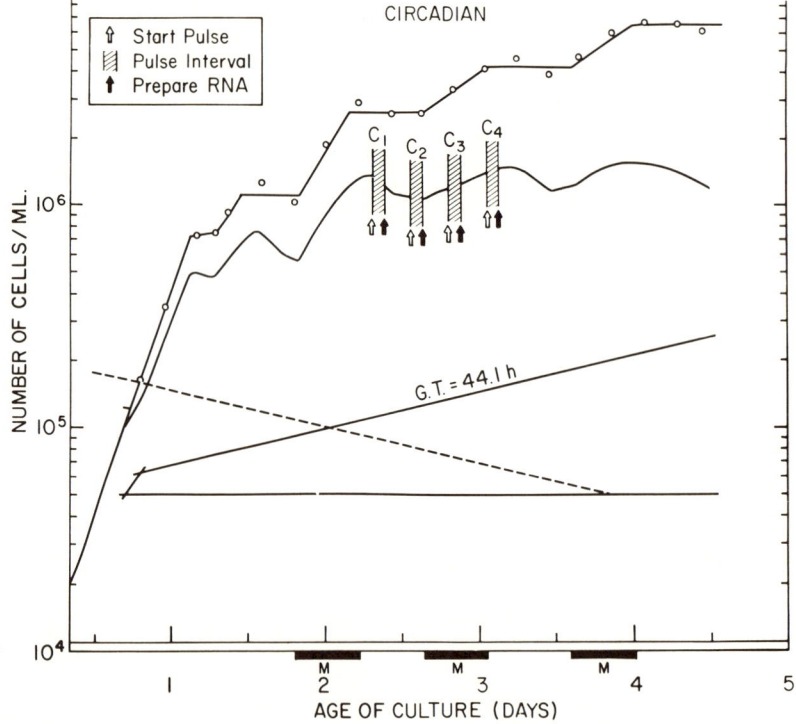

Figure 4. Protocol for ^{32}P or ^{3}H uridine labeling of Tetrahymena cells synchronized in the circadian mode. Wille and Ehret (5).

As shown in Figure 5, striking differences in hybridization capacity appear amongst the temporally unique RNA's of U and of C. Tetrahymena DNA was sheared to low molecular weight ($\approx 7.5 \times 10^5$) with the aid of a Ribi cell fractionator. It was then heat-denatured (100°C, 10 minutes) and maintained as single stranded DNA (criterion: hyperchromicity) by holding it at temperatures above 60°C until it was bound to a membrane filter (Sartorius). The latter step involved the controlled slow passage of DNA at 85°C through the filters. The filters were then air dried at room

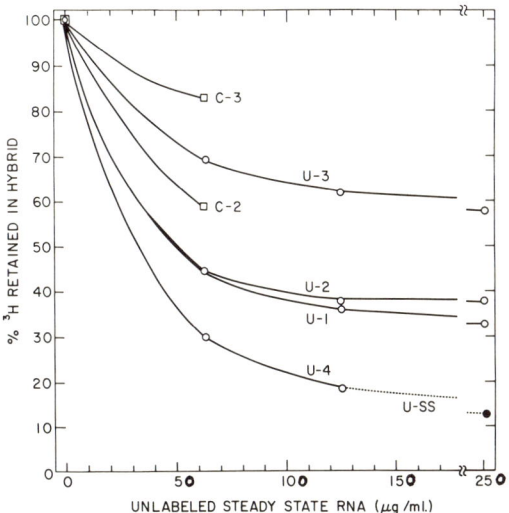

Figure 5. Hybridization competition of steady state and temporally unique RNA's from ultradian and circadian modes (see text). From Wille and Ehret (5).

temperature overnight and baken in a vacuum oven at 80°C for 2 hours. Hybridization was accomplished by annealing the DNA trapped on the filters to 2 mls of RNA in scintillation vials at 65°C. Collision kinetics were satisfied in 16 hours. After this, filters were washed, exposed to ribonuclease (5 µg/ml) for half an hour, then washed again and counted for bound activity in a scintillation spectrometer. Control filters (soaked in labeled RNA but lacking DNA) retained little activity above background.

Figure 5 shows the results of a competition experiment between labeled RNA species and unlabeled U-SS RNA. The hyperbolic function characteristic of competition between homologous species (8) is well demonstrated in the lower curve in which labeled steady state RNA competes with unlabeled steady state RNA for DNA vacancies. Because in other experiments (in progress) we distinguish between the hybridizing capacities of extremely rapidly reassociating (ERRF) DNA and slowly reassociating (SRF) DNA, we wish to emphasize here that the DNA used in this series is a composite of both classes. The failure of U-SS unlabeled RNA to compete so well against U-2 and U-3 tritium labeled molecules (middle curves, Figure 5), indicates the presence of unique species of RNA in each set. The rank order for temporally unique species is thus U-3 > U-2 > U-1 > U-4; C-3 > C-2; and U-4 ≃ U-SS. We interpret the results to indicate not only a similarity between C-3 and U-3, and between C-2 and U-2, but also the successive appearance of new messengers as the cell cycle progresses, an accumulation of "lates" and "earlys" together shortly after mid-cycle and an abrupt reduction of all classes shortly before cell division in U, or the time corresponding to it in C. Finally, we recall the endogenously rhythmical character of the cells in the population providing the C classes of RNA. There is thus underlying the endogenous circadian rhythm of cell division in Tetrahymena, (4), a more fundamental endogenous circadian rhythm of temporally unique RNA synthesis. This demonstration of temporally unique classes of RNA is borne out by other work on crosses between heterologous classes within and between U and C (5).

In conclusion, we would like to stress the basic difference between high frequency biochemical oscillations (or superpositions of such oscillations) on the one hand and the time-regulating mechanism proposed by the chronon theory on the other hand. Even within a single cell the former are described by ordinary differential equations in the concentrations of reactants and enzymes, whereas the latter process is based on sequential transcription of a linear array as the basic escapement device.

Perhaps a different but related way of looking at this state of affairs is Schrödinger's remark (9) that in many systems the observed order comes about as a result of averaging over a large number of molecules but, for particular cases of biological interest, order is inherent in the properties of single macromolecules. A chronon may be regarded as such a molecule. Its basic property, from our point of view, is the preservation of period length in spite of successive replications. Schrödinger does not discuss how the molecular order is transmitted into macroscopic systems or, as Pattee (10) puts it, how this order is expressed as a hereditary trait. Pattee's discussion, though it does not furnish a solution to the problem, is highly relevant in this context since it forces us to look at biological clocks from the perspective of reliability in the storage and transmission of hereditary information. We trust finally that we are not being unduly optimistic in expecting that our experimental demonstrations of distinctive properties of the ultradian and of the circadian-infradian modes of growth, as well as of temporally unique macromolecules associated with cell-cycle stages in synchronized populations of cells will stimulate future workers, especially in the field of high frequency oscillations, to investigate important new porperties of their systems by deriving raw materials for the <u>in vitro</u> studies from <u>synchronized</u> cells whose <u>growth mode</u> and <u>stage</u> are well-defined. From such studies will come a better understanding of the nature of the links not only between the high frequency and low frequency oscillations, but also between the ultradian and circadian-infradian modes of growth and regulation in eukaryotes.

Acknowledgements

This work was supported by the U.S. Atomic Energy Commission.

References

1. Ehret, C.F. and Trucco, E. J. Theoret. Biol. <u>15</u>, 240 (1967).
2. Ehret, C.F. and Wille, J.J. In Photobiology of Microorganisms. (Ed. Per Halldall) John Wiley (1970) pp. 369-416.
3. Szyszko, A.H., Prazak, B.L., Ehret, C.F., Eisler, W.J. and Wille, J.J. J. Protozool. <u>15</u>, 781 (1968).
4. Wille, J.J. and Ehret, C.F. J. Protozool. <u>15</u>, 785 (1968).

5. Wille, J.J. and Ehret, C.F., Manuscript in preparation.
6. DiGirolamo, A., Henshaw, E. and Hiatt, H.H. J. Mol. Biol. 8, 479 (1964).
7. Zeuthen, E. In Growth in Living Systems (Ed. M. Zarrow) Basic Books, New York (1961) pp.135-138.
8. Bolle, A., Epstein, R.H., Salser, W. and Geiduschek, E.P. J. Mol. Biol. 31, 325 (1968).
9. Schrödinger, E. What is Life? Cambridge University Press (1944).
10. Pattee, H.H. In Towards a Theoretical Biology 1. Prolegomena. (Ed. C.H. Waddington) Edinburgh University Press (1968) p.77.

Note added in proof.

The molecular hybridization parameters that were used in the experiments described above (short-term annealing kinetics, relatively low concentrations of 1 h or 2 h pulse-labeled RNAs derived from rRNA-rich cuts of sucrose gradients) favored the participation of the most highly redundant RNAs of the genome. Evidence that rarer species of RNA also have temporally characteristic properties (as shown by competition hybridization experiments) is given in two more recent papers; the one on testing the chronon theory of circadian time-keeping (1) and the other on the resolution of some component classes of complex RNA by molecular hybridization in the eukaryote Tetrahymena pyriformis (2).

1. Barnett, A., Ehret, C.F. and Wille, J.J. In Biochronometry. (Ed. M. Menaker) National Academy of Sciences, Washington (1971), p.637.

2. Barnett, A., Wille, J.J. and Ehret, C.F. Biochim. Biophys. Acta, 247, 243 (1971).

RESPIRATION DEPENDENT TYPES OF TEMPERATURE COMPENSATION
IN THE CIRCADIAN RHYTHM OF Euglena gracilis

Klaus Brinkmann

Institut für Molekularbiologie
3301-Stöckheim/Braunschweig
Germany

The unicellular organism Euglena gracilis exhibits a self-sustained circadian rhythm in dark mobility with a maximum at the middle of the physiological day (1). In cell suspensions which do not show any further cell division the rhythm persists up to three months. It is damped out under continuous bright light and reinduced by a single transition from light to darkness. As illustrated in Figure 1, it is characteristic of the rhythm to exhibit two alternative types of temperature compensation depending on the metabolic situation.

In young mixotrophic cultures, i.e. in a medium containing peptone and citrate, the length of the free-running period of the circadian rhythm increases slightly with higher temperatures. The period varies between 24 and 27 hours, indicating a temperature coefficient less than 1. A rapid temperature step does not disturb the phase. We define this type of rhythm as frequency-sensitive.

In contrast to this type, in older autotrophic cultures, the period is constant at all temperatures within the physiological range, but a rapid increase in temperature shifts the circadian phase. The extent of this shift depends only on the phase angle at which the temperature step has been applied. We define this type as phase-sensitive.

In comparing the two systems, the complementary nature of the temperature response should be noted: one oscillator is frequency-sensitive but stable in phase, whereas the other one is phase-sensitive but reestablishes exactly the same period length of 23.5 hours after being disturbed by

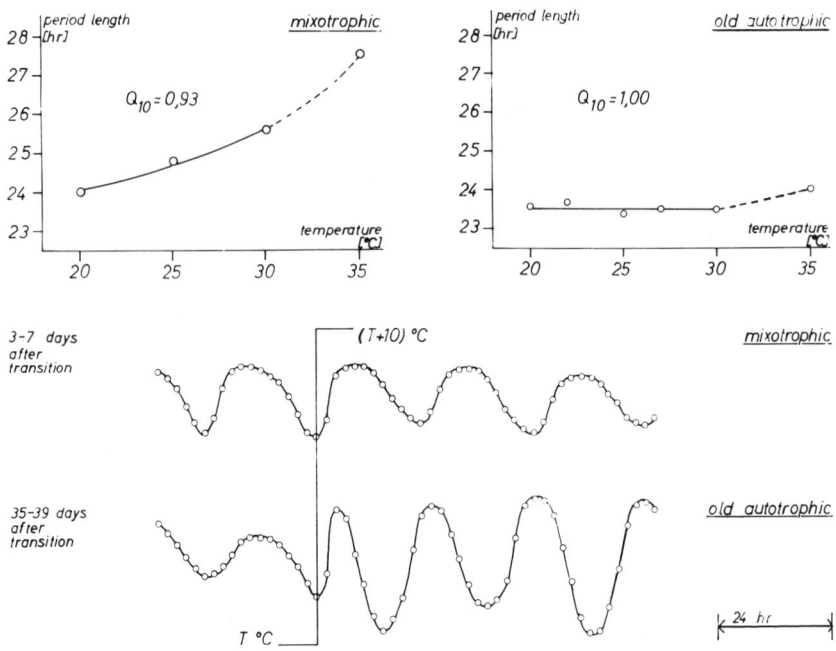

Figure 1. Types of temperature compensation in the circadian rhythm of mobility of Euglena gracilis. The top two graphs compare the free-running frequencies in young mixotrophic and old autotrophic cultures at different constant temperatures. Below, the same cultures are compared with respect to their phase reaction after input of a single temperature jump of +10°C at minimum phase. After Brinkmann (1).

any temperature jump. It is possible to convert the frequency-sensitive type into the phase-sensitive one by replacing the peptone-citrate medium by an **inorganic salt** medium without including cell division; the new type of compensation will be stabilized within 6 days after changing the medium (1).

It is the topic of this paper to state that the type of temperature compensation does not actually depend on the medium itself but is rather determined by the temperature dependency of the respiration. The temperature dependency of respiration is correlated with aging of the cells under test-conditions (see footnote 1) which might happen during

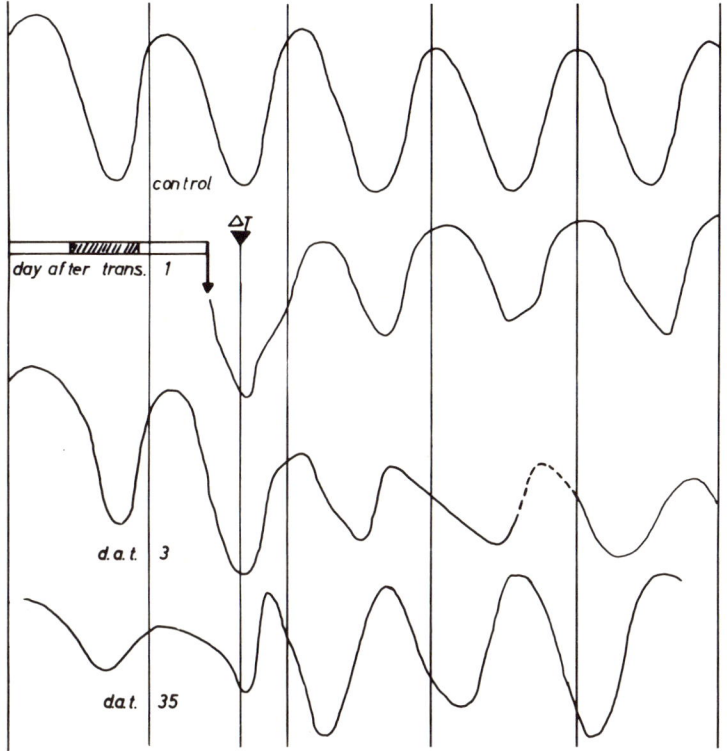

Figure 2. Phase shift of the circadian rhythm of mobility in autotrophic Euglena gracilis induced by a single temperature step of +10°C with dependence on the time after starting test conditions (see footnote 1). The numbers at left indicate the day after transition into test conditions when the temperature step was applied; the arrow indicates the temperature step.

Footnote 1. The test conditions consist of a Light-Dark rhythm of 20 min light and 100 min dark at constant temperature. The 2 hour rhythm does not affect the parameters of the circadian rhythm. The test of cell mobility is based on the sedimentation equilibrium established within each 100 min of darkness.

the medium-change experiment or when simply remaining in the same medium. As shown in Fig. 2, for a few days autotrophic cultures exhibit a phase insensitive oscillation, like that of mixotrophic cultures, if they are transferred into test conditions immediately after stopping aerobic growth. Applying a temperature step at the minimum phase of day '1' will not affect the phase. At day '3' the system shows an intermediary response (perhaps a superposition of cells which have reacted with those which have remained unaffected because the amplitude after the temperature step is decreased). Later on, full phase inversion is observed.

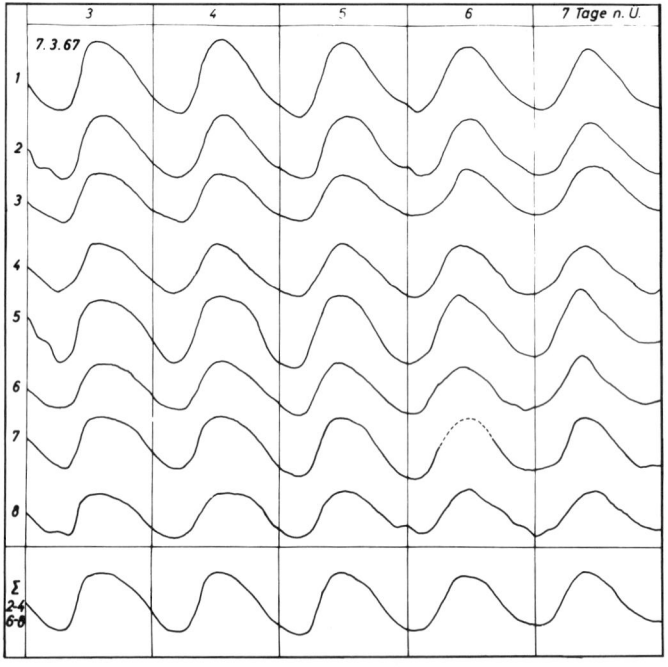

Euglena gracilis, autotroph 25°C

Figure 3. Transition from an asymmetric curve shape to a sinusoidal oscillation during the first days after starting the test conditions in autotrophic cultures. The lower curve shows the average value of the individual cultures shown above.

BIOCHEMICAL OSCILLATORS

In addition, the time at which the type of temperature compensation is changed is characterized by a transition from an asymmetric curve shape into a sinusoidal oscillation, as shown in Figure 3. The lower graph represents the mean value of the 8 individual curves above it. It should be noted that the transition is not correlated with a change in amplitude or frequency. Thus, the periodic instability of frequency reported to follow a change of the medium (1) must be a specific effect of the medium change rather than an effect essentially correlated with the induction of a new type of temperature compensation.

Looking now at what happens to the cells during the first days after the start of test conditions the clearest metabolic change is the induction of a lactic acid fermentation coupled with a leakage of pyruvate and a decrease in the pH of the medium. Figure 4 shows the increase in pyruvate on a dry weight basis. The accumulation starts immediately after the transition into the test-lighting rhythm and follows a saturation curve. The control, which remains under a daily rhythm of 12 hours light and 12 hours darkness, accumulates pyruvate only slowly.

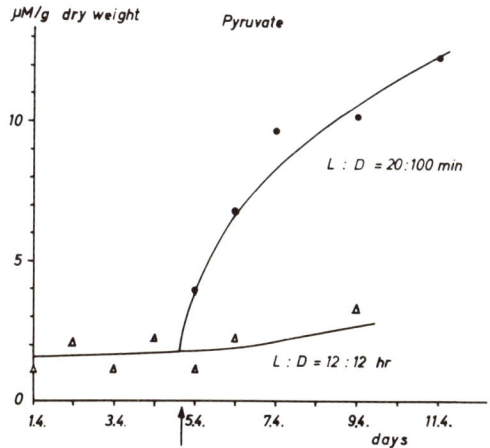

Figure 4. Accumulation of pyruvate in an autotrophic culture of Euglena gracilis during the first 6 days ('5 - 11 May') after transfer from a 12 hrs: 12 hrs Light-Dark rhythm into a constant test rhythm of 20 min light and 100 min dark. The arrow indicates time of transfer.

517

Table 1 shows the concentration of fermentation products inside and outside the cells. The data indicates a strong lactic acid fermentation with a very high lactate/pyruvate ratio. A further observation is that lactate is concentrated about 100 times more inside the cells than outside, whereas pyruvate seems to pass through the cell membrane almost unimpeded.

TABLE 1

Actual concentration	Autotrophic (27 days after transition)		
μ moles/l	cells*	medium	ratio cells/medium
lactate	99,900	913	1.09×10^2
pyruvate	120	54	2.22
ratio lactate/pyruvate	8.32×10^2	16.9	

*cell volume measured by haematocritt centrifugation

Table 1. Actual concentrations in μ moles/l of pyruvate and lactate inside and outside the cells of a stagnating autotrophic culture of Euglena gracilis 27 days after transfer into a constant rhythm of 20 min light and 100 min dark at a constant temperature of $20^\circ C$.; intensity of the white light, 1000 Lux.

The lactic acid fermentation suggests that the respiration may be inhibited - and therefore the respiration in both systems has been compared. In Figure 5, endogenous oxygen consumption, measured with a Clark-type electrode, is plotted against pH. The left-hand plot represents respiration in a growing culture and shows a characteristic pH-dependence and a temperature coefficient of about 1.9. The right-hand plot shows respiration in an old autotrophic culture which has been under circadian test for a long time. The pH-dependence has been changed but more important is the large increase of the temperature coefficient in the old cultures.

Comparing the two types at room temperature and low pH

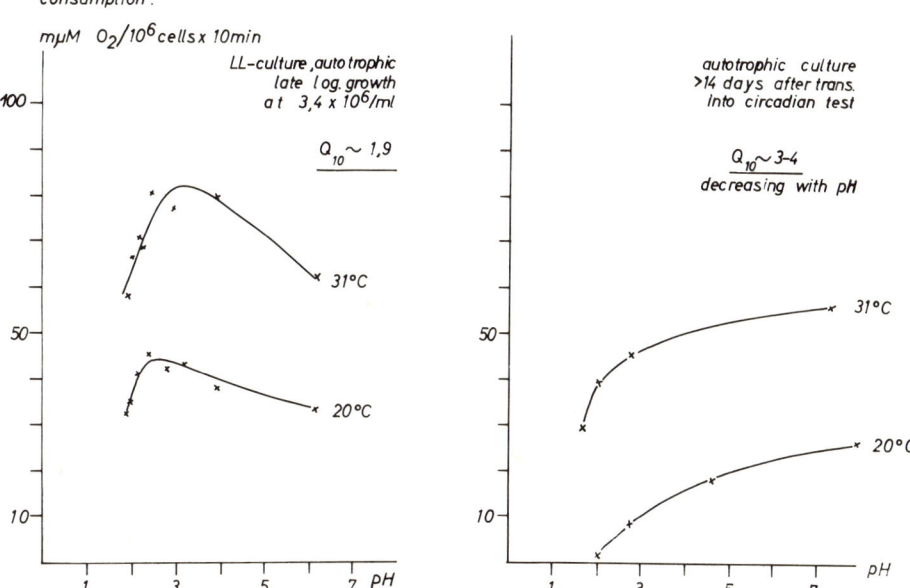

Figure 5. Endogenous respiration of autotrophic cultures of Euglena gracilis showing dependence on pH at different temperatures in the late state of growth of a continuous light culture reared at 20°C (left hand) and 14 days after transfer into the test rhythm consisting of repeating 20 min light and 100 min dark at 20°C (right hand). Oxygen consumption was measured with a Clark-type electrode beginning, at 20°C, 5 min after adjustment of pH and transfer into darkness. 20 min later the temperature was raised to 31°C. The values indicated for 31°C were recorded 5 min after stabilization of the higher temperature.

there is almost no respiration in the phase-sensitive type, whereas the frequency-sensitive type has a strong respiration. This difference initially misled us to suppose that the most important factor is whether respiration occurs or not. However, we find that a sudden rise in temperature very strongly increases or even reinduces respiration in old cultures, suggesting that this unusual increase in respiration is responsible for the phase jump of the circadian rhythm.

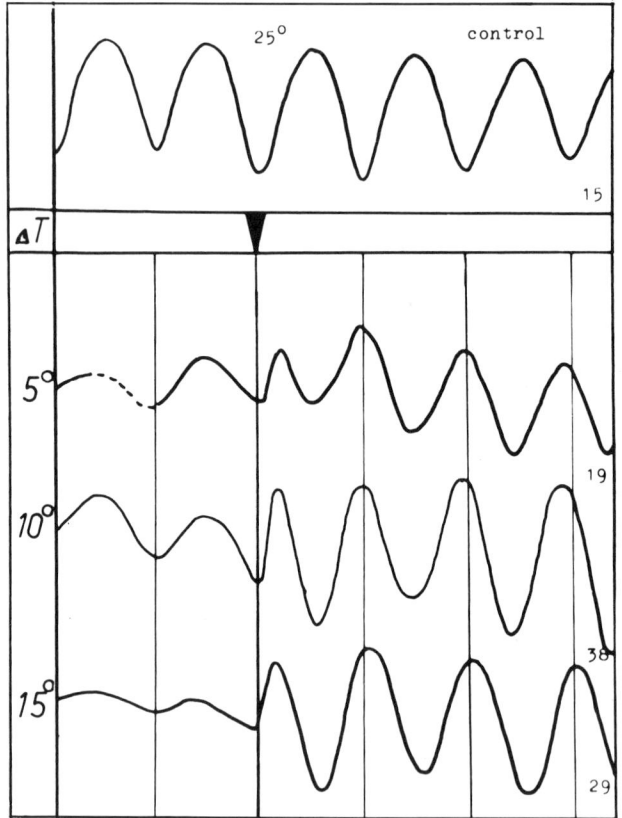

Figure 6. The inversion of the circadian phase of autotrophic *Euglena gracilis* after input of a single temperature step applied at the minimum phase of mobility. Numbers on the left are magnitude of step in °C; numbers on the right are days after transition into test conditions.

This interpretation is in good agreement with the fact that a negative temperature jump does not shift the circadian phase, as would be expected if there is already no respiration when temperature is decreased. This hypothesis is further supported by the fact that a negative temperature jump immediately stops the transient function of circadian frequency pre-induced by a positive temperature jump (1).

It seems that the circadian phase reaction does not depend on the extent of the temperature jump. The arrow in Figure 6 indicates the input time for the temperature jump and the numbers on the left indicate the extent of the jump. Although the amplitude increases with increasing temperature steps, there is only one phase reaction: an inversion of phase compared with control. This "all or nothing" response suggests that the phase shifting input is not the extent of the respiratory activity but rather its differential quotient.

We conclude that respiratory activity is involved in the control of temperature compensation of the circadian rhythm. In this context it may be mentioned that circadian rhythms are only described in organisms which have complete mitochondria, thus indicating that these organelles might have played an important role during the evolution of biological clocks.

REFERENCES

1. Brinkmann, Klaus, Planta 70, 344 (1966).

THE ROLE OF ACTIDIONE IN THE TEMPERATURE JUMP RESPONSE
OF THE CIRCADIAN RHYTHM IN Euglena gracilis

Klaus Brinkmann

Institut für Molekularbiologie,
3301-Stöckheim/Braunschweig,
Germany

In addition to the two types of temperature compensation in Euglena gracilis already presented (1), I now want to demonstrate a switch which permits the transfer of the phase-sensitive type back into the phase-insensitive type from which it evolved. This switch is caused by actidione.

It must be remembered (see Fig.1 Ref.1.) that in old autotrophic cultures of Euglena exhibiting a strong lactic acid fermentation, a sudden increase in temperature, applied at the minimum of dark mobility, shifts the phase up to complete inversion whereas the phase of young mixotrophic and strongly respiring autotrophic cultures is not affected by the same temperature jump applied at the same circadian phase.

In Figure 1, the four upper curves represent controls and the arrow indicates the temperature jump. The response is a typical transient function followed by a full inversion of phase. But if 1 or 2 µg of actidione per ml medium are added at the beginning of the temperature increase, no phase response is observed. The lower graphs show that the transient function is completely inhibited and, about 24 hours after the temperature input, a minimum of dark mobility appears instead of a maximum as in the control.

In the upper part of Figure 2 a treated and untreated culture are compared for a long time after the temperature input. The stable phase inversion of the two cultures many days after the input indicates that the effect shown in Figure 1 has not been merely transient. Another important effect shown in the lower curves is that the same agent does

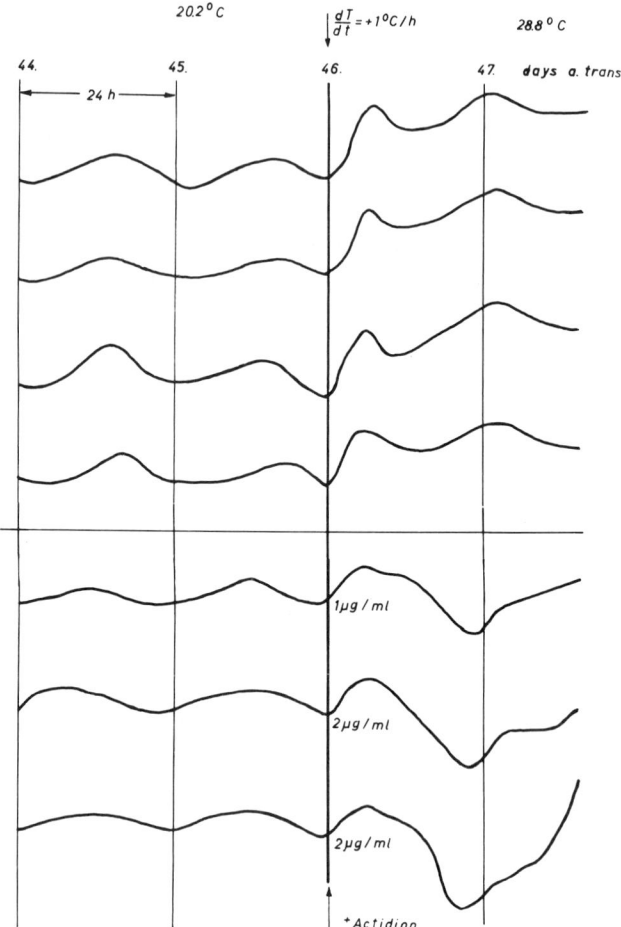

Figure 1. Phase reaction of the circadian rhythm of dark mobility in cultures of autotrophic Euglena gracilis after input of a temperature step from 20.2°C up to 28.8°C beginning with the minimum phase at the 46th day after starting the test conditions. The 4 upper curves show the reaction without actidione while the 3 lower curves show the reaction after adding different concentrations of actidione at the start of the temperature rise, indicated by an arrow. The indicated concentrations are final concentrations in the medium.

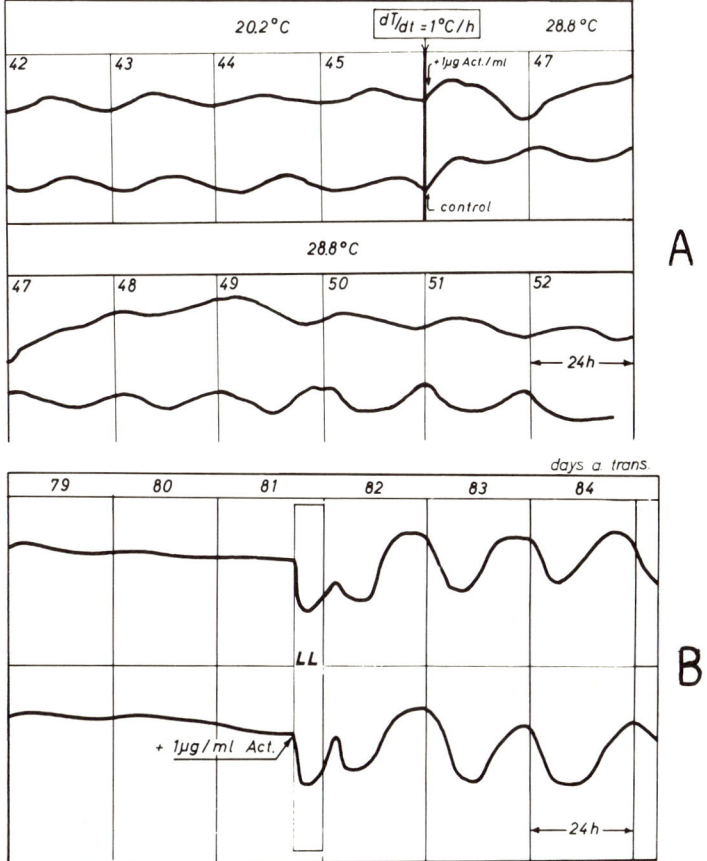

Figure 2. A: Inverse phase reactions of two cultures after input of a temperature step in the presence and absence of actidione, compared until an approach to a steady-state. B: Identical reinduction of circadian rhythms by a 6-hours light pulse in the presence and absence of actidione. Compared with their controls both cultures which have actidione exhibit a slightly decreased frequency.

not inhibit the induction and phase setting of the circadian rhythm by light pulses. The end of a 6 hr light pulse reinduces the rhythm in very old autotrophic cultures. The

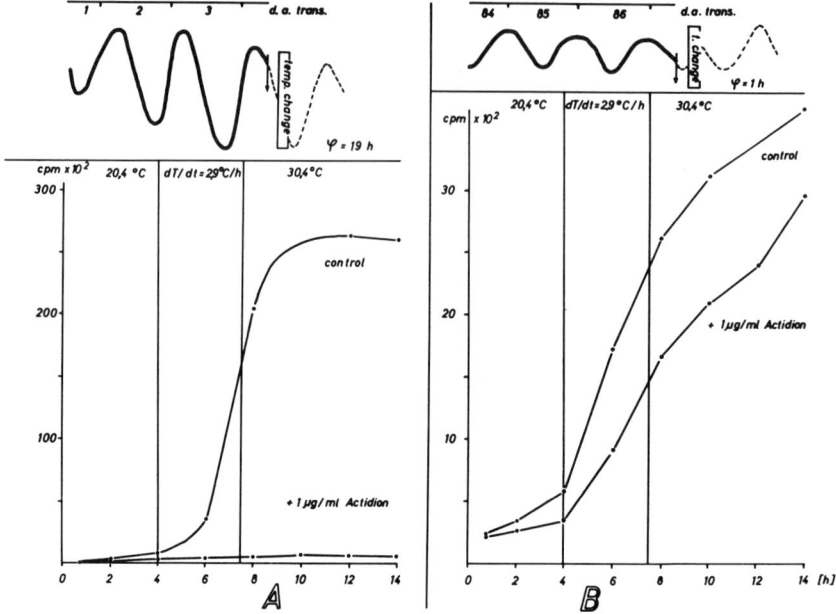

Figure 3. Inhibition of the incorporation of ^{14}c-phenylal. by actidione during a temperature rise at day '4' (left hand) and day '87' after the transfer into test conditions. The upper curves show the circadian rhythm of dark mobility up until the start of the incorporation, indicated by arrows (= time "0" on the lower schedules). Vertical lines indicate the beginning and end of the temperature rise.

phase setting is typical: 18 hrs after 'light-off' there appears a maximum of dark mobility which already represents a steady-state of oscillation (2). The same happens if actidione is added, i.e. actidione is a specific inhibitor only of phase shifting by a temperature jump and does not affect phase shifting by turning the light off.

We conclude that there must be two separated information chains; one starting from the input of a temperature signal and the other from the input of a light signal. However, it is uncertain whether the two information chains enter the oscillating network at different points or whether they end at the same trigger.

Figure 3 and Table I show that **actidione** does not act by **inhibiting** protein synthesis. In Figure 3, the incorporation of ^{14}c-phe in the presence and absence of actidione during an increase in temperature has been studied. The beginning and the end of the increase in temperature are indicated by vertical lines. The upper graphs represent the original oscillations before being prepared for incorporation; on the left, at day '4' after transition, and on the right, at day '87' after transition. It should be noted that the ordinates (in cpm) are increased 10-fold in B. It is apparent that in the young cells (A) the incorporation is ten times higher than in the fermenting old cells (B) but in the presence of actidione the incorporation by the young cells is almost completely inhibited whereas incorporation by the old cells is only slightly decreased. Furthermore, the increase in incorporation during the increase in temperature happens in old cells with actidione as well as in the young cell control without actidione. The induction of protein synthesis during the increase of temperature happens in both controls and in the old cells in the presence of actidione, whereas a temperature response occurs only in the old cells without actidione. Thus we conclude that induction of protein synthesis plays a negligible role in phase shifting the circadian rhythm.

Table I confirms that the incorporation curves may, indeed, be interpreted as indicating protein synthesis. After incorporation, a lipid fraction, a protein fraction and a fraction containing all the remaining fractions have been separated. The left part of the table shows that the percent-incorporation into the protein fraction (underlined values) have not been changed by actidione in the old cells whereas in young cells this fraction has been greatly decreased. A control experiment (right side) in which ^{14}c-pyruvate has been used for comparable incorporation into all three fractions shows again that actidione inhibits the percent-incorporation into the protein fraction only in the young cells. So we conclude that actidione does not inhibit the temperature response by inhibiting protein synthesis.

Recently Feldman (3) reported that actidione decreases the circadian frequency of Euglena gracilis. Lengthening of

TABLE I

Euglena gracilis - autotrophic

Extraction:	14c-phe				14c-pyruvate			
	4th day after trans.		87th day after trans.		4th day after trans.		87th day after trans.	
	control	+ 1 μg/ml Actidione	control	+ 1μg/ml Actidione	control	+ 1 μg/ml Actidione	control	+ 1 μg/ml Actidione
1. Acetone-extract	0.3	5.0	1.0	1.0	15.4	16.0	20.8	17.8
2. Acetone-extract	0.6	0.5	1.5	0.3	1.1	1.8	1.9	1.5
Ether-extract	---	---	0.1	---	0.2	0.1	---	---
Supernatant after tryptic digestion	84.2	69.7	85.8	87.2	40.7	24.9	41.4	42.3
Non-digested pellet	14.9	24.8	11.6	11.5	42.6	57.2	35.9	38.4

Table I. %-distribution of radioactivity in different fractions 14 hours after the start of incorporation of 14c-phe or 14c-pyruvate. For experimental conditions see Figure 3.

the circadian period could also be observed in our experiments with the effect decreasing with increasing cell age. The actidione effect reported by Feldman is most typical for young cells and is strongly correlated with an inhibition of protein synthesis, whereas the inhibition of phase reactions after temperature steps is restricted to those old cells which have become temperature-sensitive without exhibiting a correlation with an inhibition of protein synthesis. We therefore conclude that there must be different mechanisms underlying both the actidione effects.

REFERENCES

1. Brinkmann, K., This volume, p.513.
2. Schnabel, G., Planta 81, 49 (1968)
3. Feldman, J.F., Proc. Nat. Acad. Sci. U.S. 57, 1080 (1967).

SUBJECT INDEX

A

Actidione, 523
Allosteric, 177, 249
Antilimit cycles, 146, 169
Arrhythmicity, 490
Autocatalytic, 64, 101
Autonomous self-oscillatory reaction, 71, 269
Autostabilization mechanisms, 375

B

Biochemical oscillations, 127
Biological rhythms, 68
Bistability, 15

C

Calcium, 366, 443
Catabolic repression, 411, 456
Cerium ions, 63, 71, 81
Chemical oscillators, 63, 68, 71, 81, 89
 two dimensional analysis, 31
Chemical periodic reactions, 63
Chemiluminescence, 97
Chemostat, 399, 411, 453
Chronon, 488
 theory, 504
Circadian oscillation and gene action, 503
Circadian period, 529
Circadian rhythm, 173, 469, 479
 endogenous, 510
 in *Euglena gracilis*, 513, 523

Conformational changes,
 in cell membrane, 427
 in creatine kinase, 347
Control, characteristics, 227
 chemical, 227
 properties, 158
 state, 169
 strength, 151, 153
 theoretic approach, 127
Controlled state, 375, 389
Cophase, 465, 479
Creatine kinase, 347
 conformational changes, 348
Critical mass phenomenon, 331
Cross-over point, 389

D

Damped oscillations, 50, 102, 105, 115, 120, 123, 135, 243, 273, 281, 399
Damping factor and cell population, 288
Developmental age, 500
Discontinuities, 469, 489
Double, frequency, 77
 periodic, 162, 237, 273
Drosophila pseudoobscura, 461, 479
Dynamical flow, 465

E

Eigenfrequency, 68
Endogenous rhythm, 422
 circadian, 510

531

SUBJECT INDEX

Entrainment, 461
Enzyme concentration, 254
Epigenetic system,
  oscillations in, 449
Eukaryotes, 504
Excitation, biochemical cycle, 373, 384

F

Feedback, 177, 229, 399, 408
  induction loop, 450
  interactions, 41
  loop, 58
Fibrillation of heart, 329
Flip-flop, 333
Force-flux characteristics, 8
Frequency control, 278

G

β-Galactosidase, 42, 449
Genetic feedback, 42
Global, phase, 33, 39
  stable singularity, 142
Glycolytic oscillations, 41, 127, 408, 445
  component structure, 181, 253
  computer studies, 140, 157, 177, 187
  control mechanism, 197, 243, 270
  in extracts of heart muscle, 149, 170, 187
  in extracts of yeast, 177, 202, 231, 269
  in yeast cells, 177, 231, 269, 285
  net flux diagrams, 134
  substrate concentration, 253
  substrate control, 229
  theoretical studies and models, 41, 127, 149, 197
  yeast phosphofructokinase activation, 221
Glycolytic pathway, 41, 127
Glycolysis, 393

H

Harmonics, 56, 84, 86
Hill equation, 317, 326
Hill number, 43, 141, 221

I

Infradian mode, 504
Inhibitors, metabolic, 499
Inhibition, 373
Initiation, 493
Isochrons, 464, 487

L

Lactic acid, 168, 518, 523
Light, 464, 480
  pulses, 525
Light-Dark rhythm, 517
Limit cycle, 40, 58, 216
  oscillations, 142
  stable, 44, 56
Lotka model, 33, 99

M

Malonic acid, 63, 71, 81
Metamorphosis, 499
Mitochondria, 375
Mitochondrial, DNA, 426
  oscillations, 115
  oxidation, 391
Monostable systems, 19
Multi-periodic waveforms, 280
Multi-process systems, 18
Muscle contraction, 303
  kinetic model for isotonic contraction, 311
Myofibrillar oscillation, 359
  oscillations in ATPase activity, 303

O

Oscillating structures from insect muscle, 303
Oscillations,
 in glycolysis, 42, 127
 in a *Myxomycete*, 429
 in mitochondria, 285
 in mitochondrial volume, 115
 in NADH, O_2, and peroxidase, 109
 in oxygen tension, 399
 in protein conformation, 343
 in *S. cerevisiae* population, 411
 in single yeast cells, 285
 of creatine kinase activity, 347
 of cytoplasmic streaming, 437
 of light absorbance and fluorescence in plasmodium, 438
 of pyridine nucleotide and oxygen tension in cultures of *K. aerogenes*, 399
 of sodium transport, 363
Oscillator, limit cycle, 127, 462, 479
 linear, 462
 phase shift, 140, 143
 relaxation, 216, 462, 493
Oscillators, chemical, 63, 68, 71, 81, 89
 phase-shift, 140
 two component, 132
Oscillatory, dynamics, 127
 kinetics, 127, 154
 stationary states, 32, 36, 83
 synthesis of β-galactosidase, 449
Oscillatory reactions, by peroxidase, 97, 109
 space behavior, 89
Oscillatory behavior, of membranes, 7
 of muscles, 303, 311
Oscillophor, 251

P

Periodic reactions, chemical, 63
Periodicity of nuclear division, 429

Periods, 499
Perturbation, 55, 461, 470, 481
 critical, 473, 491
 in oscillatory processes, 461, 479
Phase, 221, 247, 264, 403, 462
 angle, 235, 245
 control, 235
 non-oscillatory, 78
 plane, 32
 relationship, 221, 264
 shift, 461, 468, 479
 shift oscillator, 140, 143
 shifting, 73, 273, 519
Phosphofructokinase, 149, 161, 177, 188, 197, 221, 229, 246, 253, 269
Photosensitive, 498
Physiological, activity, 374, 425
 excitation, 374
 inhibition, 373
Physiological rhythms, 172
 in *S. cerevisiae*, 419
Propranolol, 115
Protein synthesis, 527
Pulse frequency, 330

R

RNA, temporally unique, 505
Resonance frequency, 303
Response curves, 469
Reverberator, 336

S

Self-exciting, 479, 497
Short circuit current, 365
Singular, point, 297
 state, 466, 474
 unstable state, 497
Singularity, 141, 149, 153, 479, 483, 498
Siphon model, 111
Sliding filament concept, 311

SUBJECT INDEX

Space behavior of self-oscillatory chemical reactions, 89
Spontaneous oscillations, 368, 424
Stability, 11, 479, 497
State variables, 463
States, non-oscillatory, 98
  non-stationary, 8
  stationary, 12, 32, 44, 83, 149, 479, 498
  unstable, 140
Succinate, 376
  concentration burst, 389
Sustained oscillations, 113, 135, 142, 269, 281, 287, 399
Synchronization, 285, 292, 294, 399, 493, 497
  between mitochondria, 294
  between yeast cells, 281, 285
  by external force, 81, 85
  mechanism, 408
  of mitosis, 437
  zone, 84, 86
Synchronized, populations, 415, 420
  nuclear division, 429
Synchrony, 344, 359
  of cell division, 457
  oscillations, 411, 419
Systems, chemical, 10
  electrical, 10
  thermal, 10
  two dimensional, 31

T

Temperature, coefficient, 22
  compensation, 513
  dependence, 503
  effects, 76, 306, 356, 399
  response, 527
  shock, 497
Temperature jump, 290, 514, 520
  synchronization of oscillations, 290
Tetrahymena, 504
Toad bladder, 363
Trajectory, 32, 466, 473, 483
Transients, 451, 465, 469, 481, 490
Transitions, 17
Trehalose, 261, 269
Trigger, cytoplasmic, 435
Triggerable systems, 16
Triggering, of oscillations, 366
  mechanism, 414

U

Uncouplers of oxidative phosphorylation, 116, 121
Unstable, steady states, 213
  stationary states, 140

W

Wave propagation, 90, 93, 273
  during heart fibrillation, 329
Waveforms, 215, 235, 400
  double-periodic, 77, 162, 237, 273
  multi-periodic, 280
  sawtooth, 368

This book is due on the last date stamped below. Fines will be charged on all overdue books.